The Derrick Jensen Reader

The Derrick Jensen Reader

Writings on Environmental Revolution

EDITED BY LIERRE KEITH

Seven Stories Press

New York

Seven Stories Press
140 Watts Street
New York, NY 10013
www.sevenstories.com

College professors may order examination copies of Seven Stories Press titles for a free six-month trial period. To order, visit http://www.sevenstories.com/textbook or send a fax on school letterhead to (212) 226-1411.

Library of Congress Cataloging-in-Publication Data

Jensen, Derrick, 1960-
 The Derrick Jensen reader : writings on environmental revolution / Derrick Jensen, edited by Lierre Keith.
     p. cm.
 Includes bibliographical references.
 ISBN 978-1-60980-404-6 (pbk.)
 1. Environmentalism. 2. Human ecology. 3. Sustainable living. 4. Environmental degradation. 5. Environemental responsibility. I. Keith, Lierre. II. Title.
 GE195.J46 2012
 304.2--dc23

                                    2012009238

Book design by Jon Gilbert

Printed in the USA

9 8 7 6 5 4 3 2 1

Contents

Preface

The dominant culture—Western Civilization—is driven by a death urge, an urge to control, an urge to violate, an urge to destroy life. Unless it is stopped, it will kill every living being on the planet, every living being it can.

⊛ ⊛ ⊛

Fourteen years ago I wrote what has become one of my most quoted passages: "Every morning when I wake up I ask myself whether I should write, or blow up a dam. Yet I'm not always convinced I'm making the right decision. I've written books and I've been an activist. At the same time I know neither a lack of words nor a lack of activism kills salmon here in the Northwest. It is the presence of dams."

In the time since, things have gotten much worse for salmon, and for the earth. By now we all know the numbers, or we should. Two hundred species per day driven extinct, 90 percent of the large fish in the oceans extirpated, more than 98 percent of native forests destroyed, 99 percent of prairies, and on and on. Every biological indicator is going in the wrong direction. Native communities—human and nonhuman—are under assault. Just where I live, frog populations have collapsed, as have newt populations, butterfly populations, crane fly populations, dragonfly populations, banana slug populations, songbird populations. Crow populations have collapsed. Bat populations. Wooly bear populations. Harvestmen populations. Moth populations. Bumblebee and solitary bee populations. And these are just some of those I've noticed. Salmon have continued to collapse. I give salmon fifteen years. If we can bring down civilization in fifteen or twenty years, I think salmon will be, in time, fine. Much longer and they will not survive.

⊛   ⊛   ⊛

Far too many of us have forgotten, or never knew, that words can be weapons to be used in the service of our communities. Far too many of us have forgotten, or never knew, that words *should* be used as weapons in the service of our communities. For far too long too many critics and too many teachers have told us that literature should be apolitical (as though this is possible), and that even nonfiction should be "neutral" or "objective" (as though these, too, are possible). If you want to send a message, they've told us too many times, use Western Union. I once spoke with a nature writer who refused to lend his name to a campaign to protect a species about whom he had written, giving as his reason, "I'm a writer, damn it. I have to remain neutral."

When the world is being murdered, such a position is inexcusable and immoral. It's also really boring. Further, it's ignorant of literature's history, and of what it means to be a real writer. Have these people never heard of Steinbeck, Dickens, Crane, Hugo? Charlotte Perkins Gilman? Rachel Carson? Frederick Douglass? Harriet Beecher Stowe? Alexandra Kollantai? George Eliot? Katherine Burdekin? Zora Neale Hurston? Andrea Dworkin? B. Traven? Upton Sinclair? A little Tolstoy, anyone?

A few years ago I read a book that for me epitomizes the emptiness of the modern movement toward apolitical literature. It was a book called *The Diving Bell and the Butterfly*, and was a memoir of someone who had a stroke and lost the ability to voluntarily move any part of his body except to blink one eye. He eventually was able to begin communicating this way, and wrote a book. I read it because I saw so many reviews calling the book a triumph and an important piece of modern literature. I hated it, though, primarily because prior to his stroke the protagonist was a narcissistic, self-pitying, nature- and woman-hating asshole, and after the stroke, well, he was a narcissistic, self-pitying, nature- and woman-hating asshole. But the failure of that book is not why I mention it here. I mention it because as I was reading it, something kept niggling at me, and finally near the end I realized what it was. This book has essentially the same plot as Dalton Trumbo's *Johnny Got His Gun*, which is about a soldier in World War I who wakes up in a hospital and slowly realizes he was hit by a bomb, and has lost his hearing, his vision, his arms, his legs, his face. The only way he can communicate is by tapping his head against the pillow. But in this novel, written some eighty years ago, the character isn't a narcissistic self-pitying asshole, but rather becomes an anti-war messiah. The book has moved me to tears each time I've read it, and is one of the most powerful anti-war novels ever written.

That trajectory, from *Johnny Got His Gun* to *The Diving Bell and the Butterfly*, mirrors for me the degradation of too much modern literature.

I will not participate in that degradation. It is long past time we all—writers and readers alike—began working to restore and maintain a literature of justice, resistance, and revolution.

⊕ ⊕ ⊕

I recently received a note from the writer John Keeble, who twenty years ago was my teacher at Eastern Washington University, telling me he'd been following my work. He wrote, "I know you've been writing and that among some circles you're famous, which at once fills me with pride and bemuses me because I know well that nothing matters to you as much as the issues."

He's right, and he articulates well my own split feelings about what I have or have not accomplished with my work.

⊕ ⊕ ⊕

I would not be who I am and I would not write what I write without having learned from some of my elders who did not believe that writers should or can be apolitical. As one writer said to me, "We are all holding hands through time, helping to make an unbroken line of resistance." Or as another insisted, "The truth is most important. It is more important than money. It is more important than fame. It is more important than your career. It's more important than your preconceptions. Follow the truth—follow the words and ideas—wherever they lead." Yet another said, "When you lie to readers—whether this lie is something as small as an incorrect word or as large as a wrong idea; and whether this lie is explicit or by implication—you are stealing from them, stealing their time and stealing their hearts and minds." Another said, "No matter how you try, you cannot separate art from politics. So-called apolitical art supports the status quo."

But the most important thing anyone said to me, and I heard this time and again from those I respected, is this: Words matter. Art matters. Literature matters. Words, and art, and literature, can change lives, and can change history. Make sure that your words and your art and your literature move people individually and collectively in the right direction, in the direction of justice and sustainability. It is possible to create a literature that is immoral. This is something we see all the time. A literature that supports capitalism is immoral. A literature that supports patriarchy is immoral. A literature that does not resist oppression is immoral. But you do not need to follow in those traditions. You can help to create a literature of morality and resistance, as each new generation must create this literature, with the help of all those generations who came before, holding their hands for support, just as those who come after will need to hold yours.

I was taught that art can be, and, to be moral, must be, a combat discipline.

                                                   ⊠ ⊠ ⊠

If too few of us remember that words can be weapons, even fewer of us remember that as weapons, words cannot fight alone. Words can help us resist, but by themselves they do not topple dictators, they do not stop capitalism, they do not stop oppression, they do not halt species extinction, they do not stop global warming, they do not remove dams. At some point someone actually has to do something. At some point someone actually needs to physically destroy the infrastructures that allow dictators to stand, capitalism to metastasize, oppression to continue, species extinction and global warming to accelerate, dams to murder rivers.

No one can seriously believe that words alone would have stopped Hitler. Words alone wouldn't even have stopped Ted Bundy. And words alone won't stop the dominant culture from destroying life on earth.

That job is up to all of us.

                                                   ⊠ ⊠ ⊠

I think often of a question asked by a friend and mentor, "What are the largest, most pressing problems you can help to solve using the gifts that are unique to you in all the universe?"

And that is precisely where I have succeeded, and precisely where I have failed.

There are many ways my writing life could so far be considered a success far beyond anything I daydreamed when I was younger. I have written twenty books. People seem to enjoy reading them and seem to enjoy coming to my talks, both of which honor me beyond belief. Despite the truth of the old cliché about writing, that it is a terrible way to make a living and a great way to make a life, for at least the last few years I've been able to financially support myself through writing (although, once again proving the truth of this cliché, I've still never made as much as a fresh graduate in her first job after finishing at the Colorado School of Mines, where I graduated thirty years ago). More important than all of these, however, is that I have been true to my muse, and have at least attempted to tell the truth as she has presented it to me. And I have at least sometimes succeeded in articulating some of those things I know in my heart to be true, but to which I've not yet put words, and in so doing have, I hope, helped some others to articulate some of those things they may know in their hearts to be true, but to which *they* have not yet put words. And through all of this I have come to know I am living the life I was born to live, doing what I was born to do. And far more important than all of these, my work is contributing to a literature of morality and resistance, and to the formation of a culture of resistance, and the formation of a militant movement to stop this omnicidal culture from killing the planet.

This is all to the good.

But the fact remains that if we judge my work, or at this point anyone's work, by the most important standard of all, and in fact the only standard that really matters, which is the health of the planet, my work (and everyone else's) is a complete failure. Because the planet is being murdered. And my work hasn't stopped that murder. Nor has anyone else's. We haven't even slowed it down. It's embarrassing to have to explain why this is the only standard that really matters, but at this point embarrassment is the least of our problems. The health of the planet is the only standard that really matters because without a living planet nothing else is important, because nothing else exists. And also because the extirpation of two hundred species per day is profoundly immoral. Compared to this, the number of books out doesn't matter. How beautifully or poorly they are written doesn't matter. Financially supporting oneself doesn't matter. Whether one is doing what one was put here to do doesn't matter. Life itself is more important than what we create.

The health of the oceans, the forests, the rivers, the salmon, the sturgeon, the migratory songbirds, are all more important than are you or I individually, and are more important than your or my accomplishments. Their health is the measure of our success.

⊛ ⊛ ⊛

Writing books is generally considered a solitary craft. And there are ways that, at least in my case, this is true. I spend hours, sometimes days, happily alone with questions and threads and themes and words and ideas and emotions and images floating past, sometimes coalescing, sometimes not. I spend hours happily alone, asking myself what, precisely, I am trying to say; asking my muse what she wants me to say, and for help in saying it; asking trees and rivers and rocks and stars what they want me to say, and for help in saying it. I ask for dreams.

But in other ways, writing is not solitary at all. First, it is a collaboration between writer and reader (and the words themselves), as each brings offerings to the communication. And there is that matter of holding hands through time, and across species. When I write not only I am writing, but so are those who came before me, and so are the salmon and moths and bats, and so is the muse.

Social change makes no pretense of being solitary. Revolutions and other forms of social change require organization, solidarity, cooperation. And once again, this cooperation extends through time and across species, as we hold hands with Boadica, the witches, Tecumseh, Nat Turner, John Brown, Harriet Tubman, Lozen, Wesley Everest, Kartar Singh Sarabha Grewal, Maude Gonne, Eugene Debs, Erich Mühsam, Sophie Scholl, Hannah Senesh, Ken Saro-Wiwa, the Movement for the Emancipation of the Niger Delta, polar bears, wolves, chimpanzees, salmon, tuna.

Through all of my work, my hand is extended. Take it if you wish. Take the hands of all these others. Draw what strength you can from each of us, and take up where each of us has left off.

But make sure to take all of these hands into only one of yours. You will need to keep your other hand free, to make a fist or to pick up a weapon, whether that weapon is pen, paintbrush, pistol, sabot, or whatever other weapon you will need in order to use your gifts in the service of your land-base, in the service of stopping this culture from destroying life on earth.

The time for waiting is long since done. It is time to act. We are all supporting you, and counting on you. The planet needs you. Not only words matter, but so do actions, and so do you.

Listening to the Land
Conversations About Nature, Culture, and Eros

Context Books, 2002

I wrote Listening to the Land *indirectly because of Neil Evernden's extraordinary book,* The Natural Alien. *It was the first book I'd read that did not take the utilitarian worldview that's destroying the planet as a given. The book profoundly changed my life, and I wanted to interview him to help promote his work.*

When I contacted Evernden, he said, "Well, if you're going to interview me, you really should interview John Livingston. He's a more important thinker than I am." I looked into John Livingston's work and wanted to interview him, too. I started thinking it would be helpful for people to have a collection of interviews that explored the leading thinkers' ideas about environmental destruction.

It was very important to me to show the links between various forms of destruction, such as patriarchy's assault on women and civilization's assault on indigenous peoples. I also wanted to ask, "How can people commit atrocities?" which is how I ended up talking to Robert Jay Lifton. The thread that holds the book together is this culture's ubiquitous destructiveness.

We are members of the most destructive culture ever to exist. Our assault on the natural world, on indigenous and other cultures, on women, on children, on ourselves through the possibility of nuclear suicide and other means—all these are unprecedented in their magnitude and ferocity. Why do we act as we do?

I began this project because I wanted to understand our culture's pervasive destructiveness, and to know if it is possible to live another way. Perhaps, I thought, what we are doing is natural, instinctive, and no different than the expansion of bacteria on a petri dish. Or perhaps the cause of the destructiveness is more specific to our being human; our adaptability and capacity for critical thinking guarantee we will "out-compete" every other species.

But if that's the case, how do we explain the existence of the Hopi, the Inuit, the Ladakhi, the !Kung San Bushmen, and other groups of people who fashioned ways of living that were in dynamic equilibrium with their surroundings? How do we explain the continued survival until a century ago—at least twenty thousand years after humans arrived on this continent but a mere three hundred years after the arrival of Europeans—of flocks of passenger pigeons that darkened the sky for days at a time? How do we explain the possibility of humans living side-by-side with 70 million buffalo, 40 million pronghorn antelope, with grizzly, wolf, and salmon?

I determined to gain understanding from those people most likely to have answers—those environmentalists, feminists, theologians, psychologists, and indigenous peoples who have dedicated their lives to exploring and countering the destructiveness of our culture and to promulgating more peaceful ways to live. Our conversations raised yet more questions: What is the relationship between technological innovation and human misery? Is there a direct relationship between environmental destruction and other forms of oppression, such as misogyny or genocide? Given that humans are causing the greatest mass extinction in the history of the planet, can there be hope? How can we, as a culture and as individuals, rediscover our connections to ourselves, to our neighbors, to the rest of the natural world? Would these connections help us to better realize our potential as human beings?

What has emerged from this questioning is a book about listening. Early in the project, Thomas Berry said, "The universe is composed of subjects to be communed with, not objects to be exploited. Everything has its own voice." He went on to say, "Somehow we have become autistic. We don't hear the voices."

How do we remember how to listen? Would we live differently if we listened to the voices of the species we are causing to go extinct? What lessons can we learn from the one in four women within this culture who are raped within their lifetime? What do indigenous peoples worldwide have to teach us?

What would happen if each of us began listening to our own needs? Terry Tempest Williams said that "our needs as human beings are really very simple—to love and be loved, a sense of connection and compassion, a desire to be heard. Health. Family. Home. The dance, that sharing of breath, that merging with something larger than ourselves. One plus one equals three."

What will it take for each of us to remember how to dance?

There has never been a more crucial time for us to ask that question. Dave Foreman said, "All of us alive now are members of the most important generation of human beings who have ever lived, because we're determining the future, not just for a hundred years, but for a billion years. When we cut a huge limb off the tree of natural diversity, we're forever halting the evolutionary potential of that branch of life."

Not only is evolutionary potential being destroyed, but as Christopher Manes told me, "We have to realize the same forces that hack up the wilderness, put fences around it, and call this progress also shepherd us down meaningless paths, frustrate our talents, impoverish our internal lives."

I didn't understand that when in 1983 I graduated from the Colorado School of Mines with a degree in Mineral Engineering Physics. I had what it seemed everyone wants: the opportunity for a large paycheck from a not-too-meaningless job. My economic future was secure. Yet I was miserable and wondered if I were insane; thirty thousand per year, as it had been presented to me, was supposed to enable a lot of happiness.

Now I understand there are other models for happiness. There are cultures which exist without war, without rape, without child abuse. Arno Gruen said, "With the Ituris—the pygmies of the former Congo—for example, or many of the Native North or South Americans, what is of central importance is how people feel with one another, how they get along, how they can sustain and help one another." This is true of the Okanagans of British Columbia, who reject the use of aggressive force; as Jeannette Armstrong said, "To us as Okanagans, for a person to do anything in that way is unacceptable, unheard of. Not only is it unethical, but since you have so many other tools, it doesn't make sense."

The recognition that nonaggressive and synergistic ways of living exist raises a whole new series of questions: Is there a relationship between nonaggression and happiness? Why does a culture, or a person, become aggressive? What can those who have survived this aggression teach us? Is it possible to live a life based on cooperation within a culture that is essentially competitive? Where does love fit in?

I believe each of us already knows the answers to these questions. We just need to remember that we know. As Susan Griffin told me, "We are on the verge of destroying material life. One of the ways I think we can avert that is to recognize the profundity of material existence, to deeply respect our own process of knowing in the world, knowing with heart and mind and body and sensuality all at the same time. This involves a deep self-respect for that process of knowledge that has evolved over millions, even billions of years, and it involves entering that process with full passion."

It is possible—indeed imperative—for each of us to enter that "process of knowing in the world," that process of deep listening, with full passion. Linda Hogan said, "You know those moments you have when you enter a silence that's still and complete and peaceful? That's the source, the place where everything comes from. In that space, you know everything is connected, that there's an ecology of everything. In that place it is possible for people to have a change of heart, a change of thinking, a change in their way of being and living in the world." This book has helped point me to that quiet space of connection, and I can only hope it may do the same for its readers.

A Language Older Than Words

Chelsea Green Publishing, 2004

A Language Older than Words was originally supposed to address the fact that many of us experience inter-species communication on a routine basis, but almost none of us talk about it publicly. It was going to be a collection of happy interspecies communication stories, but I realized very quickly that to write a happy book about human and nonhuman relations would at this point be deeply dishonest. Further, a book purporting to show that nonhumans can think and communicate would be deeply bigoted and demeaning, just as if someone wrote a book that shows that blondes can think, or that Jews are not subhuman. It would allow the chauvinism and bigotry of the dominant culture to stand unchallenged.

Instead, I sought to answer the questions, why is it that some of us listen, and some of us don't? Why is it that some people don't care about those they exploit? Why do they exploit? I realized that before you can exploit others you have to silence them. The book exploded at that point into what it became.

There is a language older by far and deeper than words. It is the language of bodies, of body on body, wind on snow, rain on trees, wave on stone. It is the language of dream, gesture, symbol, memory. We have forgotten this language. We do not even remember that it exists.

❁ ❁ ❁

In order for us to maintain our way of living, we must, in a broad sense, tell lies to each other, and especially to ourselves. It is not necessary that the lies be particularly believable. The lies act as barriers to truth. These barriers to truth are necessary because without them many deplorable acts would become impossibilities. Truth must at all costs be avoided. When we do allow self-evident truths to percolate past our defenses and into our consciousness, they are treated like so many hand grenades rolling across the dance floor of an improbably macabre party. We try to stay out of harm's way, afraid they will go off, shatter our delusions, and leave us exposed to what we have done to the world and to ourselves, exposed as the hollow people we have become. And so we avoid these truths, these self-evident truths, and continue the dance of world destruction.

⌘ ⌘ ⌘

As is true for most children, when I was young I heard the world speak. Stars sang. Stones had preferences. Trees had bad days. Toads held lively discussions, crowed over a good day's catch. Like static on a radio, schooling and other forms of socialization began to interfere with my perception of the animate world, and for a number of years I almost believed that only humans spoke. The gap between what I experienced and what I almost believed confused me deeply. It wasn't until later that I began to understand the personal, political, social, ecological and economic implications of living in a silenced world.

This silencing is central to the workings of our culture. The staunch refusal to hear the voices of those we exploit is crucial to our domination of them. Religion, science, philosophy, politics, education, psychology, medicine, literature, linguistics, and art have all been pressed into service as tools to rationalize the silencing and degradation of women, children, other races, other cultures, the natural word and its members, our emotions, our consciences, our experiences, and our cultural and personal histories.

My own introduction to this silencing—and this is similarly true for a great percentage of children as well within many families—came at the hands (and genitals) of my father, who beat my mother, my brothers, and my sisters, and who raped my mother, my sister, and me. I can only speculate that because I was the youngest, my father somehow thought it best that instead of beating me, he would force me to watch, and listen. I remember scenes—vaguely, as from a dream or a movie—of arms flailing, of my father chasing my brother Rob around and around the house. I remember my mother pulling my father into their bedroom to absorb blows that may have otherwise landed on her children. We sat stone-faced in the kitchen, captive audience to stifled groans that escaped through walls that were just too thin.

The vagueness with which I recollect these formative images is the point here, because the worst thing my father did went beyond the hitting and the raping to the denial *that any of it ever occurred*. Not only bodies were broken, but broken also was the bedrock connection between memory and experience, between psyche and reality. His denial made sense, not only because an admission of violence would have harmed his image as a socially respected, wealthy, and deeply religious attorney, but more simply because the man who would beat his children could not speak about it honestly and continue to do it.

We became a family of amnesiacs. There's no place in the mind to sufficiently contain these experiences, and as there was effectively no way out, it would have served no purpose for us to consciously remember the atrocities. So we learned, day after day, that we could not trust our perceptions, and that we were better off not listening to our emotions. Daily we forgot, and if a memory pushed its way to the surface we forgot again. There'd be a beating, followed by brief contrition and my father asking, "Why did you make me do it?" And then? Nothing, save the inconvenient evidence: a broken door, urine-soaked underwear, a wooden room divider my brother repeatedly tore from the wall trying to pick up speed around the corner. Once these were fixed, there was nothing left to remember. So we "forgot," and the pattern continued.

This willingness to forget is the essence of silencing. When I realized that, I began to pay more attention to the "how" and the "why" of forgetting—and thus began a journey back to remembering.

What else do we forget? Do we think about nuclear devastation, or the wisdom of producing tons of plutonium, which is lethal even in microscopic doses for well over 250,000 years? Does global warming invade our dreams? In our most serious moments do we consider that industrial civilization has initiated the greatest mass extinction in the history of the planet? How often do we consider that our culture commits genocide against every indigenous culture it encounters? As one consumes the products manufactured by our culture, is s/he concerned about the atrocities that make them possible?

We don't stop these atrocities, because we don't talk about them. We don't talk about them, because we don't think about them. We don't think about them, because they're too horrific to comprehend. As trauma expert Judith Herman writes, "The ordinary response to atrocities is to banish them from consciousness. Certain violations of the social compact are too terrible to utter aloud: this is the meaning of the word *unspeakable*."

As the ecological fabric of the natural world unravels around us, perhaps it is time that we begin to speak of the unspeakable, and to listen to that which we have deemed unhearable.

A grenade rolls across the floor. Look. It won't go away.

⊛ ⊛ ⊛

Here's what I've heard about your typical slaughterhouse.[1]

The room sounds for all the world like a factory. You hear the clang of steam in pipes and the hiss of its release, the clank of steel on steel as chains pull taut, the whirr of rolling wheels on metal runners, all punctuated every thirty seconds or so by the pop of the stunner.

The rooms are always humid, and smell of grease as much as blood. The walls are often pale, the floor usually concrete. I have a picture from a slaughterhouse that will forever be etched in my mind. No matter how I try to look elsewhere, my eyes return to the newly painted chute that leads in from outside, not only because of the chute's contents, but because the color—electric blue—contrasts almost painfully with the drabness of the rest of the room.

Inside the chute, facing a blank wall stands a steer. Until the last moment he does not seem to notice when a worker places a steam-driven stunner at the ridge of his forehead. I do not know what the steer feels in those last moments, or what he thinks. The pressure of contact triggers the stunner, which shoots a retractable bolt into the brain of the steer. The steer falls, sometimes stunned, sometimes dead, sometimes screaming, and another worker climbs down to attach a chain to the creature's hind leg. Task completed, he nods, and the first worker—the one who applied the stunner—pushes a black button. There's the whine of a hoist, and the

steer dangles from a suspended rail, blood dripping red to join the coagu-
lating river on the floor.

The steer sways as wheels roll along the rail, causing the falling blood
to describe a sinusoidal curve on the way to another worker, who slits his
throat. There is barely time to follow his path before the chute door opens
and another animal is pushed in. There goes the stunner again, the hoist,
metal, steam, the grind of meshing gears. It happens again and again, like
clockwork, every half-minute.

⊛ ⊛ ⊛

We live in a world of make-believe. Think of it as a little game— the only
problem being that the repercussions are real. Bang! Bang! You're dead—
only the other person doesn't get up. My father, in order to rationalize
his behavior, had to live in a world of make-believe. He had to make us
believe that the beatings and rapes made sense, that all was as it should,
and must, be. Now, it will be obvious to everyone that my father's game of
make-believe was far from fun—it was destructive. My father rewrote the
script on a day-to-day basis, thereby making everything right—he *created*
the reality that he *required* in order to continue his behavior.

In attempting to describe the world in make-believe terms, we have forgotten what is real and what isn't. We pretend the world is silent, whereas
in reality it is filled with conversations. We pretend we are not animals,
whereas in reality the laws of ecology apply as much to us as the rest of
"God's Creation." We pretend we are at the top of a great chain of being,
although evolution is nonhierarchical.

Here's what I think: it's a sham. It's a giant game of make-believe. We
pretend that animals feel no pain, and that we have no ethical responsibility
toward them. But how do we know? We pretend that other humans—the
women who are raped, for example (a full twenty-five percent of all women
in this culture have been raped, and an additional nineteen percent have
had to fend off rape attempts)[2]—or the one hundred and fifty million children who are enslaved to make soccer balls, tennis shoes, Barbie dolls, and
the like—are happy and unaffected by it all. We pretend all is well as we
dissipate our lives in quiet desperation.

We pretend that death is an enemy although it is an integral part of life. We pretend we don't have to die, that modern medicine can cure what ails us, no matter what it is. But can modern medicine cure a dying soul?

We pretend that violence is inevitable, and in some ways it is. But can it be mitigated through better science? Rather than answer that question, most often we pretend, sheepishly, that violence doesn't exist.

Science, politics, economics, and everyday life do not exist separately from ethics. But we act like they do.

The problem is not difficult to understand: we pretend that anything we do not understand—anything that cannot be measured, quantified, and controlled—does not exist. We pretend that animals are resources to be conserved or consumed, when, in reality they have purposes entirely independent of us. It is wrong to make believe that people are nothing more than "Human Resources" to be efficiently utilized, when they (we!) too have independent existences and preferences. And it is wrong to make believe that animals are not sentient, that they do not form social communities in which members nurture, love, sustain, and grieve for each other, that they do not manifest ethical behavior.

We act like these pretenses are reasonable, but none of them are intuitive or instinctual; nor are they logically, empirically, or ethically defensible. Taken together, a way of life based on these pretenses is destroying life on this planet.

But a real world still awaits us, one that is ready to speak to us if only we would remember how to listen.

⊕ ⊕ ⊕

When I was a child, the stars saved my life. I did not die because they spoke to me.

Between the ages of seven and nine, I often crept outside at night to lie on the grass and talk to the stars. Each night I gave them memories to hold for me—memories of beatings witnessed, of rapes endured. I gave them emotions too large and sharp for me to feel. In return the stars gave me understanding. They said to me. "This is not how it is supposed to be. This is not your fault. You will survive. *We* love you. You are good."

I cannot overstress the importance of this message. Had I never known an alternative existed—had I believed that the cruelty I witnessed and suffered was natural or inevitable—I would have died.

My parents divorced during my early teens. It was a bitter divorce in which my father used judges, attorneys, psychologists, and most of all money, with the same fury and relentlessness with which he had once used fists, feet, and genitals. The stars continued to foster me, speaking softly whenever I chose to listen.

Time passed, I grew older. I went to college, received a degree in physics, and on my own read a fair amount of psychology. I came to a new understanding of my place in the world. It had not been the stars that saved me, but my own mind. My earlier thesis—that the stars cared for me, spoke to me, held me—made no physical sense. Stars are inanimate. They don't *say* anything. They can't, and they certainly couldn't care about me. And even if they had cared there remained the problem of distance. How could a star a thousand light-years away respond to my emotional needs in a timely fashion? It became clear that some part of my own psyche had known precisely the words I needed to hear in order to endure, and had projected those words onto the stars. It was a pretty neat trick on the part of my unconscious, and this projection business seemed a wonderful adaptive mechanism for surviving in a world that I had come to recognize as largely insensate, with the exception of its supreme tenant—humankind.

⊛   ⊛   ⊛

I've often wished that I could have been in the room when Descartes came up with his famous quip, "I think, therefore I am." I would have put my arm around his shoulder and gently tapped, or I would have punched him in the nose, or I might have taken his hands in mine, kissed him full on the lips, and said, "René, my friend, don't you *feel* anything?"

I used to believe that Descartes' most famous statement was arbitrary. Why hadn't he said, "I love, therefore I am," or "I breathe, therefore I have lungs," or "I defecate, therefore I must have eaten," or "I feel the weight of the quill on my fingers and rejoice in the fact that I am alive, therefore I must be"? Later I grew to see even these statements as superfluous; for anyone living in the real world, life *is*: existence itself is wondrously sufficient proof of its own existence.

I no longer see Descartes' statement as arbitrary. It is representative of our culture's narcissism. This narcissism leads to a disturbing disrespect for direct experience and a negation of the body.

Descartes had been attempting to find one point of certainty in the universe, to find some piece of information he could trust. He stated, "I suppose, then, that all the things that I see are false; I persuade myself that nothing has ever existed of all that my fallacious memory represents to me. I consider that I possess no senses; I imagine that body, figure, extension, movement and place are but the fictions of my mind. What then can be esteemed as true? I was persuaded that there was nothing in all the world."[3] Estranged from all of life, Descartes thought that everything was a dream, and he the dreamer.

You may have played this game, too. During tenth grade I occasionally bedeviled a friend of mine by saying, "Jon, the entire world doesn't exist. You'll be glad to know that includes you. You are nothing more than a figment of my imagination. Because you don't exist, everything you do is a result of my having willed it." Since Jon was a good friend, and because we were high school sophomores, his response was a fairly straightforward sock in the arm. I then countered by smiling and saying, "I willed you to do that." He'd throw a couple more jabs for good measure, and then we'd go to the gym and shoot baskets.

I guess Descartes didn't have a close friend with Jon's good sensibilities. So, instead of going to play basketball, he found himself pushing his philosophy of narcissism to its logical, albeit empty, conclusion. He realized that since he was thinking his thoughts—because he was doubting the existence of the universe—then he must exist to be doing the doubting. "I think, therefore I am." So far, so good. But as Descartes continued his line of reasoning, the world congealed for him into two groups, the thinker, in this case Descartes (or more precisely his disembodied thought processes), and that which he thought (i.e., everything and everyone else). He who matters, and that which doesn't.

Had Descartes stopped there, the response by other philosophers would probably have been similar to Jon's: a violent backlash at having been philosophized out of subjective existence. But he didn't. He and many other philosophers eventually agreed that subjective personhood should certainly be granted to all of them, as well as to others with political, economic, or military power, while they decided that just as certainly it should not be granted to those who could not speak, or at least those whose voices they chose not to hear.

The latter group of course included women: "Let the woman learn in silence with all subjection. But I suffer not a woman to teach, nor to usurp authority over the man, but to be in silence."[4] It also included Africans, because they were "extremely ugly and loathsome, if one may give the name of Men to such Animals,"[5] and because "when they speak they fart with their tongues in their mouths."[6] But the bottom line was that these thinkers thought it was "a great pity that such creatures as they be should enjoy so sweet a country."[7] The subjective persons—those who actually existed—set out immediately to rectify this situation by exterminating these "creatures" and appropriating their land. The same logic was used to deal with Native Americans, who also occupied land the Europeans wanted.[8] It was ethical to steal their land because they were "animals who do not feel reason, but are ruled by their passions," and who "were born for forced labor."[9] It included non-Christians, whose poor choice of religion meant they were not fully human, and so could be enslaved. It included children born to non-Christians, whose poor choice of parents meant they too were not fully human, and so too could be enslaved. The definition of those precluded from being fully subjective and rational beings included *anyone* whom those in power wished to exploit.

Regarding the world of nonhumans (i.e., "animals") we find a contemporary of Descartes who reported that "scientists administered beatings to dogs with perfect indifference and made fun of those who pitied the creatures as if they felt pain. They said the animals were clocks; that the cries they emitted when struck were only the noise of a little spring that had been touched, but that the whole body was without feeling. They nailed the poor animals up on boards by their four paws to vivisect them to see the circulation of the blood, which was a great subject of controversy."[10]

Searching for certainty, René Descartes became the father of modern science and philosophy. Even if his philosophy were not such an easy justification for exploitation his search was fatally flawed before it began. Because life is uncertain, and because we die, the only way Descartes could gain the certainty he sought was in the world of abstraction. By substituting the illusion of disembodied thought for experience (disembodied thought being, of course, not possible for anyone with a body), by substituting mathematical equations for living relations, and most importantly by substituting control, or the attempt to control, for the full participation in the wild and unpredictable process of living, Descartes became the prototypical modern man. He also established the single most important rule of Western philosophy: if it doesn't fit the model, it doesn't exist.

Welcome to industrial civilization.

✾ ✾ ✾

I do not know what my father was thinking or feeling during those days and nights of violence when I was young. I do not know what was in his heart or mind as he cocked his fist to strike my sister, or as he lunged across the table at my brother, or as he stood beside my bed and unzipped his pants. Throughout my childhood an unarticulated question hung in the air, then settled deep in my bones, not to be defined or spoken until it had worked its way back to the surface many years later: If his violence isn't making him happy, why is he doing it?

I will never know what my father was feeling or thinking during those moments. For him, at least consciously, the moments don't exist. To this day and despite all of the evidence, he continues to deny his acts of violence. This is often the first response to the undeniable evidence of an awful truth; one simply denies it. This is true whether the evidence pertains to a father's rape of his children, the murder of millions of Jews or scores of millions of indigenous peoples, or the destruction of life on the planet.

I would imagine this denial of evidence is often unconscious. My father is not the only person in my family whose recollection for those years is unaccountable. As he leapt across the table, do you know what I did? I continued eating, because that is what you do at the table, and because I did not want to be noticed. I ate, but I do not know what I felt or thought as I brought the sandwich to my mouth, or the spoonful of stew, or the bean soup.

I do not know how I arrived at it, but I do know that I had a deal with my unconscious, a deal that has been made in one form or another by nearly everyone living in our culture. Because I was spared the beatings,

I pretended—*pretended* is not the right word, perhaps it would be more accurate to say I *made believe* because the process became in time virtually transparent—that if I did not consciously acknowledge the abuse, it would not be visited directly on me. I believed that if I focused on my own moment-to-moment survival—on remaining motionless on the couch, or forcing beans down a too-tight throat—then my already untenable situation would get no worse.

My father's first visit to my bedroom did not abrogate the deal. It couldn't because without the deal I could not have survived the violence he did to me, just as I'm sure that without a similar deal, that removed *him* from *his* own experience, my father could not have perpetrated the violence. In order to maintain the illusion that if I ignored the abuse I would be spared the worst of it—in order to maintain the illusion of control in an uncontrollably painful situation, or simply to stay alive, even if I had to divorce myself from my emotions and bodily sensations—the events in my bedroom necessarily did not happen. His body behind mine, his penis between my legs, these sensations and images slipped in and out of my mind as easily and quickly as he slipped into and out of my room.

⊕ ⊕ ⊕

It's probably best if you don't believe a word I say.

What I wrote about my father beating and raping us simply isn't true. I was not only wrong, I was lying. My childhood was nothing like that, because if it had been, I couldn't have survived. No one could survive that. So the truth not only *is*, but especially *must be* that my father never chased Rob around the house, and my mother and sisters never threw pans and glasses of water on him trying to make him stop. That would all have been just too implausible. Oh, he may have gotten a little out of control when he spanked one or the other of us, but he never beat anyone to the ground, then kicked her again and again. And rape? Out of the question. The constant insomnia, the incessant nightmares, the painful and itching anus, all these had their origin in some source other than my father. The same was true for my nightly ritual of searching my room, and later, barricading my door. Doesn't every child have a terror of someone catching him asleep?

I do not remember—I *specifically* do not remember—sitting at the table for dinner early one summer evening, and I do not remember my father asking my brother where he was the night before. I don't recollect if my brother said he went to an amusement park. But if my brother *had* said that, my father would never have asked him how much it cost to get in. And most certainly if my brother *had* said an amount, in response to this question that was never asked, my father would not have lunged at him across the table, not even if my brother's answer was incorrect, meaning my brother had not gone to the amusement park but instead perhaps to a bar. Food would not have scattered. My brother would not have made a break for the door, only to be cut off by the bottleneck at the refrigerator. My father would never have called him a cocksucking asshole stupid fuck, nor would he have begun to pummel him. My sisters would not have screamed, and my mother would not have clutched at my father's back. My brother would not have broken free only to stumble, fall, and get kicked in the kidneys. None of this happened. None of it could have happened. I swear to you. My brother could not have made it to his feet, and made it out the door and to his car, a pink Camaro, if you can believe *that*. My brother would not have locked the doors, and even if he had it would never have occurred to my father to kick in the side of the car. And even if by some strange chance all this did happen, I can tell you for certain that I do not remember continuing to sit at the table, a seven-year-old trying desperately not to be noticed, trying to disappear.

I can tell you for certain also that I was never, even as a young child, awakened and summoned to the living room to watch someone get beaten. This did not happen daily, weekly, or even monthly. And even had the beatings occurred—which I need to reassure both you and me that they did not—they could never have been made into such a spectacle. Who could endure such a thing? And who could perpetrate it? I have no recollection of sitting frozen on the couch, eyes directly forward, feeling more than seeing my siblings near to me, none of us touching, none of us moving, none of us making a sound, each of us simultaneously absent and preternaturally present, hyperaware of every one of my father's movements. I do not remember my father's leg frozen in mid-kick, nor can I see his face closed off with fury. I recollect nothing of this. Because it didn't happen. My brother doesn't have epilepsy, and if he does it could not have been caused by blows to the head. My sister never wakes up screaming that someone is in her room, in her bed. She never fears that someone will step out from behind a door to hit her, or to push her onto a bed. The smell of alcohol on a lover's breath does not terrify me, because my father did not drink. And even if he had, he would never have become drunk. And even if he would have become drunk, he would never have entered my room.

And the worst of it all is that even if he would have, I would never have remembered a thing.

Do not believe a word that I write in this book, about my father, about the culture, about anything. It's much better that way.

⊛ ⊛ ⊛

A study of Holocaust survivors by the psychologists Allport, Bruner, and Jandorf revealed a pattern of active resistance to unpleasant ideas and an acute unwillingness to face the seriousness of the situation.[11] As late as 1936, many Jews who had been fortunate enough to leave Germany continued to return on business trips. Others simply stayed at home, escaping on weekends into the countryside so they would not have to think about their experiences. One survivor recollected that his orchestra did not miss a beat in the Mozart piece they were playing as they pretended not to notice the smoke from the synagogue being burned next door.

And what do we make of the good German citizens who stood by? By what means did they suppress their own experiences and their own consciences in order to participate or (similarly) not resist? How did they distract themselves from the grenade that slowly rolled across the floor?

Think for a moment about the figure I gave earlier: twenty-five percent of all women in this culture are raped during their lifetimes. One out of four. Next, think for a moment about the number of children beaten, or of the one hundred and fifty million children—*one hundred and fifty million*—enslaved, carrying bricks, chained to looms, chained to beds.[12] If you were not one of the women raped, if you were not one of the children beaten, if you were not one of the children enslaved, these numbers probably don't mean very much to you. This is understandable. Consider your own life, and the ways you deny your own experience, the ways you have to deaden your own empathies to get through the day.

We live our lives, grateful that things aren't worse than they are. But there has to be a threshold beyond which we can no longer ignore the destructiveness of our way of living. What is that threshold? One in two women raped? Every woman raped? 500 million children enslaved? 750 million? A billion? All of them? The disappearance of flocks of passenger pigeons so large they darkened the sky for days at a time? The death of salmon runs so thick that it was impossible to dip an oar without "striking a silvery back"? The collapse of earthworm populations?

This deal by which we adapt ourselves to the receiving, witnessing, and committing of violence by refusing to perceive its effects on ourselves and on others is ubiquitous. And it is a bad deal. As RD. Laing has written about our culture, "The condition of alienation, of being asleep, of being unconscious, of being out of one's mind, is the condition of the normal man. Society highly values its normal man. It educates children to lose themselves and to become absurd, and thus to be normal. Normal men have killed perhaps 100,000,000 of their fellow normal men in the last fifty years."

The question still hangs heavy in the air: If our behavior is not making us happy, why do we act this way?

☙ ☙ ☙

The zoologist and philosopher Neil Evernden tells the familiar story of how we silence the world. During the nineteenth century, many vivisectionists routinely severed the vocal cords before operating on an animal. This meant that during the experiment the animals could not scream (referred to in the literature as emitting "high-pitched vocalization"). By cutting the vocal cords experimenters simultaneously denied reality—by pretending a silent animal feels no pain—and they affirmed it by implicitly acknowledging that the animal's cries would have told them what they already knew, that the creature was a sentient, feeling (and, during the vivisection, tortured) being.

As Evernden comments, "The rite of passage into the scientific," or, I would add, modern, "way of being centered on the ability to apply the knife to the vocal cords, not just of the dog on the table, but of life itself. Inwardly he [the modern human being] must be able to sever the cords of his own consciousness. Outwardly, the effect must be the destruction of the larynx of the biosphere, an action essential to the transformation of the world into a material object."[13] This is no less true for our relations with fellow humans.

If we are to survive, we must learn a new way to live, or relearn an old way. There have existed, and for the time being still exist, many cultures whose members refuse to cut the vocal cords of the planet, and refuse to enter into the deadening deal which we daily accept as part of living. It is perhaps significant that prior to contact with Western Civilization many of these cultures did not have rape, nor did they have child abuse (the Okanagans of what is now British Columbia, to provide just one example, had neither word nor concept in their language corresponding to the abuse of a child. They did have a word corresponding to the violation of a woman: literally translated it means "someone looked at me in a way I don't like"[14]). It is perhaps significant as well that these cultures did not drive the passenger pigeon to extinction, nor the salmon, the wood bison, the sea mink, the Labrador heath hen, the Eskimo curlew, the Taipei tree frog. Would that we could say the same. It is perhaps significant that members of these cultures listen attentively (as though their lives depend on it, which of course they do) to what plants, animals, rocks, rivers, and stars have to say, and that these cultures have been able to do what we can only dream of, which is to live in dynamic equilibrium with the rest of the world.

The task ahead of us is awesome, to meet human needs without imperiling life on the planet.

⊛   ⊛   ⊛

Every morning when I wake up I ask myself whether I should write or blow up a dam. Every day I tell myself I should continue to write. Yet I'm not always convinced I'm making the right decision. I've written books and I've been an activist. At the same time I know neither a lack of words nor a lack of activism kills salmon here in the Northwest. It is the presence of dams.

Anyone who lives in this region and who knows anything about salmon knows the dams must go. And anyone who knows anything about politics knows the dams will probably stay. Scientists study, politicians and businesspeople lie and delay, bureaucrats hold sham public hearings, activists write letters and press releases, I write books and articles, and still the salmon die. It's a cozy relationship for all of us but the salmon.

I don't like it. I do not wish to merely describe the horrors that characterize our culture; I want to stop them. Sometimes it seems to me terribly self-indulgent to write, to shuffle magnetically-charged particles on a hard drive, when day after day it's business as usual. Other times it seems even worse, as if the flow of words were not merely self-indulgent, but an act of avoidance. I could be blowing up dams. I could be destroying the equipment used to deforest our planet. I could be physically stopping perpetrators of abuse. How many social critics, I often wonder, how many writers, really want to stop the cycle, bring down this culture of death? How many have found a way to make a comfortable living while comforting themselves with beautiful descriptions of nature and the occasional outburst of righteous indignation?

The world is drowning in a sea of words, and I add to the deluge, then hope that I can sleep that night, secure in the knowledge that I have "done my part." Sometimes I don't know how we all live with ourselves. What can I say that will give sufficient honor to the dead, the extirpated, the beaten, the raped, the little children? I don't know.

In the ten minutes I have stared at this computer screen, trying to fashion a conclusion to this section, more than sixty women have been beaten by their partners, and twelve children have been killed or injured by their parents or guardians. At least one species of plant or animal has been permanently eradicated from the face of Earth, and approximately a square mile of the planet has been deforested.

In the time it took me to write this last sentence, another woman was beaten by her lover.

Ⓜ Ⓜ Ⓜ

My mother has often stated she wishes my father were dead. This seems reasonable to me, not only because of the pain he caused her and her children, but also because it would stop at its source the rolling wave of pain he leaves in his wake.

My own wish for him would be that he live in the full understanding of the damage he has caused. Better minds than my own have pointed out that this is the psychic meaning underlying the Christian notion of Hell. Remove Hell from its literal interpretation, which trivializes the profound psychic content in order to create yet one more means to control people ("Give up your land-based religion and accept Jesus Christ as your personal savior or you'll roast in hell"), and what remains is precisely what those like my father—those who would destroy—lack, which is an honest appreciation of their actions. Another way to say this is that for someone who is destructive, for someone who is controlling, for someone who is civilized (and in more general terms, for *anyone*), Hell is the too-late realization that everything and everyone are interdependent. This realization is our only salvation.

ⓒ ⓒ ⓒ

Most everyone I know speaks openly about the clear and present collapse of industrial civilization in one-on-one conversations. This is true not only for my friends, who are mainly writers, activists of one sort or another, or revolutionaries, but also for people I encounter on buses or planes, in airports, and so on. It's also true that almost without exception the people I know most intimately speak of the certainty that unless it is stopped, our culture will destroy every living being on the planet. Once again, they say this only in private.

I once stood behind a woman and her little boy in line to board an airplane. He looked up at her. "What if the plane crashes?"

"Shhh," she said, "we don't *talk* about that."

The price of admission to public discourse is an optimistic denial pushed to absurd lengths. I live less than three miles from the Spokane River, which begins about forty miles east of here as it flows out of Lake Coeur d'Alene. Lake Coeur d'Alene, one of the most beautiful lakes in the world, is also one of the most polluted with heavy metals. There are days when more than a million pounds of lead drains into the lake from mine tailings on the South Fork of the Coeur d'Alene River. Hundreds of migrating tundra swans die here each year from lead poisoning as they feed in contaminated wetlands. Some of the highest blood-lead levels ever recorded in human beings were from children in this area. Yet just last summer the *Spokesman-Review*, the paper of record for the region, wrote that concern over this pollution is unnecessary because, in their words, "there are no human bodies lining the Spokane River."

The resemblance between this behavior—a steadfast refusal to acknowledge physical reality—and my own denial as a child is frightening. I see myself at the kitchen table, bringing the spoon to my mouth with a mechanical precision that would have made Descartes proud. I see my father by my bed, a dark figure on a background nearly as dark, and in that one so-brief instant of awful recognition, I feel my consciousness slip away—*This cannot happen. This cannot happen to me*—quickly, like running footfalls down a distant corridor, or like the last bit of water sucking down an open pipe. As my consciousness disappears, so, too—poof—does my father. Poof, no more father, no more rape. Poof, no more clearcuts, no more lead, no more crash. Suddenly for all our claims to rationality we are, each and every one of us, as much out of our minds as we are out of our bodies. Poof.

Just today a friend told me she used to date a man who hunted. She hated the fact that he killed.

"Do you eat meat?" I asked.

"Yes," she replied, "but I don't have to see them die."

A moment's pause, and she added, "I can't believe I just said that." Another pause, and we both laughed.

Poof.

A few weeks ago I participated in a conference of about twenty-five environmentalists and small farmers. For three days we attempted to name values we hold in common and in opposition to each other. The purpose was to begin a dialogue between these two beleaguered groups, which may lead to better working relationships as we both try to stop the destruction of family farms and farming communities by transnational agribusiness corporations.

One of our exercises was to pretend that the year was 2018, and that somehow our culture had undergone a revolution in values such that we were now living sustainably. We wrote what we believed sustainable communities and farms would look and feel and smell like, what technologies would be used, and so on.

I don't know whether it broke my heart more to perform the exercise or for the group to share the results. This was due in part to the fact that no mention was made, either in the setup of the exercise or in the answers, of the nearly insuperable physical difficulties we face—for example, the fact that those in power control guns, tanks, airplanes, biological and nuclear weapons, as well as all major media outlets, and have shown themselves time and again more than eager to use these various tools to destroy any perceived threat. Nor did anyone mention the probably unconscious and certainly irrational imperative that drives all of the destruction. My discomfort arose primarily because even when we spoke of technology; no one mentioned the crash. We spoke much of "appropriate" or "friendly" technologies, but we did not define either one, nor did we mention how or why people would implement these technologies. Finally I could hold back no longer.

"Everyone here knows industrial civilization isn't sustainable. We all know that any technology that relies on the use of non-renewables is by definition not sustainable. We also know that by definition, any technology or activity that damages any other community—human or nonhuman—isn't sustainable. Finally, everyone here knows that there's no way within the next twenty years we'll make a transition to a technologically sustainable culture. The best we can hope for is that we begin to throttle down our overblown technology, to bring ourselves to a soft landing instead of a full crash."

Everyone seemed to agree, and it came clear to me that while these thoughts had probably occurred to nearly everyone in the room, no one else had been willing to speak until someone broke the ice. Poof.

இ இ இ

When dams were erected on the Columbia, salmon battered themselves against the concrete, trying to return home. I expect no less from us. We too must hurl ourselves against and through the literal and metaphorical concrete that contains and constrains us, that keeps us from talking about what is most important to us, that keeps us from living the way our bones know we can, that bars us from our home. It only takes one person to bring down a dam.

இ இ இ

For years I've been haunted by a fantasy involving someone like Jesus. This person—woman or man, it doesn't matter— comes into a community and talks about love. She, or he, tells people they should treat each other with respect, and that this respect must extend to humans and non-humans alike. A crowd gathers as this person says they should do unto others as they'd have others do unto them, and they begin to murmur quietly as they hear that they should share with each other everything they own. The discomfort of especially the crowd's children grows more noticeable as this stranger tells them they should love each other, love the land. (He or she says nothing about loving the enemy.) The children hide giggles behind their hands, and now even the adults bite the insides of their cheeks. Finally, after much hesitation, one of the community members responds, "Friend, we respect what you have to say, and thank you for telling us, but can't you tell us something we don't already know?" The stranger looks closely, and seeing the obvious well-being of the people, realizes that her (or his) words are redundant. The stranger merges into the community, and all continue with the dailiness of their lives.

The reality of our Judeo-Christian culture is of course far different. A primary purpose of Judeo-Christianity has not been to move us toward a community where the teachings of someone like Jesus—simple and necessary suggestions for how to get along with each other—are made manifest in all aspects of life, but instead to provide a theological framework for a system of exploitation. Easy as this is to say, not many people say it (at least in public). It is more convenient for exploiter and exploited alike to pretend their parasitic relationship is Natural, ordained by God. It is easier to believe in a logic that leads directly from original sin to totalitarianism— *Because human beings are selfish, evil creatures, they must be controlled; therefore might, guided by an all-seeing God as interpreted by an elite priesthood, makes right*—than it is to take responsibility for one's own actions, and to fight for egalitarianism. It is easier to listen to the voice of God than it is to listen to the voice of one's conscience, suffering, and outrage. And it is easier to follow the well-worn yet faulty logic leading once again from original sin this time to apocalypse—*Because human beings are evil, and have sinned, they must die. All beings on earth die. Therefore, all beings on Earth must be evil, and must have sinned Death is the flower of sin. To avoid death requires the annihilation of evil: therefore, all things on Earth must be annihilated*—far easier than it is to accept one's death as natural. It is all so easy, so sanctimonious, to shift responsibility for your own choices and their consequences onto the divine plan of some invisible God.

If you feel like raping a woman, don't just do it; have your God decree that under some circumstances such behavior is not only acceptable, but righteous, your God-given right: "And seest among the captives a beautiful woman, and hast a desire unto her then thou shalt bring her home to thine house. . . . If thou have no delight in her, then thou shalt let her go wither she will; but thou shall not sell her for money, thou shalt not make merchandise of her, because thou hast humbled her."[15]

Rape alone is problematic as a method of social domination, in that it only temporarily provides the rapist control; to extend this over time, to permanently "make merchandise of her," you must have your God issue a series of decrees hemming women in, binding them to you as your property. Have your God say (while hoping no one notices your own lips moving) that because Eve listened to the serpent—remember, my father never beat anyone who didn't have it coming—every woman "shall welcome [her] husband's affections, and he shall be [her] master."[16] Have your God say any woman who has sexual intercourse freely will be put to death (her body not being her own). Any man, too, who has sex with another man's wife shall die, because he has diddled with another man's property, although no punishment shall be meted out on the man who has intercourse with an unmarried woman (but we'll put her to death for good measure). Small wonder that one of the daily Orthodox prayers reads: "I thank thee, O Lord, that thou has not created me a woman."[17]

⊕ ⊕ ⊕

It's not unheard of for old trees—big pines, firs, and cedars a thousand years old—to scream audibly when they're cut down. I've heard from loggers that the screams are disturbing at first, but as with anything else, you get used to it.

We've had a long time to get used to the screams. Just as our civilization's expansion is marked by a widening circle of genocide, so too forests and all of their inhabitants precede us. Deserts dog our heels.

The need to deforest started in what used to be the Fertile Crescent of the Middle East, Mesopotamia. The land was fecund, as land so often is before we get our hands on it. Cedar forests stretched so far that no one knew their true size, and sunlight never penetrated far enough to touch the humus that has long since baked, crumbled, and blown away. Forty-seven hundred years ago Gilgamish, ruler of Uruk, a city near the Euphrates River, decided to make a name for himself by building a great city Armed with "mighty adzes" and more importantly with a justification—the promise of "a name that endures"—that would allow him and his cronies to deafen themselves to "the sad song of the cedars" as they cut them down, Gilgamish entered the forest, briefly reflected on its beauty, vanquished its protector, and took what he needed.

There goes the neighborhood! It's not unlike the times my father found fault with one of us—he was right, end of conversation. So too the transformation of wild nature to usable resource marked the end of our conversation with wild nature. The rest has been a steady journey to an all-too-familiar destination, one devoid of life.[18]

The story of this journey is as monotonous in its own terrible way as the story of our culture's genocidal practices, which is not surprising, considering, as we shall eventually see, that they spring from the same hollow impulses. Soon after Gilgamish was history (i.e., dead), the ruler Gudea of the nearby city of Lagash took up the mantle, and built his own city, cutting trees to build temples, and once again, to build a name. Name after name rulers are recorded, building up like silt in streams from the eroded hillsides they left in their paths. And nations, too, rise with the fall of forests and fall when they are gone. Troy, Greece, Lebanon, Rome, Sicily, the trees were cut for the greater good, for ships, for commerce, for this reason or that. Always a reason, always deforestation. France, Germany, Britain, the United States, a sandy thread of dead and dying forests that leads to South America, Siberia, Southeast Asia, and now back to my own home, where the last of the American forests fall.

It is not possible to commit deforestation, or any other mass atrocity—mass murder, genocide, mass rape, the pervasive abuse of women or children, institutionalized animal abuse, imprisonment, wage slavery, systematic impoverishment, ecocide—without first convincing yourself and others that what you're doing is beneficial. You must have, as Dr. Robert Jay Lifton has put it, a "claim to virtue." You must be convinced—as the Nazis were convinced that the elimination of the Jews would allow the Aryan race to thrive; as the founders of Judeo-Christianity were convinced their misogynist laws were handed down not from their own collective unconscious but from the God they could not admit they created; as my father was convinced he was not beating his son but teaching him diligence, respect, or even spelling; as politicians, scientists, and business leaders today are convinced they're not destroying life on earth but "developing natural resources"—that you are performing a service for humankind.

Forests have fallen as surely to these claims to virtue as they have to axes, saws, and feller bunchers. By looking at the successive claims used to rationalize the deforestation of *this* continent, perhaps we can begin to see not only the transparent stupidity of them but further still to the motives that underlie the destruction.

Early European accounts of this continent's opulence border on the unbelievable.[19] Time and again we read of "goodly woods, full of Deere, Conies, Hares, and Fowle, even in the middest of Summer, in incredible abundance," of islands "as completely covered with birds, which nest there, as a field is covered with grass," of rivers so full of salmon that "at night one is unable to sleep, so great is the noise they make," of lobsters "in such plenty that they are used for bait to catch the Codd fish." Early Europeans describe towering forests of cedars, with an understory of grapes and berries that stained the legs and bellies of their horses. They describe rivers so thick with fish that they "could be taken not only with a net but in baskets let down [and weighted with] a stone." They describe birds in flocks so large they darkened the sky for days at a time and so dense that "a single shot from an old muzzle-loader into a flock of these curlews [Eskimo curlews, made extinct by our culture] brought down 28 birds."

The early Europeans faced much the same problem we face today: their lofty goals required the destruction of these forests and all life in them, but they couldn't do it without at least some justification. The first two claims to virtue were the intertwining goals of Christianizing the natives and making a profit. These embodied a bizarre yet efficient exchange in which, as Captain John Chester succinctly put it, the natives gained "the knowledge of our faith" while the Europeans acquired "such riches as the country hath."[20] Both the natives and the "ritches"—including the forests of New England—were quickly cut down.

Soon the claim to Christianization was dropped, and the rationalization became "Manifest Destiny," the tenet that the territorial expansion of the United States was not only inevitable, but divinely ordained. Thus it was God and not man who ordered the land's original inhabitants be removed, who ordered the destruction of hundreds of human cultures and the killing or dispossession of tens of millions of human beings, who ordered the slaughter of 60 million buffalo and 20 million pronghorn antelope to make life tougher. Thus it was God and not man who ordered that the native forests of the Midwest be felled by the ax.

Manifest Destiny as a claim to virtue soon evolved back into the ideal of making money. An enterprise was deemed as good as it was profitable, while domination and control remained safely unspoken. The forests of the Northwest were described by a corporate spokesperson as "a rich heiress waiting to be appropriated and enjoyed."[21] To be honest, not even this claim was new in any meaningful sense, but a mere recycling of the words of our Judeo-Christian fathers—"And seest among the captives a beautiful woman, and hast a desire unto her, then thou shalt bring her home to thine house"—with a substitution of trees for women, whipsaws for penises, and the immutable laws of economics for the immutable laws of God.

That brings us to today. As the effects of industrial forestry on this continent become increasingly clear—fisheries vanish, biodiversity goes monotone, communities fall apart, and rich biomes become tree farms—corporate profitability loses its effectiveness as a claim to virtue. Another claim—jobs—has arisen, but this has no ring of truth in an era of automation, downsizing, and the Asian lumber mill. The search for a different justification begins anew.

Recognizing that the forests of this country are in a state of ecological collapse, the timber industry and the politicians and the governmental agencies that serve it have begun to claim the way to improve the health of these massively over cut forests is, unsurprisingly enough, to cut them down. The government has provided, in the words of one of the industry's Senators, "exemptions from environmental laws for logging needed to improve forest health." The Forest Service has disallowed citizens from purchasing federal timber sales to leave the trees standing, because "then the trees won't get cut down." A clearcut is then rationalized by declaring that "while insect and disease populations are currently at endemic levels, there is a potential for spruce bark beetle populations to reach epidemic proportions." In other words, we must cut these admittedly healthy trees because they might get sick someday. The timber transnational corporation Boise Cascade has run advertisements likening clearcuts to smallpox vaccinations.

It's all insane. It doesn't take a cognitive giant to see that if logging were "needed to improve forest health" there'd be no need to exempt it from environmental laws. The most difficult and disturbing task is to understand how and why, after millenia of deforestation, the destroyers and defenders alike accept each new, ephemeral, transparently false claim to virtue at face value. One reason, of course, is that the pattern itself is horrifying, too terrible to think about. A second reason is that if we allow ourselves to recognize the pattern and fully internalize its implications we would have to change it. And so we propagate, or at least permit the myths. It's called passing the buck.

Rational discussion presupposes rational motivations, yet claims to virtue are always attempts to place rational masks over non-rational urges. This means that to focus on the claims without broadening the debate so that it includes a consideration of the underlying urges is to be irrational and ultimately to fall into the same pattern of destructiveness. Another way to say this is that while the claims themselves possess the veneer of rationality the process is not rational, and cannot be resolved by rational discussion. It can seem rational, but only within a severely distorted, non-rational framework—and then only so long as one doesn't question the framework itself.

Take the doctors at Auschwitz. As has been made clear by Lifton, the physicians working there would not have been effective cogs in the Nazi machine without first being quite certain they acted in the best interests of the world, and even in some cases of the Jews themselves. Some exhibited genuine concern for the well-being of the Jews, but only within the strict confines of the Auschwitz reality. In other words, while refusing to question the justice, sanity, or humanity of working prisoners to death or gassing them in assembly-line fashion, and refusing to question the abysmal conditions under which prisoners were housed, they often did what little was left to alleviate suffering.

One of the most common ways they did this was by preventing out-breaks of typhus, tuberculosis, and other communicable diseases by injecting patients with phenol. Children, adults who had long been on the medical block, and others who were ill or had the potential to become ill were selected for injection. The physician or technician filled the syringe from the phenol bottle and thrust the needle into the heart of the patient, emptying the contents of the syringe. Most patients fell dead almost immediately, although some lived for seconds or even minutes. Just like the Forest Service and timber companies, these physicians were pre-empting outbreaks by killing their patients. This could be rationalized by saying that dead and burned prisoners were no longer infectious risks to the living. Rationale aside, it was murder.

Paradoxically, the way out from these destructive frames of mind is to step in—experience, not thought or rationalization is the only cure-all. Instead of hiding behind notions of racial purity or pretending to prevent epidemics, notice that at this moment I am lifting this boy's arm. He is six. His skin is pale. His eyes lock on mine: he is terrified. I am inserting the needle between his fourth and fifth ribs. It slides in easily. He winces, stifles a sob. I depress the plunger. He stiffens, and before he can fall off the stool my attendant carries him to the back door. The attendant returns, and ushers a woman through the front door. She takes her place on the stool. I begin to lift her left arm. Her eyes, too, lock on mine. I realize, in that instant, that I am the last thing she will ever see.

Trust experience. Descartes' inversion of what is to be believed makes no sense to me, not to any of my senses. Thought divorced from expe-rience is nonsense. I know from my own childhood this divorce can be essential to survival, but paradoxically, it is this same divorce on the part of perpetrators that gives rise to these awful claims to virtue.

I was not at Auschwitz. I wasn't there for the first clearcut, or the second, or the thousandth. I can read about death camps, and I can read about the forest that was turned into this book, but if the story is to mean anything to me, if it is to change my life, it must lead back to my own experience. And it does. From a rocky knoll not far from my home, the knoll where the coyotes came to remind me of our deal, I can see clearcuts white on green on a snowy winter's day. Entering the region's forests I am sure to encounter more stumps, slash piles, and dead hillsides than trees ancient enough to scream as they go down. There's a place I know near Spokane—by no means unique—where clearcuts wrap around a mountain, drop into a valley, climb a nearby ridge, and cut a swath deep into the next watershed. Recently I walked those clearcuts, past whitened slash piles of wood cut a dozen years ago and past the green limbs of this year's cut, and in ten consecutive miles I never once came within twenty yards of a live tree.

Do not, however, for a moment believe me. I could be conjuring this clearcut as easily as Gilgamish conjured his claim to virtue. Go look for yourself. Listen to the wind pick up soil that has baked and crumbled. Listen to distant trees speak as they sway, and listen, finally, if you can hear them above the whine of the chainsaws, to their screams as they are felled.

❀ ❀ ❀

The relationship between Judeo-Christianity and exploitation is not so straightforward as I may have made it seem. It would not have been enough for the religion's founders to have simply made up a God and, as with a hand puppet, put words into the Deity's mouth. To become deaf to the clamoring of one's conscience and to the pleas of victims requires more than fabrication: it requires a belief stronger than experience, unshakable by the spilling of blood.

This doesn't mean that those who exploit don't consciously fabricate or lie, for they clearly do. Like the layers of an onion, under the first lie is another, and under that another, and they all make you cry.

In 1939, Germany invaded Poland. Hitler's justification was that Polish troops had already attacked. This was a simple and conscious lie. Beneath and behind this lie, pushing it into action, was the notion that in order to fulfill its destiny the Aryan "race" needed room to expand. I don't know to what degree Hitler believed this; already we are sliding out of the realm of the conscious lie and into the realm of semiconscious justification for unspoken (and often unspeakable) urges. Pursuing this further we find the belief that Aryans were superior to Poles or other Eastern Europeans, and so more deserving of land. Beneath this belief there were undoubtedly others equally absurd, nesting like so many Russian dolls.

Another example: The United States Forest Service regularly uses the presence of armilleria root rot as an excuse to cut trees. This is often a conscious lie: examination by plant pathologists routinely reveals armilleria at or beneath endemic levels. In any case, armilleria is a secondary pathogen, which means disturbances, such as cutting trees, actually increase its prevalence in the remaining roots. So the conscious lie becomes self-fulfilling, and the rot becomes pandemic. And then comes another lie, the only way to improve the health of the forest is to cut it down. Beneath all of these lies is the notion that we can manage the landscape without destroying it. And underneath this? The God-given mandate, the evolutionarily ordained duty, the economic policy, all driven by the notion that we are not normal citizens of this planet, that instead we are the most important creatures—really the only ones who matter—on Earth.

Back to Judeo-Christianity, and the relationship between this religion and exploitation. Many of the men who drafted the Bible, and the men who later helped shape the Christian worldview probably believed, sincerely, that it was their God-given right to rape a woman and their holy duty to silence all women. Don't take my word for this. Let the fathers speak for themselves. Tertullian stated that women were "the devil's gateway," and was in agreement with Ambrose that all evil stems from women.[22] Origen, the father of the Alexandrian church, so hated the flesh that he castrated himself to become, in the words of St. Matthew, one of the "eunuchs for the kingdom of heaven."[23] He stated, "What is seen with the eyes of the creator is masculine, and not feminine, for God does not stoop to look upon what is feminine and of the flesh."

Examples of Christian misogyny are legion: St. Chrysotom said, "What else is woman but a foe to friendship, an inescapable punishment, a necessary evil, a natural temptation, a desirable calamity; a domestic danger, a delectable detriment, an evil of nature." St. Thomas Aquinas, "The voice of a woman is an invitation to lust, and therefore must not be heard in church." St. Augustine got right to the point: "I know nothing which brings the manly mind down from the heights more than a woman's caresses."

This notion that sexuality and women—in fact the earth and all direct experience—bring men down from what is considered most important—the heights to which manly men may attain—is a central theme of our culture.[24] The denigration of the flesh is essential to science, where a body is considered "nothing but a statue or a machine," and where direct experience is considered "mere anecdotal evidence" or noise to be ignored while we search for the real signal. And of course, it is fundamental to Christianity: I recently came across a paragraph in the book *Porneia: On Desire and the Body in Antiquity*, which details some of the methods used by early Christians to try to control the erotic. It states: "Ammonius used to burn his body with a red-hot iron every time [he felt sexual desire]. Pachon shut himself in a hyenas den, hoping to die sooner than yield, and then he held an asp against his genital organs. Evagrius spent many nights in a frozen well. Phioromus wore irons. One hermit agreed one night to take in a woman who was lost in the desert. He left his light burning all night and burned his fingers on it to remind himself of eternal punishment. A monk who had treasured the memory of a very beautiful woman, when he heard that she was dead, went and dippled his coat in her decomposed body, and lived with this smell to help him fight his constant thoughts of beauty." Clearly, these men suffered.

It would be comforting, as always, to believe these are the words of a few sad men. We would, as always, be wrong. Origen, St. Augustine, St. Thomas Aquinas were influential men. And they articulated—as did Descartes, Bacon, Hitler—deep cultural urges in ways that obviously resonated with a great many people.

Take Martin Luther, who wrote, "I would have such venomous, syphilitic whores broken on the wheel and flayed because one cannot estimate the harm such filthy whores do to young men who are so wretchedly mined and whose blood is contaminated before they have achieved full manhood." If his message hadn't resonated, would Luther be known as the father of a church? Would any of these men be esteemed, even beatified? No, their words would have been ignored instead of being translated into action.

By the time Luther gave voice to his hatred, in the sixteenth century, women had been getting burned at the stake for nearly seven hundred years, since the council of Salzburg in 799 C.E. approved the torture of witches. Of course the legalized murder of women goes much farther back. Millions of women—up to twenty percent in many communities—were tortured and killed on the pretense that they were witches, and that they had committed crimes against men.

The extremely influential and popular tome *Malleus Maleficarum*—which in 180 years went through thirty-five editions in four languages—

detailed many of these crimes. Women were murdered, for example, because they did "marvellous things with regard to male organs." It goes on to tell us that women "collect male organs in great numbers, as many as twenty or thirty members together, and put them in a bird's nest, or shut them up in a box, where they move themselves like members, and eat oats and corn, as has been seen by many and is a matter of common report." According to the women's executioners, these women copulated with devils, traveled on clouds, stole milk, rode atop he-goats or broomsticks and so should be tortured and killed—remember my father never beat anyone without good reason.

The belief that men own women continues to permeate our culture. Even today politicians and others cite the Bible frequently to support their view. Recently the local newspaper ran a multi-page profile of a young woman who, according to the profile's first sentence, is a "role model for the entire city."[25] This is not only because as a Christian she decided to abstain from sex until she married, but because she decided until that time never to be alone with a man. When the man who eventually became her husband asked her father, a Christian pastor, for permission to court his daughter, her father refused. The suitor took her acquiescence to her father's wishes as a positive sign: "I knew that how she respected and honored her father was how she would respect and honor me." He asked again in six months, and this time her father agreed to turn his daughter over to the clearly like-minded suitor. After the article ran, the newspaper received many letters praising this woman and her family. Not a single letter commented on the woman's ownership by successive males.

More recently, a drunken woman was raped by two men after having danced an impromptu striptease at a party. The newspaper responded with an editorial entitled, "Act provocatively and you provoke," with the subtitle, "Sometimes, women lead men into temptation."[26] The primary author of the editorial was a self-described fundamentalist Christian. The editorial's last line is, "If a woman demands the right to be a promiscuous fool . . . she shouldn't expect society to embrace her as a victim when she gets burned." The newspaper also published a rebuttal by a solitary dissenting member of the editorial board. Although his editorial did not ascribe blame to the woman (hedging bets, however, by stating that "ultimate accountability for this woman's behavior occurred when she was raped by the men she teased"), he did state that "because two men couldn't control themselves, a rape occurred," implying that rape is a crime of sexual passion, not violence and subjugation. Taken together, the editorials, both written by men, reinforce the Biblical notion that a woman's body is not her own, to do with as she pleases, and also that for men, rape is desirable, a temptation.

How does this come about? How is it that Jesus was a fair feminist for his day—treating women with a deference remarkable for his immersion in a deeply patriarchal culture—yet the religion that bears his name shows no such respect? How is it that pagans, Jews, Muslims, heretical Christians, Indians, Africans, Polynesians, Asians, women, men, children, salmon, forests, have been murdered by the millions in the name of a man who said that people should love their neighbors and love their enemies?

At least part of the answer is that the words of Jesus are ultimately irrelevant to the course of historical Christianity. Far more important to this course are deeply hidden urges that grope not only for expression but also for sufficient pretext to allow fulfillment without acknowledgment, and hence without accountability.

Few people would be senseless enough to believe that my father beat my sister because she found dead puppies in the swimming pool, or for any reason other than those emanating from my father's damaged psyche. He blamed it on the dead puppies simply to confuse us all, himself especially, and to drown out the horrific experience of beating his own child.

Few would be ignorant enough to believe Hitler's justifications for murdering millions of people in Europe and Africa. We can see—probably more easily than he—through and beyond his words to his intent, made deathly clear in the showers at Treblinka and on the stone cold killing fields of the Soviet Union.

By the same token, it isn't difficult to honestly evaluate the insanity and hatred that characterized the witch trials, that conceptualized, then created a suitable political, social, philosophical, and theological context for the torture and murder of women because they were alleged to "do marvelous things with regard to male organs."

How then do we so blind ourselves to the same impulses that surround us today, that are central to, and propel our culture? Do you think today's destruction of the salmon is so much less than last century's destruction of the passenger pigeon? Do you think the enslavement of 150,000,000 children is so much less than the race-based slavery of not-so-long-ago? Is the ongoing genocide of indigenous peoples so much less than the Final Solution, so much less than Manifest Destiny? It is safe to speak of Hitler because he is dead, and because you and I were not there to participate. My father is safely out of my life. None of this emerged in the midst of the beatings.

This fear and hatred of life, shape shifter that it is, stays always one step ahead of our discernment, slipping each time we nearly understand it to faster, more efficient ways to control and then destroy the objects of our hatred, and with them ultimately ourselves.

The patriarchal family gives rise to a patriarchal God, who can be internalized to wield Fatherly control even when the father is absent. When threats wear thin the patriarchal God sends a Son to prove His love. My father always knew exactly how far to push with violence before relenting to confuse with signs of affection, and to get us to agree that our suffering, compared to his own, was nothing. So, too, with Christianity. And now what? Christianity—by now entirely divorced from the teachings of its nominal founder—inevitably gives way to science, an infinitely stronger tool to control and destroy not only humans but the entire planet.

Those who wish to destroy will do so. It really is that simple. Remove the words, and the acts are there. Beatings, rapes, enslavement, sanctified murders in autos-da-fé, or industrialized death-dealing with Zyklon B, chainsaws, driftnets, mink coats, time cards, clocks, protein drinks, satellite surveillance systems, and the soul-murder of lives wasted in quiet desperation.

In the beginning is the urge. In the people who would destroy it is always there. Like poisoned water, it is heavy; like poisoned water, it is ungraspable; like poisoned water, it always seeks for cracks to seep through, to exploit, to wear away, to open; like poisoned water it emerges, and when the vessel breaks, as so often it does, like poisoned water it comes out raging its mantra of death.

⬚ ⬚ ⬚

Make no mistake, our economic system can do no other than destroy everything it encounters. That's what happens when you convert living beings to cash. That conversion, from living trees to lumber, schools of cod to fish sticks, and onward to numbers on a ledger, is the central process of our economic system. Psychologically, it is the central process of our enculturation; we are most handsomely rewarded in direct relation to the manner in which we can help increase the Gross National Product.

It's unavoidable: so long as we value money more highly than living beings and more highly than relationships, we will continue to see living beings as resources, and convert them to cash; objectifying, killing, extirpating. This is true whether we're talking about fish, fur-bearing mammals, Indians, day-laborers, and so on. If monetary value is attached to something it will be exploited until it's gone. This story is oft-repeated and oft-ignored. Take the great auk, also called the spearbill in tribute to its massive bill, and called by the Spanish and Portuguese *pinguin*, which means the fat one, in reference to the soft jumpsuit of blubber that enveloped it. This flightless bird was common throughout Europe, existing side-by-side with humans as far south as the Mediterranean coast of France. By the year 900, the great auk was no longer perceived as a neighbor; it had become a commodity. It was slaughtered commercially for the oil derived from its fat, and for its soft elastic feathers. By the mid-seventeenth century, hyperexploitation had killed all but one of the great auk nesting sites in Europe, and that was destroyed before 1800.[27]

In North America, too, humans coexisted with great auks for thousands of years, perhaps thousands of human generations. But they didn't develop an economics requiring the objectification of all others, and so the relationship continued. Humans smoked auk meat to eat through winter; they ate their eggs; they rendered fat into oil which they stored in sacks made from the birds' inflated gullets; they dried the contents of eggs, then ground them into flour from which they made winter pudding. Humans did all this, season after season, generation after generation, causing no appreciable harm to the birds. I do not know what these humans gave to great auks in return, but I would stake any hope I have for continued human existence on the belief that the humans gave something back to these stately black birds, with their powerful lungs and wings made for diving and undersea propulsion. Perhaps all they gave back was the right for them to be.

The earliest description we have of a North American encounter between

Europeans and great auks ends, as these encounters always do, in tragedy for the natives: "Our two barcques were sent off to the island to procure some of the birds, whose numbers were so great as to be incredible. . . . In less than half-an-hour our two barcques were laden with them as if laden with stones." The next year another chronicler noted, "This Island is so exceedingly full of birds that all the ships of France might load a cargo of them without anyone noticing that any had been removed." Having been noticed by members of our culture, the fate of the great auk was sealed.

They were slaughtered for their meat, which was sold. They were slaughtered for their oil, which was sold. They were slaughtered for their feathers, which were sold. Their eggs were taken for markets in Boston and New York. Wrote an Englishman: "These Penguins are as big as geese and . . . they multiply so infinitely upon certain flat islands that men drive them from hence upon a board into their boats by the hundreds at a time, as if God had made the innocence of so poor a creature to become such an abundant instrument in the sustenation of man."

At last, around the turn of the nineteenth century, bans were placed upon the killing of remnant auk populations. The bans, being as nominal as environmental restrictions are today, were of course ignored, and the last known rookery was destroyed in 1802. But one colony, a tiny one of perhaps 100 individuals, remained, near Iceland. Word of this colony finally reached Europe, and collectors quickly offered a local merchant high prices for eggs. By 1843, most of the birds were gone, and on June 3, 1844, three fishermen killed the last two auks, and smashed the last auk egg.

It would be easy for me to hate that local merchant and his three hirelings for what they did to the world in general, and to me in particular, when they eradicated these creatures. But as with Chivington, Hitler, Descartes, Bacon, the authors of the Bible, "free market" economist Milton Friedman, and so on ad nauseam, these men were not alone. They had, and continue to have, an entire culture for company. A bureaucrat with the Canadian Department of Fisheries and the Ocean stated the matter perfectly. His honesty is frightening: "No matter how many there may have been, the Great Auk had to go. They must have consumed thousands of tons of marine life that commercial fish stocks depend on. There wasn't room for them in any properly managed fishery. Personally, I think we ought to be grateful to the old timers for handling the problem for us."

Any being that sparks economic interest is doomed. Eskimo curlews, passenger pigeons, puffins, teals, plovers, all these and more were exterminated or diminished by the insatiable lust for killing that our economics both rationalizes and rewards.

Sea mink, exterminated for their fur. Beavers, decimated. Wolverines. Fisher, marten, otter. Buffalo, wood bison, pronghorn antelope. Salmon: "A ball could not have been fired into the water without striking a salmon." Cod: "So thick by the shore that we hardly have been able to row a boat through them." Halibut. Herring: "I have seen 600 barrels taken in one sweep of a seine net. Often sufficient salt cannot be procured to save them and they are used as manure." Capelin: "We would stand up to our knees in a regular soup of them, scooping them out with buckets and filling the wagons until the horses could scarcely haul them off the beaches. You would sink to your ankles in the sand, it was that spongy with capelin eggs. We took all we needed for bait and for to manure the gardens, and it was like we'd never touched them at all, they was so plenty."[28]

You or I could catch all the fish we could ever eat, cut all the trees we could ever use, kill all the animals whose skins we could wear, and we still would not destroy the planet. Or rather, we could kill all that is given to us only so willingly as we give back. What the hell use would it be for me to overfish West Medical Lake, where just tonight I caught my dinner? Why would I possibly take every fish? They would rot. It makes more sense to leave them so I can come back next week or next year, or never. Why should I stop them from living out their lives in their own manner?

Right now in the Bering Sea forty-five trawlers, each larger than a football field, drop nets thousands of yards long and catch up to 80 tons of fish per day. These ships can remain at sea for months, catching sea lions, seals, pollock, whales, halibut: anything that crosses their paths. Most of what they catch is not worth any money, so it is simply shredded and dumped back in the ocean. If none of the eighty tons of fish could be converted to cash, no sane people would ever want to kill so many, which is itself powerful support for the thesis that our economic system makes us crazy, or at least manifests prior insanity, or both.

But money doesn't rot. It doesn't swim away to live another day. It doesn't fight back. It doesn't disappear to the bottom of the ocean. It doesn't get eaten by other fish.

Like the Christian heaven far from Earth, money perfectly manifests the desires of our culture. It is safe. It neither lives, dies, nor rots. It is exempt from experience. It is meaningless and abstract. By valuing abstraction over living beings, we seal not only our own fate, but the fates of all those we encounter.

⊛ ⊛ ⊛

I attended a workshop where we performed an exercise entitled "Peace-making and Voluntary Simplicity." We sat in a large circle, candles burning in the center of the room, each person speaking in turn as a "talking stick"—a piece of wood with a feather on one end dangling from a leather thong—was passed hand to hand. As the stick made its way around, I considered what I was going to do or say. My first inclination was to not touch the stick: the person in charge of the exercise was not traditional Indian, and had the night before shown herself willing to exploit indigenous traditions. My second inclination was to simply tell the truth, that I was uncomfortable with our unauthorized use of a symbol belonging to a tradition that has explicitly declared itself off-limits to us.

As the stick came closer, I found myself increasingly agitated, at least as much by what was being said as by the cultural appropriation. Person after person stepped close to the edge of outrage, then stopped to turn their anger and shame regarding our culture on themselves: "Sometimes I find myself getting angry at the heads of corporations or at politicians who design and implement murderous policies. But then I always have to realize that I am part of the problem, because I, too, drive a car. I realize that most of all I need to have compassion for politicians. They must suffer, simply being who they are."

What about compassion for the murdered? The comments around the circle took me back a few years to a panel discussion I heard at an environmental law conference. The panelists were Buddhists, addressing much the same topic, and saying much the same thing. There was talk of compassion for wounded wretches who wound us all, of taking pleasure in the dailiness of our lives, of living simply, but not much talk about how to slow or stop the destruction. Afterwards, a woman from the audience stood to ask her question: "Everything you say makes perfect sense, but what do you do if you are standing in front of someone who is aiming a machine gun at a group of children, or is holding a chainsaw in front of a tree?"

This is the point at which virtually all of our environmental philoso-phizing falls apart. It is the central question of our time: what are sane and appropriate responses to insanely destructive behavior? In many ways it is the only question of our time. Future generations will judge us according to our answers. So often, environmentalists and others working to slow the destruction are capable of plainly describing the problems (Who wouldn't be? The problems are neither subtle nor cognitively challenging), yet when faced with the emotionally daunting task of fashioning a response to these clear and clearly insoluble problems, we generally suffer a failure of nerve and imagination. Gandhi wrote a letter to Hitler asking him to stop com-mitting atrocities, and was mystified that it didn't work. I continue to write letters to the editor pointing out untruths, and continue to be surprised each time the newspaper publishes its next absurdity. At least I've stopped writing to politicians.

It is desperately true that we each need to look inside, to make ourselves right—as a poet friend of mine writes, "The Old One says you must put your house in order before you can have guests"—but it's also true that because we are embedded in and dependent upon this planet, and because we owe the planet our lives for having given us life, and because (one hopes) a deep spring of love lies hidden within us, this making ourselves right, this inner work, if it is to mean anything at all, must of necessity lead us to effective action, to actions arising from the love and responsibility we feel toward our neighbors.

The members of the panel on Buddhism blew it. Each in turn stated that the most important thing is to have compassion for the killer, to see the Buddha-nature in each of us. That was a very fine, enlightened position, I thought, but one that helps neither the children nor the trees, nor for that matter the murderers. Nor, in fact, does it help the bystander. Enlighten-ment as rationalization for inaction. Pacifism as pathology. As Shakespeare so accurately put it, "Conscience doth make cowards of us all."

I mentioned this to George, who has been a Buddhist since his early teens. George's response was even more direct than mine. "That's bullshit," he said. "There's a story that the Buddha killed someone who was going to later be a mass murderer. He did it so that he, instead of the murderer, could take on the bad karma caused by killing. And also, presumably, to save the innocent lives. The appropriate response is to stop the murderer by any means possible, as mindfully and compassionately as you can. If you must use force do so, and if you must kill, do that, too, the whole time being fully aware of the implications of what you're doing."

I related to George a story I once heard of a samurai whose master had been killed, and so who was bound to track down the murderer. For years he followed him, until finally he cornered the man in a room. The samurai raised his sword, and from terror the other man spat in the samurai's face. The samurai held the sword poised, shaking now with anger. Finally he sheathed his sword, wiped his face, and walked away. He could not kill the man in that moment, because had he done so it would have been for the wrong reason.

The stick was close now, only two people away. I didn't know what to do. I thought of a conversation I'd had with Jeannette and one of the Maoris. I said that I feel bad whenever I drive, because I'm adding to global warming. The Maori nodded agreement. So did Jeannette. Then she added fervently, "But you didn't set up the system. Do what you can, but don't identify with the problem. If you internalize what is not yours, you fight not only them but yourself as well. Take responsibility only for that which you're responsible—your own thoughts and actions. You didn't make the car culture, you didn't set up factory farming. Do what you can to shut those things down."

The stick came to me. I took it, despite my earlier misgivings, and suddenly calm, said, "There can be no real peace when living with someone who has already declared war, no peace but capitulation. And even that, as we see around us, doesn't lead to further peace but to further degradation and exploitation. We're responsible not only for what we do, but also for what is in our power to stop. Before we can speak of peace, we have to speak honestly of the war already going on, and we have to speak honestly of stopping, by any and all means possible, those who have declared war on the world, and on all of us. Those who destroy won't stop because we live peacefully, and they won't stop because we ask nicely. There is one and only one language they understand, and everyone here knows what it is. Yet we don't speak of it openly."

I took a breath, then continued, "I have to be honest here. During the reading last night I told you of my childhood, but I didn't tell you this: If I were once again a child, with only the options open to me as a child—in other words no running away to fend for myself—but also knowing what I know now of the futility of trying to talk my father out of his violence, I know I would have killed him. How else do I protect the innocent, the little boy who was me? Pacifist as I am—I've never been in a fistfight, nor even shouted in anger—I still would have killed him. And I don't think that would have been wrong."

I looked at the faces around the room: some people were stunned, some looked away, a few disappeared behind a mask of impassivity, many looked intrigued, and quite a few nodded, eyes fierce with solidarity of understanding. I continued, "The point is that we're all in a room with a cannibal, with a mass murderer, and we need to figure out what to do about it."

⊛　⊛　⊛

Viktor Frankl died yesterday. Although most famous for his book *Man's Search For Meaning*, in which he described his experiences as a prisoner at Auschwitz, and articulated his understanding that those who found meaning in their lives and in their suffering were better able to survive the horrors of the camp, I mention him because of something he said toward the end of his life: "There are only two human races—the race of the decent and the race of the indecent people."

He is right, of course. To restate this in terms of this book's exploration: there are those who listen, and those who do not; those who value life, and those who do not; those who do not destroy, and those who do. The indigenous author Jack Forbes describes those who would destroy as suffering from a literal illness, a virulent and contagious disease he calls *wétiko*, or cannibal sickness, because those so afflicted consume the lives of others—human and nonhuman—for private purpose or profit, and do so with no giving back of their own lives.

There are those who are well, and those who are sick. The distinction really is that stark. Attending to this distinction leads again to the central question of our time, restated: How can those of us who are well learn to respond effectively to those who are not? How can the decent respond to the indecent? If we fail to appreciate and answer this question, those who destroy will in the end cause the cessation of life on this planet, or at least as much of it as they can. The finitude of the planet guarantees that running away is no longer a sufficient response. Those who destroy must be stopped. The question: How?

▢   ▢   ▢

In the 1930s, anthropologist Ruth Benedict tried to discover why some cultures are "good," to use her word, and some are not. She noticed that members of some cultures were generally "surly and nasty"—words she and her assistant Abraham Maslow recognized as unscientific—while members of other cultures were almost invariably "nice."[29]

Benedict is of course not the only person to have made this distinction. The psychologist Erich Fromm found that cultures fell, sometimes easily, into distinct categories such as "life-affirmative," or "destructive." The Zuni Pueblos, Semangs, Mbutus, and others that he placed in the former category are extraordinary for the way in which they contrast with our own culture. "There is a minimum of hostility, violence, or cruelty among people, no harsh punishment, hardly any crime, and the institution of war is absent or plays an exceedingly small role. Children are treated with kindness, there is no severe corporal punishment; women are in general considered equal to men, or at least not exploited or humiliated; there is a generally permissive and affirmative attitude toward sex. There is little envy, covetousness, greed, and exploitativeness. There is also little competition and individualism and a great deal of cooperation; personal property is only in things that are used. There is a general attitude of trust and confidence, not only in others but particularly in nature; a general prevalence of good humor, and a relative absence of depressive moods."

Readers may more closely recognize our own culture in Fromm's description of the Dobus, Kwaikutl, Aztecs, and others he put into the category of "destructive." These cultures, he said, are "characterized by much interpersonal violence, destructiveness, aggression, and cruelty, both within the tribe and against others, a pleasure in war, maliciousness, and treachery. The whole atmosphere of life is one of hostility, tension, and fear. Usually there is a great deal of competition, great emphasis on private property (if not in material things then in symbols), strict hierarchies, and a considerable amount of war-making."

Fromm also defined a third category, "nondestructive-aggressive societies," which included Samoans, Crows, Ainus, and others who are "by no means permeated by destructiveness or cruelty or by exaggerated suspiciousness, but do not have the kind of gentleness and trust which is characteristic of the . . . [life-affirmative] societies."

Given the ubiquity of this culture's destructiveness as well as its technological capacity, there has never been a more important time to ask Ruth Benedict's question: Why are some cultures "good" and others not?

Benedict found that good cultures, which she began to call "secure," or "low aggression," or "high synergy cultures," could not be differentiated from "surly and nasty" cultures on the basis of race, geography, climate, size, wealth, poverty, complexity, matrilineality, patrilineality, house size, the absence or presence of polygamy, and so on. More research revealed to her one simple and commonsensical rule separating aggressive from non-aggressive cultures, a rule that has so far evaded implementation by our culture: the social forms and institutions of non-aggressive cultures positively reinforce acts that benefit the group as a whole while negatively reinforcing acts (and eliminating goals) that harm some members of the group.

The social forms of aggressive cultures, on the other hand, reward actions that emphasize individual gain, even or especially when that gain harms others in the community. A primary and sometimes all-consuming goal of members of these cultures is to come out ahead in their "dog eat dog" world.

It all comes down to how a culture handles wealth. If a culture manages it through what Benedict called a "siphon system," whereby wealth is constantly siphoned from rich to poor, the society as a whole and its members as individuals will be, for obvious reasons, secure. They will not need to hoard wealth. Since this generosity is manifested not only monetarily but in all aspects of life, they will also not need to act out their now-nonexistent insecurities in other ways. On the other hand, if a culture uses a "funnel system," in which those who accumulate wealth are esteemed, the result is that "the advantage of one individual becomes a victory over another, and the majority who are not victorious must shift as they can." For reasons that should again be obvious, such social forms foster insecurity and aggression, both personal and cultural.

⊛ ⊛ ⊛

A few years ago I had the opportunity to ask Grey Reynolds, second-in-command of the Forest Service, "If we discover that industrial forestry is incompatible with biodiversity, what then?" The question was, of course, absurd; I mentioned it that day to a high school jumper I was coaching, who said, "What a stupid question! Everyone knows they're incompatible." Reynolds' non-answer that evening unintentionally validated the teen's response: "What do you want us to do, live in mud huts?"

Pointing out that the needs of mass production are counter to the requirements of a good culture and incompatible with long-term survival doesn't mean I don't like hot showers, baseball, good books, or Beethoven. I wish that the items we produce—the good ones, at least—were separable from the larger processes; I wish we could have hot showers without building dams and nuclear power plants.

On some level, of course that is possible. It wouldn't take long to rig up a system to heat water on my woodstove, then pour it into a reservoir that releases water over my head when I pull a cord. But where do I get the metal and glass for the woodstove? Where do I get the cord, or the reservoir? Where do I get the wood? We seem to have painted ourselves into a corner.

As Lewis Mumford observed, our choices have been grossly limited: "On the terms imposed by technocratic society, there is no hope for mankind except by 'going with' its plans for accelerated technological progress, even though man's vital organs will all be cannibalized in order to prolong the mega-machine's meaningless existence."[30] All is not lost, though, as he also remarked, "But for those of us who have thrown off the myth of the machine, the next move is ours: for the gates of the technocratic prison will open automatically, despite their rusty hinges, as soon as we choose to walk out."

I think he's a bit optimistic. Although it's as possible as it is imperative to throw off the myth of the machine, it's not quite so simple to throw off the machine itself. The modern economy is a complicated web, sticky in every thread, and to disentangle oneself personally is difficult, requiring knowledge (much of it long lost), forethought, effort, vigilance, and access to land. For an entire community to disentangle itself from that web may be well-nigh impossible, given the modern economy's interconnected nature as well as the overpopulation, resource depletion, and environmental degradation that comes with civilization.

Food exemplifies the difficulty of withdrawing from the modern economy, because you can't live without it, and because not many people produce all of their own. And if it's uncommon for a modern person to be food self-reliant, it is almost unheard of for a community to supply all of its own food. That is only recently the case; merely one hundred and fifty years ago here in Spokane, the natives lived self-reliantly and sustainably. One of their staples, for example, was salmon. During the massive runs, people placed boxes tinder falls over which salmon leapt on their way to spawn and die. Some of the salmon fell into the boxes; these, the people who lived here ate, or dried, to eat later. Salmon and human communities coexisted, and could presumably have done so indefinitely. Now, even had the salmon not been killed by the dams that destroyed the Columbia as a free river, there are too many people here in Spokane—300,000 in the county—for the salmon to have supported all of us over the longterm. Nor can the community take other food from the river; signs near the Spokane River warn that its fish—native and introduced trout—are contaminated with PCBs.

I recently had dinner with George. We did not eat fish. Instead we ate at a wonderful Vietnamese restaurant. I had lemongrass chicken with chile, and George had stir-fried vegetables. Both meals were excellent, and both consisted of foods originating far from Spokane. Although we didn't ask the cook where the chicken and other foodstuffs came from, it isn't difficult to construct an entirely plausible scenario. Here it is: the chicken was raised on a factory farm in Arkansas. The factory is owned by Tyson Foods, which supplies one-quarter of this nation's chickens and sends them as far away as Japan. The chicken was fed corn from Nebraska and grain from Kansas. One of seventeen million chickens processed by Tyson that week, this bird was frozen and put onto a truck made by PACCAR. The truck was made from plastics manufactured in Texas, steel milled in Japan from ore mined in Australia and chromium from South Africa, and aluminium processed in the United States from bauxite mined in Jamaica. The parts were assembled in Mexico. As this truck, with its cargo of frozen chickens, made its way toward Spokane, it burned fuel refined in Texas, Oklahoma, California, and Washington from oil originating beneath Saudi Arabia, Venezuela, Mexico, Texas, and Alaska. All this, and I have chickens outside my door.

The making of the vegetarian dish was no less complex. The broccoli in George's stir-fried vegetables was grown in Mexico. The field was fertilized with, among other things, ammonium nitrate from the United States, phosphorous mined and processed by Freeport McMoRan from deposits in Florida, and potassium from potash deposits in Sasketchewan. This potash was processed by any one of the multinational mining, oil, and chemical companies: Texasgulf, Swift, PPG Industries, RTZ, or Noranda. The pesticides we ingested are equally cosmopolitan.

Another company associated with nearly every facet of our meal was AKZO, which has 350 facilities in 50 countries. The meal utilized many of their 10,000 chemical products: chicken vaccines that enable Tyson to keep their operations relatively disease free; automobile coatings; chemicals used in many steps of the agricultural and manufacturing processes, and so on.

This was truly an international meal, and not merely because we ate at a Vietnamese restaurant. The simple pleasure of eating a fine meal is tied to processes involving literally thousands of people working for many companies in numerous countries, manifesting the intricate and interconnected nature of the global economy, which runs like a well-oiled machine.[31]

The processes behind the meal manifest not only the complexity of the modern economy's web, but also its destructiveness. Our meal was tied inescapably to pernicious activities across the globe: Tyson Food's monopoly and "virulently antiunion" attitudes, the unspeakable cruelty and debasement of factory farming, and water pollution in Arkansas; loss of topsoil and the depletion of the Oglala aquifer in Nebraska and Kansas; the indescribable immiseration and debasement of labor exploitation in Mexico; air pollution in Japan; toxic mining wastes in Australia, South Africa, and Jamaica; chemical pollution from refineries in four states, and degradation from oil exploration and extraction in four countries; soil toxification, the poisoning of groundwater, more labor exploitation, and the poisoning of agricultural workers in Mexico; air, water, and ground pollution in the United States and Canada, and so on. The food, the cruelty, the pollution, the exploitation, the debasement—all are tied together in this convoluted web that is the modern economy.

The point is not to confess George's and my own particular hypocrisy, nor to explicitly condemn Tyson, PACCAR, or Freeport McMoRan, although Freeport McMoRan is the single most polluting company in the United States, but instead to point out the interconnectedness of the modern economy and the ubiquity of the destruction it causes. The same exercise could be performed for the clothes we wear (sweat shops in Burma's military dictatorship, cotton pesticides, polypropylene petrochemicals), the houses we live in (formaldehyde in plywood, deforestation, extinction of fish and wildlife), other consumer products (40,000 American workers killed on the job each year), or any other activity that vibrates the strings of the web.

If our emphasis on production requires that resources be funneled toward producers, which seems self-evident; if the funneling of resources toward the already wealthy is a characteristic of a culture in which, as Ruth Benedict observed, "the advantage of one individual becomes a victory over another, and the majority who are not victorious must shift as they can"; if this funneling is also a cause of widespread inequality and insecurity, then it makes sense that our hyperemphasis on production leads to hypermilitarism. The rich have to protect what they've got, and take what they don't. An emphasis on production requires an emphasis on private ownership requires a means to protect this ownership requires, in the end, murder.

You may say it's crazy to suggest that hot showers are predicated on dams, nuclear power plants, hydrogen bombs, and napalm. I'd say it's even crazier to think we've built these things if they aren't necessary for hot showers.

Although it seems clear to me that the two are linked—that is, hot showers, computers, vaccinations, major league baseball games, and compact disks of Mozart on one hand are tied inextricably to global warming, evolutionary meltdown, ubiquitous genocide, institutionalized cruelty to nonhumans, immiseration of the majority ("who must shift as they can"), high rates of incarceration, and NASA space probes on the other (not to mention the designated hitter rule)—it doesn't really matter whether they are or not. Pretend for a moment that they are. Are you going to argue that compact disks are worth genocide? Or to take a "more difficult" dilemma, are you going to suggest that the wonders of modern medicine (available to the few) are worth the immiseration of the majority? To state these trade-offs are fair, as Grey Reynolds seemed to be suggesting, would immediately show that one is not fit to be a member of a functioning community. It would suggest that one has become deafened to the sufferings of others, and to one's own conscience.

Now pretend that they are not linked. We can have hot showers and email and a computer that plays chess without having any of the negative characteristics of our culture. This leads immediately to an even more difficult question: in that case, why the hell the ubiquitous genocide, the mass rapes, the biological meltdown?

The primary link is not causal, in that my hot shower does not lead causally to the showers at Treblinka, but familial, in that my own shower and the other are distant cousins. Both ultimately spring from the same ancestor, which is the need for control, and a willingness to deafen oneself to all other considerations. I'm not talking about the simple act of heating water to pour over oneself: I'm talking about the systematic bending of others—human and nonhuman, animate and "inanimate"—to our will. There is obviously a difference between me taking a shower, and Jews being killed in gas chambers. And there is obviously a difference between hot showers and napalm. I've not said they're identical: they're kissing cousins. One seemingly benign—at least so long as we ignore the death of the salmon from dams, the irradiation of the region from nuclear power plants, the draw-down of the Spokane aquifer from wells, the toxication of the landscape caused by the production of metal and plastic used in plumbing, and so on—and the other obviously malevolent, but not without its uses, as those in power are only too aware.

⊗ ⊗ ⊗

Early on I asked what level of violence it will take to make the destructiveness of our way of living obvious. That's not a fair question, because the relationship between the perpetration or even receipt of violence and the awareness of that violence is not linear. As violence becomes more ubiquitous it also becomes more transparent. If one woman out of four is raped within her lifetime, and two out of five are either raped or fend off rape attempts, is the condition of having been sexually assaulted now normal? Is the condition of being capable of, or even having committed, sexual assault also now normal? Add to this the isolating effects of trauma—the erection of internal walls to keep a dangerous world at bay—and the way these walls later facilitate—at least make possible and in some cases make inevitable—the committing of further violence (emotional, spiritual, and physical), and the result can be no other than a constantly expanding sphere of traumatized, isolated individuals. Those on the inside—the already traumatized—will consume those at the frontier (the new children, the newly contacted indigenous peoples, the newly discovered reserves of exploitable human and nonhuman resources), never once seeing the damage they cause nor the isolation they engender. To see the damage would be to revisit their original insult. If they did that, where would they be, and what would they do?

Monkeys are made permanently insane by their artificial removal from the social embeddedness in which they evolved. These creatures were too fearful to interact normally with other beings, and became the best "monster mothers" the scientists could devise. Isolation leads to psychopathology. Isolated from the rest of nature, isolated from each other by walls of fear, isolated from our own bodies, and isolated most of all from our own horrifying experience, is it any wonder that we are all crazy?

If we have become so inured to the coercion that engulfs, forms, and deforms us that we no longer perceive it for the aberration it is, how much more is this true for our ignorance of the trauma that characterizes our way of life? Salmon are going extinct? Pass the toast, man, I'm hungry. A quarter of a million dead in Iraq? Damnit, I'm gonna be late for work. If coercion is our habitat, then trauma is the food we daily take into our bodies.

I spoke with Dr. Judith Herman, one of the world's experts on the effects of psychological trauma. I asked her about the relationship between atrocity and silence.

She said, "Atrocities are actions so horrifying they go beyond words. For people who witness or experience atrocities, there is a kind of silencing that comes from not knowing how to put these experiences into speech.

At the same time, atrocities are the crimes perpetrators most want to hide. This creates a powerful convergence of interest: no one wants to speak about them. No one wants to remember them. Everyone wants to pretend they didn't happen."

I asked her about a line she once wrote: "In order to escape accountability the perpetrator does everything in his power to promote forgetting."

"This is something with which we are all familiar. It seems that the more extreme the crimes, the more determined the efforts to deny the crimes happened. So we have, for example, almost a hundred years after the fact, an active and apparently state-sponsored effort on the part of the Turkish government to deny there was ever an Armenian genocide. We still have a whole industry of Holocaust denial. I just came back from Bosnia where, because there hasn't been an effective medium for truth telling and for establishing a record of what happened, you have the nationalist governmental entities continuing to insist that ethnic cleansing didn't happen, that the various war crimes and atrocities committed in that war simply didn't occur."

"How does this happen?"

"On the most blatant level, it's a matter of denying the crimes took place. Whether it's genocide, military aggression, rape, wife beating, or child abuse, the same dynamic plays itself out, beginning with an indignant, almost rageful denial, and the suggestion that the person bringing forward the information—whether it's the victim or another informant— is lying, crazy, malicious, or has been put up to it by someone else. Then of course there are a number of fallback positions to which perpetrators can retreat if the evidence is so overwhelming and irrefutable it cannot be ignored, or rather, suppressed. This, too, is something we're familiar with: the whole raft of predictable rationalizations used to excuse everything from rape to genocide: the victim exaggerates; the victim enjoyed it; the victim provoked or otherwise brought it on herself; the victim wasn't really harmed; and even if some slight damage has been done, it's now time to forget the past and get on with our lives: in the interests of preserving peace—or in the case of domestic violence, preserving family harmony— we need to draw a veil over these matters. The incidents should never be discussed, and preferably should be forgotten altogether."

I asked her a question that has bothered, entranced, and terrified me since childhood: "To what degree do perpetrators and their apologists believe their own claims? Did my father really believe he wasn't beating us?"

Her response made clear what I understood as a child: that I'll never know. She responded, "Do perpetrators believe their own lies? I have no idea, and I don't have much trust in those who claim they do. Certainly we in the mental health profession don't have a clue when it comes to what goes on in the hearts and minds of perpetrators of either political atrocities or sexual and domestic crimes.

"For one thing, we don't get to know them very well. They aren't interested in being studied—by and large they don't volunteer—so we study them when they're caught. But when they're caught, they tell us whatever they think we want to hear.

"This leads to a couple of problems. The first is that we have to wend our way through lies and obfuscation to attempt to discover what's really going on. The second problem is even larger and more difficult. Most of the psychological literature on perpetrators is based on studies of convicted or reported offenders, which represents a very small and skewed, unrepresentative group. If you're talking about rape, for example, since the reporting rates are, by even the most generous estimates, under twenty percent, you lose eighty percent of the perpetrators off the top. Your sample is reduced further by the rates at which arrests are made, charges are filed, convictions are obtained, and so forth, which means convicted offenders represent about one percent of all perpetrators. Now, if your odds of being caught and convicted of rape are basically one in one hundred, you have to be extremely inept to become a convicted rapist. Thus the folks we are normally able to study look fairly pathetic, and often have a fair amount of psychopathology and violence in their own histories. But they're not representative of your ordinary, garden-variety rapist or torturer, or the person who gets recruited to go on an ethnic cleansing spree. We don't know much about these people. The one thing victims say most often is that these people look normal, and that nobody would have believed it about them. That was true even of Nazi war criminals. From a psychiatric point of view, these people didn't look particularly disturbed. In some ways that's the scariest thing of all."

"Given the misogyny, genocide, and ecocide endemic in our culture," I said, "I wonder how much of that normality is only seeming."

"If you're part of a predatory and militaristic culture, then to behave in a predatory and exploitative way is not deviant, per se. Of course there are rules as to who, if you want to use these terms, might be a legitimate victim, a person who may be attacked with impunity. And most perpetrators are exquisitely sensitive to these rules."

I wondered out loud, "Why is our behavior so predatory? What are the common factors among predatory cultures?"

"It's interesting," she responded. "The anthropologist Peggy Reeves Sanday looked at data from over a hundred cultures as to the prevalence of rape, and divided them into high- or low-rape cultures. She found that high-rape cultures are highly militarized and sex-segregated. There is a lot of difference in status between men and women. The care of children is devalued and delegated to subordinate females. She also found that the creation myths of high-rape cultures recognize only a male deity rather than a female deity or a couple. When you think about it, that is rather bizarre. It would be an understandable mistake to think women make babies all by themselves, but it's preposterous to think men do that alone. So you've got to have a fairly elaborate and counterintuitive mythmaking machine in order to fabricate a creation myth that recognizes only a male deity. There was another interesting finding, which is that high-rape cultures had recent experiences—meaning in the last few hundred years—of famine or migration. That is to say, they had not reached a stable adaptation to their ecological niche. Sadly enough, when you tally these risk factors, you realize you've pretty much described our culture."

⌖ ⌖ ⌖

If words alone could bring down our culture, I would write them. If actions by themselves would stop the atrocities, I would commit them. If a change of heart would bring back the salmon, I would change my heart again and again and again.

It is not enough at this point (necessary, as they say, but not sufficient) to merely right ourselves from trauma, to dismantle the walls we've so laboriously and necessarily constructed to constrict our broken hearts, and then to try to pick up the shredded and scattered fragments of our experience to reassemble like a precious vase that wont quite go back together no matter how we try, or better, like the lifeless body of a loved one who is never coming back.

My dog was going blind, a dog with whom I had travelled for so many years. Then she was hit by a train. Focused only on me, her friend, farther along the way, she never knew what hit her but merely tumbled off the tracks into the tall weeds by the side. I ran to her, found her, picked her up and could not believe she was dead until I saw the gash that split her, back to belly, white flesh that never had time to bleed. There was nothing I could do except hurl sobs at my stupidity for taking a blind dog for a walk on railroad tracks, and to wish that just this once I could go back to before and this time do it right.

In our case, too, seeing the mistake after it is done is not enough. Nor is wishing it away. Nor, especially in this case, is grieving.

When I took her for that walk, I found myself wandering far ahead of this ancient, arthritic, overweight pup. I worried about a train, but not overmuch, because we were in the midst of a long straightaway, at the far end of which, beyond the dog, was a tight curve. When I saw a train round this corner, I began to run toward the dog. Not because I needed to run, but for the joy of running. Then I saw the train had more speed than I had anticipated, and I ran faster than I ever had. And then I saw her tumble into space.

What do you do—what do you feel—when you see destruction rushing down steel rails toward someone you love, and you see that nothing you can do will stop it? You may be a fast runner, but you cannot outrun a train.

I have read that while every culture has invented comedy, tragedies are the unique invention of civilizations. A hero, doomed, stupid, blind to his own faults, falls quickly—or more accurately inexorably—toward a fate he can neither comprehend nor avoid.

I can try to make right those parts inside of me which should never have been made wrong, and I can grieve losses both inside and out, and I can try for all my life to improve my relationships with those around me, human and nonhuman alike. I can even accept that the oncoming train will most likely crush us—or rather continue to crush us—and will stop only under the weight of our bodies and the gumming of its gears by our flesh.

But I will not give up. I know in my bones what it is like to sit stone-faced and frozen in the face of inevitable evil, and I know in my flesh what it is to lie down and take it. I know also what it is like to resist. I know that I am no longer a child, faced with only the options of a child. I know that I am now an adult, and I know that it is at long last time I began to act like one. It is time for me to fight back.

The Culture of Make Believe

Chelsea Green Publishing, 2004

The Culture of Make Believe *built on Language, but not intentionally. At first it was supposed to be a five-page introduction to an encyclopedia of hate groups. I began with the question, "What's a hate group?" and the answer became a book.*

If you go just by the numbers, you see that the biggest racist, segregationist, organization in the country is not the KKK, but the US judicial and penal system. It has achieved segregation of African American males on a scale the KKK can only dream of. Culture is an exploration of many of the ways hatred can be manifested.

As I wrote in the book, hatred felt long enough stops feeling like hatred; it feels like economics, or religion, or science, or government, or just the way things are.

At first glance, particularly if that glance is cursory, hate groups may seem to have little in common with the culture at large. We're told that hate groups, while on the rise, remain an aberration, and stand in opposition to everything we hold dear. This line of thought holds that while racism may at one time have reigned supreme in this country, we're well on our way to being color-blind and widely tolerant, with the exception of a few white-robed buffoons who chant "White Power" in mush-mouthed accents. No longer, for example, do we allow black men to be lynched with impunity. When a black man was recently dragged to death behind a pickup those guilty faced not only lethal injection, but the scorn of an entire nation.

But the relationship between hate groups and the whole culture isn't so simple as that first glance would lead us to suspect. To attempt to really understand hate groups is to begin to burrow into the culture's soft white underbelly, and to confront painful truths about who we are, what we believe, and how we act.

⊛   ⊛   ⊛

How, exactly, would you define a hate group? The obvious answer—a group of people who promote hate—is slippery. For example, most people would agree that the Ku Klux Klan (KKK) is a hate group, the grandaddy of American racist organizations. But literature from the Knights of the Ku Klux Klan states explicitly that the KKK "is not a hate group, but we are a LOVE group. We are a love group because we LOVE America and we LOVE our people." The literature continues, "We don't want those who are only looking for an outlet for their hatred. Hatred never accomplishes anything. We feel terrible for those who have been victims of non-white crime and anti-white discrimination, but turning your life over to hatred isn't the answer. It will only cause self destruction. On the other hand if you have a deep sense of love for your white brothers and sisters and truly desire for them to have a better life, then your efforts to awaken them to the plot to destroy western Christian civilization will be fruitful. God will bless your efforts which are based on love."[1]

The point seems elementary, but bears stating explicitly: either the KKK isn't a hate group, or we shouldn't blindly rely on a group's self-description. Choosing the former—relying on the group's own rhetoric—means devaluing the definition of a hate group to mean only those groups whose members aren't sophisticated enough to mask their messages of hate behind claims to virtue. Choosing the latter leads to another question: on what *should* we rely? The group's actions? If so, then which actions? Murder? If so, are some murders more important than others? Does the race, gender, or sexual preference of the victim play into the discussion? Or is the simple number of killed more important? How about the motivation of the killers?

The man in Texas who was dragged to death behind a car was killed because he was black. This seems a hate crime, pure and simple. But what if the killers were more sophisticated than these happened to be, and their rhetoric was not then to be trusted? Had these killers masked their motives with a robbery, would it still be a hate crime? What if they honestly believed their motivation was fiscal, yet simply chose, for whatever reason, to rob only or primarily black people, or only or primarily poor people? And say, then, that most of these robbery victims just happened to end up dying? Would these be hate crimes? Of course you could also say it doesn't really matter: economics or hate, the victim is just as dead.

⊗ ⊗ ⊗

Most of the people convicted of Crimes Against Humanity at the main Nuremberg trials for major war criminals in 1946 were precisely those we would have expected: Hermann Goerring, director of the Luftwaffe and founder of the Gestapo; Ernst Kaltenbrunner, head of the Reich Security Office and second in command of the SS after Himmler; Hans Frank, governor of occupied Poland; and others directly responsible for the committing of hate crimes on a continental scale. But one person—Julius Streicher—was convicted and hanged for the crime of running a newspaper.

Of course there's more to it than this. As one of the prosecutors presented it at Nuremberg: "It may be that this defendant is less directly involved in the physical commission of crimes against Jews. The submission of the prosecution is that his crime is no less the worse for that reason. No government in the world . . . could have embarked upon and put into effect a policy of mass extermination without having a people who would back them and support them. It was to the task of educating people, producing murderers, educating and poisoning them with hate, that Streicher set himself. In the early days he was preaching persecution. As persecution took place he preached extermination and annihilation. . . . [T]hese crimes . . . could never have happened had it not been for him and for those like him. Without him, the Kaltenbrunners, the Himmlers . . . would have had nobody to carry out their orders."[2]

⊗ ⊗ ⊗

Perhaps it's not murder as such that defines a hate group, but the attempt to inspire terror in a specific category of victim. That was one purpose, to choose an obvious example, of the Night Riders of the KKK's nineteenth-century incarnation. They visited the homes of black people with the specific intent of, among many other things, terrorizing them into not using their recently-gained right to vote.

Yet here, once again, the definition slips from our grasp. If the definition of a terrorist is anyone who wishes to create terror in a specific category of victim, with the purpose of altering the behavior of the members of that category, does this then mean that anyone who supports imprisonment and especially the death penalty as deterrents to crime is by definition a terrorist (the same question could be asked, then, of anyone who spanks or threatens to spank a child)? Clearly the stated purpose is to terrify a specific group of people into changing their behavior. That's what deterrence is. And given the rates at which blacks, Latinos, American Indians, and so on are imprisoned (and on death row) it could be then argued that much of the United States judicial and penal systems combine to constitute a giant racist, terrorist organization. Simply looking at the numbers, it becomes clear that the judicial and penal systems have achieved the segregation of black males—into prisons—on a scale the KKK and their puny brethren could only dream of.

⊛   ⊛   ⊛

The purpose here is not to blur distinctions between the KKK and the United States government. Clearly there are differences. But what *are* the differences, and what are the similarities?

There are many important distinctions: for one thing, if the KKK had the full resources of the state, they would presumably work hard to lock up the other two out of three young black males who've thus far been able to avoid the judicial system.

Another important distinction: the judicial system is part of the government—*our* government—and when we speak of hate groups we're normally referring to groups acting in opposition to the government, and in opposition to the will of the ultimate governors, the people.

Or are we? To state that governments, and in fact entire peoples, cannot be hate groups pushes us into an absurdity: it would mean that while the American Nazi Party with its paltry membership *is* a hate group, the Third Reich, with its death camps, military aggression, slavery, race-based murder, and other crimes against humanity was not. Given the awesome power of the state to inflict violence and terror on its own and other citizens (think of the Soviet Union), it seems unwise to arbitrarily exclude nations from consideration as hate groups.

But there is another argument against categorizing the judicial and penal systems as hate groups: they're imprisoning only *criminals*, people who've done something to deserve imprisonment. Ostensibly they're not targeting specific races or classes, but statistics as well as racial profiling policies—such as routinely stopping motorists for the crime of DWB, Driving While Black—put the lie to this. All that said, we need to admit that the judicial and penal systems exist to protect all of society. But this argument leads us just as quickly into difficulty: most extralegal lynchings were precipitated by some offense—real or imagined—on the part of the victim, and even the modern KKK states that it, too, is protecting society, in this case against "the plot to destroy western Christian civilization." Don't forget that the Nazi government stated it acted defensively when with due process of law it ordered Jews segregated into concentration camps, or prisons.

Nothing is so simple as at first glance it seems.

⊗ ⊗ ⊗

I have to admit I'm glad I was born white. Very rarely have I received stares of hatred based on my race. When I was a teenager I spent a summer at the University of Southern California, and often wandered off-campus into Watts, where I'd see black men playing basketball. At first I was scared to join them. I was white, they were black, and I'd been told by my television that these men could be dangerous. Finally I asked someone in my dorm—a black man—if he thought it would be okay. I will never forget the pained look he gave me before he said, "You grew up in a white community, didn't you? You don't know anything." Then he turned my insult into a joke. He started laughing and said, "So long as they're holding a basketball, they can't pull a gun on you." After that we began running early mornings through Watts. He cautioned me, playfully poking fun at my racist fears, "Make sure you don't lag behind me. A white man chasing a black man just won't fly here."

I went to play basketball many nights. We had fun. The only night I got worried was during a marathon one-on-one game (to one hundred) with a fellow who kept dashing to his bag to grab what he called "pep pills." I got scared. I'd been warned on TV, after all, about black youths and drugs. He could become violent at any moment. He could try to hook me on something. Finally he asked if I wanted any, and held out his hand. They were jelly beans.

But there've been times I have been hated for my race. Not often, but a few times playing basketball with black men the sport has failed to erase the racial tension, and I've been elbowed a bit harder than necessary, been cursed when I've done nothing to deserve it. A few times among Indians I've received hard stares appropriate of nothing I've said. Other times I've blundered and received stares for insulting them as ignorantly as I insulted my friend at U.S.C.

But the most hateful stares I've received because of my race have been from whites. I was in New Zealand with my friend Jeannette Armstrong, an Okanagan Indian. She was doing some work with Maori friends, and I was there to help. At least one of the Maori disliked me intensely for my race, but on the main they were among the most welcoming people I've ever met. As I went to leave, a couple of them gave me a beautiful black shirt with a Maori flag on the front, with the caption *Maori Rangotiri-tanga—Maori Sovereignty*. I was as proud to wear it as they were to give it to me.

I had never before understood the phrase *nigger lover*, nor the contempt with which such race traitors are regarded. But now at least to some small degree I did. White men and women looked at my face, stared at my shirt, then moved their gaze back up to meet mine, their lips set and downturned, jaws tight, cheeks hard, eyes angry. I knew if given the opportunity, at least some of them would have been eager to hurt me.

I was on my way out of the country. I experienced the hatred only for the few hours it took to get to the airport and fly across New Zealand. After that, on the international flight and now at home, the shirt is no affront to those who don't know what it means.

It's a cliché to say one *cannot imagine* something, but the truth is that it's impossible for me to fathom what it would be like to receive those stares not for a shirt I can remove, but for skin I cannot. I don't know what it would do to my heart to always be so noticed, and so unwelcome, no matter where I went, to be met with contempt or worse every day of the year. I think it wouldn't be very difficult for me to internalize those stares, and to begin even to hate myself.

There's another reason I'm glad I'm white. I enjoy going into the prison to teach, but I enjoy even more the fact that I get to come home when I'm done. My students don't have that luxury.

Although blacks make up only twelve and a half percent of the population of this country,[3] they account for more than forty-five percent of prison inmates sentenced to more than one year.[4] The United States imprisons black men more than six times more frequently than it does whites,[5] and four times more than South Africa did during apartheid.[6] Over thirty percent of this nation's African American males between the ages of twenty and twenty-nine are under criminal justice supervision—awaiting trial, in jails or prisons, or on probation or parole. In some areas the figures are even worse: in Washington, DC, forty-two percent of black men between the ages of eighteen and thirty-five are under criminal justice supervision, and in Baltimore it's *fifty-six* percent. More than half. During the first two years California's three-strikes law was in effect, forty-three percent of those sentenced for a third strike were black, even though blacks account for only seven percent of the population and twenty percent of those arrested. Hispanics, Asian-Americans, and Native Americans are also grossly overrepresented in the prison system.[7]

I teach four classes at the prison. Three of them are in "level four," or maximum security. One is in "level one," or minimum security. Out of about sixty level four students, six are white. Of six students in level one, only one isn't.

I'm glad I was born white.

⊛ ⊛ ⊛

I'm glad I was born a male.

There's something interesting about the rate at which men in prison are raped: it's lower than the rate at which women are raped in the culture at large. Most studies suggest that about twenty-five percent of women in this culture are raped during their lifetimes, and another nineteen percent have to fend off rape attempts.[8] What this means, I suppose among many other things, is that you could say that for women who live in our culture, rape is, as is true of imprisoned males, "a fact of life." Some guy goes to prison, and everyone thinks: "Oh, shit, he's going to get raped." But everyday, women walk down the streets, or stay in their homes, and face that same possibility.

⊛   ⊛   ⊛

I don't know what to make of the fact that when I do a quick Alta Vista search of the Internet using the keyword "rape," I see far more pornography than any other single category of site, e.g., rape crisis hot lines, support groups, scholarly analyses, histories, news, and so on. Pornography makes up more than a third of the total sites. And remember, the keyword here was *rape*, not *sex, body, nude,* or even *vagina, penis, dick, or pussy*; we're talking action, not anatomy.

I visited some of these sites. Leaving aside the more obvious and routine treatment of women as objects to be invaded ("Feeling a little sneaky? Take a tour through a house with live hidden cams. Watch unsuspecting victims get caught! Shower cam. Inside her toilet cam!") I was struck by the sheer number of depictions of outright violence against women accompanied by a correspondingly violent—and sorry to be naïve, but disrespectful—ambience. "Nasty little breeders." "You command the action of these young sluts. Your wish is there [sic] command." "Look at those dirty little Asian sluts." "Fuck this Asian teen in every hole." And there were images of women tied, being struck, having jars or feet or things I couldn't quite figure out what they were being put into their vaginas. Asian teens being "fucked in every hole."

The point here is not to express outrage at the depictions—though that would be easy enough to do—but to point out once again how slippery is our notion of hate. I strongly suspect that if the photos were not of women but instead of members of a "protected class"—imagine, for example, sites with literally tens of thousands of pictures of black men bound because they're black, with captions of "You command the action of these young bucks," or white men gagged because they're white, with captions labeling them "dirty little breeders"—the sites would be categorized as sites promoting hate. The organizations that monitor hate groups would watch these sites closely. But even the most comprehensive hatewatch sites—for example the extraordinary "Hate Directory," which monitors even such obscure sites as American Christian Nationalists CyberMinistries Sodomy Information Center, Grendel's White Power Video Games, and Why Christians Suck—do not count these as hate sites.[9] But truth be told, I've yet to encounter at any of the racist sites 1/100th the crudity or overt violence manifest at these. This is not to say the racist sites aren't hateful, but to point out an obvious blind spot.

This all leads to a slew of questions. The first and most obvious is, why are sites depicting (and even reveling in) violence against women not counted as hate material? The "protected class" argument doesn't actually work in the case of the Hate Directory, because the Directory's "Criterion Statement" reads: "Included are Internet sites of individuals and groups that, in the opinion of the author, advocate violence against, separation from, defamation of, deception about, or hostility toward others based upon race, religion, ethnicity, gender or sexual orientation."[10] The second and in many ways more important and profound question is, why are some forms of hate so transparent to us? The third and perhaps most troubling question is, how many more of these "invisible" forms of hate are there?

And here are two more questions: What *are* these invisible forms of hate? And why are they invisible?

❀ ❀ ❀

It really isn't possible to talk about hate without talking about children as objects of hatred. I know that age isn't a protected class under the Violent Crime Control and Law Enforcement Act of 1994, but bear with me.

Each year an estimated 20,000 Mexican children disappear, many for use as mules to transport drugs inside their bodies, others taken for the harvest of their organs, to be transplanted into children in the United States.[11]

Worldwide, entire economies have been founded specifically on the sexual trade in children. Eight hundred thousand Thai girls and boys work as prostitutes[12] (A brochure distributed in England advertising a Thai resort reads "If you can suck it, use it, eat it, feel it, taste it, abuse it or see it, then it's available in this resort that truly never sleeps").[13] Nearly all of them are enslaved or indentured. A good portion have received death sentences from HIV. There are 1.5 to 2 million child prostitutes in India[14] (those in Bombay, for example, are often held in cages; fifty cents buys half an hour of sex with a twelve-year-old).[15] Five hundred thousand child prostitutes work in Brazil[16] (a child of thirty-five pounds is considered "good" in many mining towns[17]). There are 200,000 child prostitutes in or from Nepal (most of these girls are kidnapped, sold for between forty and a thousand dollars, "broken in" through a process of rapes and beatings, and then rented out up to thirty-five times per night for one to two dollars per man). Between 100,000 and 300,000 children work the sex trade in the United States[18] (one study of US survivors of prostitution found that seventy-eight percent were victims of rape by pimps and buyers an average of forty-nine times per year; eighty-four percent were the victims of aggravated assault; forty-nine percent had been kidnapped and transported across state lines; fifty-three percent were victims of sexual abuse and torture; and twenty-seven percent had been mutilated).[19] On average, a child prostitute services more than two thousand men per year.[20] At least a million new girls per year are forced into prostitution.[21]

Kids are not, of course, injured only through sexual exploitation. A half million children die every year as a direct result of debt repayment by the third world—by the colonies—and eleven million children die annually from easily treatable diseases. This latter has been called by the World Health Organization Director-General "a silent genocide."[22]

This is not counting the children who are simply beaten. In 1993, approximately 614,000 American children were physically abused, 300,000 were sexually abused, 532,000 were emotionally abused, 507,000 were physically neglected, and 585,000 were emotionally neglected. 565,000 of these children were killed or seriously injured. That's just in the United States.[23]

So here's the question: Do all these numbers—or more precisely the reality behind these numbers—imply that we hate children? Perhaps the answer would be more evident if we simply invert the question: "Do we value children?"

The answer, of course, is yes. One to two dollars per fuck, unless we happen to be in the Philippines, in which case it will cost us six dollars to have sex with a six-year-old.[24]

So let me put the question another way: was slavery in the United States based on hatred of the Africans, or was it based on economics?

The problem we have in answering (or even asking) these questions comes from the fact that hatred felt long enough and deeply enough no longer feels like hatred. It feels like economics, or religion, or tradition, or simply the way things are. Rape is not a hate crime because our hatred of women is transparent. Child prostitution is not a hate crime for the same reason that beating a child is not a hate crime, because our hatred of children is transparent. The economic murder of children (or creating the economic conditions for their slavery as prostitutes) is not a hate crime because we've held this hatred long enough to enshrine it into our macro-economic policies.

If we did not hate children, we would not cause or even allow them to be destroyed by any of these means. And if we do not love even our children, what, precisely, can we truly say that we love?

⊛　⊛　⊛

Property is the central organizing feature of our culture. The protection or sanctity of private property—or at least the private property of those in power—informs nearly every decision made by the rulers, and certainly informs the great moral debates of recent history. The question of property was central, of course, to the slave debates of the 1840s and 1850s. In his classic study, *Abolition of Negro Slavery*, Professor of Political Law Thomas Roderick Dew used as his first argument against emancipation, "We take it for granted that the right of the owner to his slave is to be respected, and consequently that he is not required to emancipate him, unless his full value is paid by the state."[25] In *his* defense of slavery, William Harper begins a description of a hypothetical Utopia with the phrase, "Let us suppose a state of society in which all shall have property. . . ."[26] The point is that slaves were property—no more and certainly no less—and a nearly insurmountable philosophical, political, and practical difficulty in even *talking* about their emancipation was the enormous cost of compensating the traffickers in humans for their property. Dew threw out a figure of $100 million in 1832 dollars for the slaves in just Virginia, which compares far too closely with the assessed value of all of the houses and lands in that state at the time: $206 million.[27] In all of the debate, even the most fervent abolitionists objected merely to *humans* as property: it was as unthinkable then as it is now to discuss the morality of property itself.

Property has *always* been the central consideration of the United States government, but it has become even moreso over time. Between the signing of the Declaration of Independence in 1776, to provide just one obvious, and in some ways silly, example (silly because all of the terms are seemingly obvious yet in fact nearly impossible to adequately define) and the passage of the Fourteenth Amendment to the US Constitution in 1868, the unalienable rights with which men are self-evidently endowed by their Creator, and which may not be abridged by the State, changed from "Life, Liberty, and the pursuit of Happiness," to "life, liberty, or property." The Fourteenth Amendment, passed during the KKK's maiden reign of terror ostensibly to protect the rights of blacks from racist state governments, has been used far more often to protect the rights to property: of the Fourteenth Amendment cases brought before the Supreme Court between 1890 and 1910, only nineteen dealt with the rights of blacks, while two hundred and eighty-eight dealt with the rights of corporations.[28]

On the other side of the Atlantic, private property even informed something so open-and-shut as the debates about the safety of chimney sweeps in England: attempts to mandate larger flues to facilitate their cleaning by adults or machines were rejected for decades on the grounds that, in the words of Lord Sydney Smith, one of those opposing the mandate, they "could not be carried into execution without great injury to property."[29] Smith said this after listing five pages of horrors inflicted upon climber boys,[30] and was explicit that the essential reason for opposing the mandate was that property is worth more than life, stating that "it was quite right to throw out the bill for prohibiting the sweeping of chimneys by boys— because humanity is a modern invention; and there are many chimneys in old houses which cannot possibly be swept in any other manner."[31] Influential economist David Ricardo consistently and steadfastly refused to speak out in favor of the climbing boys (although he spoke out on nearly every subject related to the economy); the reason can be inferred from the premise constant to his writings, which was that legislature (including that to protect child laborers) must not be allowed to infringe on the rights of property owners.[32]

I have two questions. The first has to do with David Ricardo. In graduate school I studied economics. My macroeconomics textbook listed Ricardo as one of the three most important classical economists (his work, along with Adam Smith's and John Stuart Mills's, "dominated" the period).[33] Here's the first question: What sort of a culture would value the thoughts of someone who perceives property rights as more important than the health and safety of children (or of anyone, for that matter), and most especially, what sort of a culture would value this person's thoughts highly enough to continue to teach them (with no mention of their inhumanity) a couple of hundred years later? Even more so, what sort of culture would enact policies that actualize this sort of inhumanity?

The second question has to do with me, and with this book. Why am I exploring private property? What is the relationship, if any, between the sanctity of private property and hate? I keep asking myself why a discussion of hate seems to keep leading me to talk about property. On one hand it seems absurd. Hate is an emotion. The sanctity of private property is a belief system. End of discussion. No connection. But there is *something* there. The chimney sweeps. The Middle Passage. Each of these horrors occurs because people—the owners and their employees—value *things*, value money, value property, over living beings. But that still doesn't lead me to the relationship between these atrocities and hate.

Or does it? Is this a case of hatred—hatred of life itself, perhaps?—having been felt long enough that it no longer feels like hate, but tradition—in this case the sacrality of private property?

So far I've been speaking with a straight face about democracy—and the elegancies of life—being made possible for the owning class through the efforts of the enslaved, and I've been quoting without disagreement defenders of this politics and luxury. But maybe now it's time to begin asking some more fundamental questions. What sort of democracy can be based explicitly on the misery of others? And what sort of people would desire and claim a luxury which has as its cost the hopes and lives of, as John Henry Hammond put it, a race of slaves? I include myself in this question, as I sit now in front of my computer made in Thailand, wearing a sweatshirt and a sweater made in sweatshops in Korea (at least the shirt I'm wearing today was made in the United States by a company that does not exploit—too much—its workers). I don't *feel* like I hate the workers in

Thailand and Korea. Truth be told, I feel *nothing* toward them. I don't even know who they are. How, then, do I support their suffering, and in the case of the workers in Thailand, their deaths? It's too easy for me to simply blame it all on our economic system, or to call that system one current in the cultural river of hate. It's too sanctimonious to invoke the defense of the good Germans: I did not know, and to the degree that I did, I was merely following orders. Or just getting along.

The problem isn't property and ownership *as such*. You don't have to own people to misuse them. All you have to do is see them as means to ends, to see the world through the lens of utilitarianism, or instrumentality. Writer John Keeble points out that corporations and hate groups are branches from the same tree, different forms of the same cultural imperative: to rob the world of its subjectivity, to turn everyone and everything into objects.

Here's a new question, or maybe a new version of an old question. Does someone who objectifies the world—perceives the living planet and its members as objects to be used—hate the world, and hate life? Or is it more true that for these people the world (or children, or blacks, or women, or whatever category they (or is it we?) wish to objectify) simply doesn't exist except instrumentally? I have to admit that it seems like a stretch to say that those who objectify the world hate life, but given that our culture clearly objectifies the world, and given that it is rapidly destroying life on the planet, maybe it's not such a stretch after all.

⊛   ⊛   ⊛

One primary argument in favor of chattel slavery was philanthropic: far from being something to be abolished, the argument went, slavery was a positive good, beneficial not only to the masters but most especially to the slaves. Not only did contact (albeit involuntary) with Euro-Americans raise Africans (Indians, Chinese, and so on) out of their squalid life of savagery, but because slaves were expensive they were treated, on the main, better than their free counterparts, the wage slaves.

Early European accounts of their contact with Africans often reveals revulsion on the part of the Europeans (who are strangely silent on what the Africans thought of them). For example, Pyard de Laval wrote in 1610,

"They eat . . . as do dogs . . . they live . . . like animals."[34] In 1616 the Reverend Terry disparagingly commented on the fact that the Khoikhoi of southern Africa dressed in skins, calling them, "Beasts in the skins of men rather than men in the skins of beasts."[35] Frederick Andersen Bolling called them "the most hideous folk that can be found in the world,"[36] and Wouter Schouten got right to the point when he said, "truly they more resemble the unreasoning beasts than reasonable man, living on earth such a miserable and pitiful life, having no knowledge of GOD nor of what leads to their salvation."[37]

Fortunately for the Africans, the Europeans were more than willing to provide salvation, in the form of slavery, which, as William Harper pointed out, "has done more to elevate a degraded race in the scale of humanity; to tame the savage; to civilize the barbarous; to soften the ferocious; to enlighten the ignorant, and to spread the blessings of christianity [no caps] among the heathen, than all the missionaries that philanthropy and religion have ever sent forth."[38] Harper asked rhetorically, "Can there be a doubt of the immense benefit which has been conferred on the race, by transplanting them from their native, dark, and barbarous regions, to the American Continent and Islands?"[39]

I thought of the argument put forward now about how global trade is supposed to benefit members of nonindustrialized nations, and wondered how much things have really changed in the last one hundred and seventy years. Substitute the words *capitalism*, *industrialization*, or *free markets*, for *slavery* and I could find these quotes tomorrow in the newspaper. I wondered also how many of the things I am told are good for me actually are, and how many are nothing more than rationales for exploitation: sweet words to keep me from perceiving my own predicament. It made me wonder also something even worse: what things are done to me and to others in ways I cannot even begin to perceive? Do those in power have my best interests at heart? Did the enslavers really believe their own rhetoric? I'm not sure I want to know the answers.

Although "slavery educates, refines, and moralizes the masses," according to George Fitzugh, by "bringing them into continual intercourse with masters of superior minds, information, and morality,"[40] there was only so much—because of the raw material with which the masters had to work—that slaveowners could hope to accomplish. Without more or less constant supervision, Africans almost always reverted to their prior or natural state of shiftlessness: "We have already seen that the principle of idleness triumphed over the desire for accumulation among the savages of North and South America, among the African nations, among the blacks of St. Domingo, &c, and nothing but the strong arm of authority could overcome its operation," wrote Thomas Roderick Dew.[41] The reason for the triumph of idleness was that, according to Dew, "In dealing with a negro we must remember that we are dealing with a being possessing the form and strength of a man, but the intellect only of a child."[42]

Harper, too, noted that "Slaves are perpetual children,"[43] who needed protection, and Fitzhugh provided the answer as to how this protection should be accomplished: "To protect the weak, we must first enslave them, and this slavery must be either political and legal, or social; the latter, including the condition of wives, apprentices, inmates of poor houses, idiots, lunatics, children, sailors, soldiers and domestic slaves. Those latter classes . . . require masters of some kind, whose will and discretion shall stand as a law to them, who shall be entitled to their labor, and bound to provide for them."[44]

I have spent the past several hours now thinking about the notion that masters "shall be entitled to their labor," and at the risk of overstating, it seems to me that entitlement is key to nearly all atrocity, and that any threat to perceived entitlement will provoke hatred.

The man who flayed the cat presumably felt that his employers were entitled to the cat's skin. Europeans felt that they were (and are) entitled to the land of North and South America. Slave owners clearly felt they were entitled to the labor (and the lives) of their slaves, not only in partial payment for protecting slaves from their own idleness, but also simply as a return on their capital investment. Owners of nonhuman capital today feel they, too, are entitled to the "surplus return on labor," as economists put it, as part of their reward for furnishing jobs, and to provide a return on *their* investment in capital. Rapists act on the belief that they are entitled to their victims' bodies. Americans act as though we are entitled to consume the majority of the world's resources, and to change the world's climate. All industrialized humans act like we're entitled to anything we want on this planet.

Nietzsche wrote, "One does not hate so long as one despises."[45] There seems to me a pretty clear relationship between feeling entitled to exploit someone and despising her or him. My dictionary defines *entitle* as "to qualify (a person) to do something: to give a claim to; to give a right to demand or receive."[46] It comes from the Latin *in-titulus*, to give a title, meaning to honor or dignify with a title. By right of my title as a white man, I have a claim to a black man's labor. Any black man's. By right of my title of a man, I have a claim to a woman's body. Any woman's. By right of having enough money to invest in capital, I have a claim to "surplus" of other peoples' labor. By right of having enough money to buy the rights to land, I have a claim to all of the resources it holds. My dictionary defines *despise* as "to look down upon, to scorn; to disdain; to have a low opinion of; to regard as contemptible." It comes from the Latin *de-specere*, to look down.[47] If I am above, I have claims upon those I look down upon. I am entitled to take from those I despise.

From the perspective of those who are entitled, the problems begin when those they despise do not go along with—and have the power and wherewithal to not go along with—the perceived entitlement. That's where Nietzsche's statement comes in, and that's where hatred of the sort I'm trying to get at in this book becomes manifest. Several times in this book I have commented that hatred felt long and deeply enough no longer feels like hatred, but more like tradition, economics, religion, what have you. It is when those traditions are challenged, when the entitlement is threatened, when the masks of religion, economics, and so on are pulled away that hate transforms from its more seemingly sophisticated, "normal," chronic state—where those exploited are looked down upon, or despised— to a more acute and obvious manifestation. Hate becomes more perceptible when it is no longer normalized. Another way to say all of this is that if the rhetoric of superiority works to maintain the entitlement, hatred and direct physical force remain underground. But when that rhetoric begins to fail, force—and hatred—waits in the wings, ready to explode.

⊛ ⊛ ⊛

Let me put this another way. Pretend that you were raised to believe that blacks—niggers would be more precise in this formulation—really are like children, but strong. And pretend that niggers working for whites is simply part of the day-to-day experience of living. You do not question it any more than you question breathing, eating, or sleeping. It is simply a fact of life: Whites own niggers, niggers work for whites.

Now pretend that someone from the outside begins to tell you that what you are doing is wrong. This outsider knows nothing of the life you live and that your father and his father have lived. To your knowledge, this outsider has never walked the fields and actually watched the slaves work, has never gone over the figures to see that your farm wouldn't be viable without these slaves, and doesn't know the slaves well enough to know that they, too, could not survive without the things you provide for them. Pretend that your slaves listen to this outsider, and because of this, your relationship with them begins to deteriorate, even to the point that you begin to lose money. If it were me—had I been raised under these circumstances and with those beliefs—I think it possible that, once I got over my initial shock at the temerity of this outsider meddling in something that is none of his or her business, I would have become angry, and perhaps, eventually, felt outrage toward this interloper who was threatening to ruin my way of life. Raised in those circumstances, it would have taken more courage than most of us have, I think, to admit that one's way of life is based on exploitation, and to gracefully begin to live a different way.

It's easy enough at this remove to simply say that slaveholders were immoral, and that members of the KKK and other hate groups were a bunch of stupid bigots with whom we have nothing in common. But are you sure? Try this. What if, instead of owning people, we're talking about owning land. Someone tells you that no matter how much you paid to purchase title to some piece of land, the land itself does not belong to you. No longer may you do whatever you wish with it. You may not cut the trees on it. You may not build on it. You may not run a bulldozer over it to put in a driveway. All of those activities are immoral, because they're based on your exploitation of a living thing: in this case, the land. Did you ask the land if it wants you to build on it? Do you care what the land thinks? But the land can't think, you say. Ah, but that's just what *you* think. It is how you were taught to think. Let's say, further, that your livelihood and your way of life are based on working this land—the outsiders call it exploiting—and that if the outsiders have their way, you'll be out of business. Again and again they tell you that you are a bad person, a stupid bigot, because you refuse to see that your way of life is based on the exploitation of something you don't perceive as having any rights-or sentience-to begin with.

Angry yet?

Then how about this? Outsiders take away your computer because the process of manufacturing the hard drive killed women in Thailand. They take your clothes because they were made in sweatshops, your meat because it was factory-farmed, your cheap vegetables because the agricorporations that provided them drove family farmers out of business (or maybe because lettuce doesn't like to be factory-farmed: "lettuce prefers diversity," say the outsiders), and your coffee because its production destroys rain forests, decimates migratory songbird populations, and drives African, Asian, and South and Central American subsistence farmers off their land. They take your car because of global warming, and your wedding ring because mining exploits workers and destroys landscapes and communities. They take your TV, microwave, and refrigerator because, hell, they take the whole damn electrical grid because the generation of electricity is, they say, so environmentally expensive (dams kill salmon, coal plants strip the tops off mountains and generate acid rain, wind generators kill birds, and let's not *even talk* about nukes). Imagine if outsiders wanted to take away all these things—without your consent—because they had determined, without your input, that all of these things are exploitative and immoral. Imagine that these outsiders actually began to succeed in taking away these parts of your life which you see as fundamental. I'd imagine you'd be pretty pissed. Maybe you'd start to hate the assholes doing this to you, and maybe if enough other people who were pissed off had already formed an organization to fight these people who were trying to destroy your life—I could easily see you asking, "What do these people have against me anyway?"—maybe you'd even put on white robes and funny hats, and maybe you'd even get a little rough with a few of them, if that was what it took to stop them from destroying your way of life.

⊛ ⊛ ⊛

What if, even without outside agitation, the slaves began to revolt? They began to upset—to violate—the natural order of things. Would you find it incomprehensible that slaves could be so ungrateful as to not appreciate your providing them food, clothing, and shelter? Would you try to teach them, at first with words, and then through other means, that their thinking was in error? Would you try to put your life back in order through any means possible? Would you find it hard to hear what your slaves-or rather (though not from your perspective) the people whom you had enslaved-were trying to tell you?

Now, what if the land itself were speaking to you? What if creatures opted for extinction rather than submit to the type of world you're trying to order around you? What if the planet were changing its climate in an attempt to get you to stop enslaving it? What if the planet and its inhabitants were doing everything they could to tell you that they do not like what you're doing, that they do not want to live this way? Would you be able to hear them?

I didn't think so.

⊕   ⊕   ⊕

I just came across the transcript of a nationally syndicated commercial radio talk show hosted by Tom Leykis. The program began with Leykis reading letters from male listeners who said that women who've been sexually molested "put out" more. He commented, "If you believe that what guys are saying is true—they've had this experience that they're with these chicks who're all messed up in the head because they were abused, molested, whatever—and their [the men's] experience has been that the sex is really good—is there something unethical, is there something wrong with trying to determine that? I mean, as it is, you'll try to determine if a woman is easy, right? You'll try to determine if she'll put out for other reasons. You'll try to determine if you can, you know, give a woman some booze and see if she'll put out, or smoke pot to get her to put out, or buy her dinner or buy her drinks or buy whatever to get her to put out. . . . So that makes you wonder is everything fair game? If you think that a woman's more likely to put out or more likely to be good in bed because she has a history of abuse, is it wrong to try to find that out and then to go for the gold?"

Following this came a commercial consisting of music with a recorded female voiceover: "Roll me over and do me. Do me. Turn me over and do me. Do me. Throw me down and do me. Do me." A female caller said, "I agree with you . . . about women who've been molested. I was molested. . . . And the reason women put out like that when they've been molested is because they have a fear that men won't accept them any other way because that was basically how they were taught or how they were brought—up that this is what you do."

"So it's easier to get lucky with a woman who's been molested."

"It is . . . because their self-esteem is destroyed and so they don't have any more respect for themselves than that. I went through years of it."

"So," he replied, "it might be a good idea for a guy to try to find out if a woman's been molested."

Later, Leykis had male and female callers on at the same time. The man said, "If she's grown up consenting with it and it's turned her into some nymphomaniac, then who am I not—you know, what am I gonna do?"

The woman responded that she thought it was wrong to target women who'd been abused.

Leykis asked why.

"Because I think if you're targeting that and the girl has low self-esteem and the reason that she's going to perform for you in that way is because she's looking for some kind of healing—"

Leykis interrupted to say that most women have low self-esteem. The two argued this point for a while, before he said, "The woman who is the most insecure is the one you target. We, on this show, have talked about a theory and I've taught a class on the air called 'Leykis 101.' It's a ratio. It's a mathematical formula and you express it in the form of a fraction and the way it works is this . . . the number on top you want to be as high as possible. You want a woman who's the most attractive you can find. That's the top half of the equation. Bottom half of the equation? The lowest possible self-esteem. And that produces a ratio. The higher the ratio, the more you should go for it."

A little later in the show, another woman said that playing on a woman's weakness in order to get laid is the "same thing as going out and playing on a kid's innocence when you molest them."

Leykis responded, "I don't think it is and you want to know something? All men do that. All men do it. Now, we find different weaknesses. . . . We find out all kinds of weaknesses you have and that's how we get in."

"Don't you think it's a little cruel?"

"It's not cruel," he said. "It's reality. This is how you do it. Men want to get laid. We're not here to get to know you."

⊛ ⊛ ⊛

It took me a long time to understand why so many people hate, or at the very least are repulsed by homosexuals. It's always seemed to me that the most reasonable action for someone disturbed by homosexuals is simply not to date one. But I realized several years ago a reason for the vehemence. The realization came, oddly enough, when I was listening to a military spokesperson outlining reasons gays should not be allowed into the military (I've always believed, by the way, that homosexuals *should* be prohibited from the military, but then again, I've always believed so should hetero- and bisexuals, and the transgendered, for that matter). His main reason was that it would destroy the chain of command. "Can you imagine what would happen," he asked, obviously rhetorically, "if a private had sex with a captain? The captain would no longer have any authority." It was clear from the context his point was not intimacy, but rather penetration. If the inferior penetrated the superior, the superior would no longer be such. I suddenly understood what a lot of feminists have been saying for many years, that within our system, sexuality is an act of power, with a *fucker,* and a *fucked.* And in a culture where men consider themselves superior to women, men fucking women is a sign that all is well with the world. Feminist author Catharine MacKinnon put this as succinctly as possible: "Man fucks woman, subject verb object." Any deviation from that behavior—and homosexuality is certainly not the only way to deviate—chips away at the illusions that it's natural for men to rule women and that it's also natural to manifest this rule through sex (as opposed to "using" sex for pleasure and communication, if I may venture a wild and crazy idea). Recognizing that our most intimate relationships need not be based on power points toward an even more dangerous possibility, that *all* of our relationships can have as their basis pleasure and communication: that they can consist of communion.

⊛ ⊛ ⊛

Our culture's last chance for redemption came when European explorers encountered North America. The Middle East—cradle of civilization—had been deforested, its people had been enslaved. So, too, in Europe, where "those who were born to obey," as Aristotle put it, were killed or enslaved first by the Greeks, then the Romans, then by their former indigenous brothers who had themselves donned the iron armor—both physical and emotional—of civilization. The people, forests, and wildlife of France, Germany, the British Isles—all fell to sword and ax.

North America presented a chance for a fresh start. Here was the chance to do it right this time (or better, to take the lessons back home and attempt to integrate them there). The continent was lush, redolent, ludicrously rich in life. Fish: "Cods are so thick by the shore that we hardly have been able to row a boat through them."[48] Mammals: Martens, now endangered, and a once-in-a-lifetime sight even for those who love the wild, were "a creature with which the whole country abounds, and is of all others most easily entrapped by the furrier."[49] Birds: In the sixteenth-century, a fisherman making a raid on an island rookery (to use the bird flesh for bait to catch fish), found "so great an abundance of all kinds [of seabirds] so that all my crew and myself having cut clubs for ourselves, killed so great a number . . . that we were unable to carry them away. And aside from these the number of those which were spared and which rose into the air, made a cloud so thick that the rays of the sun could scarcely penetrate it."[50] Unbroken forests stretched from the Atlantic to the Mississippi. Trees five hundred or a thousand years old—members of the type of natural communities seen now only in tiny remnants preserved essentially as museums, displays more sad than hopeful, small samples of what was and what could have been—covered what is now Manhattan, what is now Brooklyn, what is now Boston, what is now Washington, DC, what is now Baltimore, what is now Richmond, Charleston, Atlanta, Knoxville, Cincinnati, Cleveland, Chicago, and everywhere in between.

A decade ago I drove from Fort Providence to Fort Simpson, along the Mackenzie River in the far north of Canada, one hundred and eighty miles of wild forest. The trees are small, because the summers are short, the winters long and sharp. But there were trees. There were no clear-cuts, no cities, nothing but one lonely gravel road cutting into the flesh of the forest. I was moved, both by the beauty of what I saw and by the idea of the beauty of long-gone forests that I will never see. I was moved also because I knew that the forests just to the south were already being clear-cut by transnational timber companies. The Europeans could have learned from the generosity of the land, and from the generosity of the peoples who had lived here for millennia. The natives lived far differently from the Europeans. Central American and Caribbean expert Carl Sauer commented on the Arawak of that region: "The tropical idyll of the accounts of Columbus and Peter Martyr was largely true. The people suffered no want. They took care of their plantings, were dextrous at fishing and were bold canoeists and swimmers. They designed attractive houses and kept them clean. They found aesthetic expression in woodworking. They had leisure to enjoy diversion in ball games, dances, and music. They lived in peace and amity."[51] Everywhere in the Americas, Europeans were met cordially by those they deemed savages (defined denotatively, removed from the word's pejorative connotations, these people *were* actually savages: "uncivilized, undomesticated, wild," coming from the Middle English *sauvage,* from the French *salvage,* from the medieval Latin *salvaticus,* alteration of Latin *silvaticus,* "of the woods, wild," from *silva,* "a wood." The savages of North America were as lush, rich with life, and generous as the forests that were their home). This fact has been mentioned by everyone from Amerigo Vespucci in South America in 1502 (the Indians "swam out to receive us . . . with as much confidence as if we had been friends for years") to Henri Cartier up in Canada in 1535 (the Indians "as freely and familiarly came to our boats without any fear, as if we had ever been brought up together").[52] William Brandon lays out an impressive list of possible reasons for the differences in attitudes between the indigenous and the civilized, in his book *New Worlds For Old: Reports from the New World and their effect on the development of social thought in Europe, 1500- 1800:* "Religion rather than business being the principal business; living to live rather than to get; belonging rather than belongings as a reigning value; apparent rarity of enforced civil or military service . . . group ownership of land and wealth, and consequent tendencies toward individual cooperation rather than competition, and apparent rarity of the police and lawsuits necessary to regulate individual possession."[53]

In many cases the conquerors could not have survived to conquer had not the natives first been wildly generous, not only with their food but also in instruction on how to live on the land. Edmund S. Morgan commented that the Indians near Roanoke "welcomed the visitors, and the Indians gave freely of their supplies to the English, who had lost most of their own when the *Tyger* grounded. By the time the colonists were settled, it was too late to plant corn, and they seem to have been helpless when it came to living off the land. They did not know the herbs and roots and berries of the country. They could not or would not catch fish in any quantity, because they did not know how to make weirs. And when the Indians showed them, they were slow learners: they were unable even to repair those that the Indians made for them. Nor did they show any disposition for agriculture. . . . The English, for lack of seed, lack of skill, or lack of will, grew nothing for themselves, even when the new planting season came round again."[54] Had the Indians desired to kill the invaders—whom they thought at the time were merely visiting—they needn't have cut their throats, much as many of us may later have wished they had. All they would have had to do to avert the genocide perpetrated against them would have been to let the newcomers fend for themselves. Morgan states that "the Indians . . . could have done the English in simply by deserting them."[55]

For ten years I lived in Spokane, Washington. The Spokane River runs through the center of the city. Often I walked to stand or sit by Spokane Falls—actually a number of falls in close succession—or what is left of them after the erection of a series of dams decades ago. The falls are still beautiful—liquid thunder in the swell of spring thaw, settling to a chatter come the hundred-degree days of late July and August—but often I wondered how much more beautiful they must have been before. I've seen photographs taken in the nineteenth century, but blurry washed-out sepia does not sufficiently replicate the experience of standing next to a rolling explosion of water, ice forming at the edges of liquid, cold spray finding its way into your mouth and onto the end of your nose, the force of the sound leaning into your belly like an open hand that pushes you back while beckoning you forward. A photograph cannot do that. I want to know what the old falls *felt* like. James Glover, one of the white founders of the city of Spokane, saw the falls and felt perhaps the same thing, but in him it turned over into something else. He said, and these words—words eerily similar to the words Tom Leykis said about getting laid—have stayed with me ever since I first read them on the wall of a museum in Spokane, "I was enchanted, overwhelmed with the beauty and grandeur of everything I saw. It lay just as nature had made it, with nothing to mar its virgin glory. I determined that I would possess it."

The Europeans did not come to North America to get to know the land, nor to get to know its inhabitants. They did not come to get to know themselves. They did not even come here to get laid. They came to possess, which I suppose was what Tom Leykis was really after as well *(possess:* "to have and hold as property; to enter into and control firmly; to insinuate one's gestalt into another, as in possession by a devil; to dominate; to have sexual intercourse with": from the Middle French *possesser,* to take possession of; from Latin *possessus;* from *potis,* having the power; more at potent: "having or wielding force, authority, or influence; able to copulate, usually said of a male: Middle English, from the Latin *potis,* having the power, akin to Gothic *brilfatbs,* bridegroom; Greek *posis,* husband; Sanskrit *pati,* master). They came here, if we are bluntly honest, to enslave or exterminate the inhabitants—and we have to admit that they (or rather we) have done an extraordinary job. They came also to enslave the land, to yoke it to their own purpose, and ultimately to remove from it everything of monetary value. This behavior, of course, continues to this day.

Do Tom Leykis and his listeners hate women? If not, precisely what behavior would be considered hateful? I'm looking for a specific threshold at which it will finally be acceptable to use the word *hate.* Similarly, do Europeans—Euro-Americans, whites, nonsavages, members of the dominant culture, the civilized: whatever you want to call us—hate the continent and its original inhabitants? If not, why does it seem so? Once again if not, what threshold must be finally passed before we will begin to use that word?

If, on the other hand, we decide to use that heavily loaded word—*hate*—and to say that members of civilization as a whole hate this continent and its members, we are led immediately to many more questions, such as: Can we hate without even knowing it? If we can, what does that feel like? When I wrote earlier that hatred felt long enough and deeply enough no longer feels like hatred, but like economics or religion or tradition, I was referring to someone else's hate, certainly not my own. What does it feel like—I now have to ask myself—to swim in this river of cultural hatred? Does it feel congruent, a simple unexamined life? Does it feel self-righteous? Does the feeling resemble pride, as you—or I—reap the social rewards of going with the flow? If we really are swimming in a river of hatred that extends to the land, to the inhabitants of the land, to women, what are we going to do about it?

⊛ ⊛ ⊛

Narcissistic individuals must ultimately be disappointed, and must then always displace onto others the blame for their disappointment. This is often, but not always, true. There is at least one condition-and, to be sure, this happens all the time-under which those who are narcissistic will accept blame, and in fact will act with all speed and diligence to correct their mistake.

The mistake, of course, is weakness, also known as empathy, compassion, communion, love, relationship, or humanity. More generally, the mistake that can be acknowledged and rectified is that of a failure to objectify. More generally still, the mistake can be known as a failure to be narcissistic enough, the failure consisting of acknowledging the other's uniqueness and existence as a subject. In practice, this weakness finds its way into the world as a lack of will sufficient to annihilate one's enemies.

Failure to eradicate their enemies was, to go back to the cradle of our civilization, a huge problem among the Israelites. God warned them time and again not to make covenants with those He delivered unto them: those they were supposed to exterminate and whose land they were to take. The deal was pretty clear, and it's just as clearly a deal we still adhere to: Give up your humanity and dissolve all interconnection with others, and you will receive power beyond your most insane dreams. Here's God's part of the bargain (and if you're an atheist or otherwise a humanist, just substitute for *God* the Market, Science, Technology, Capitalism, Free Enterprise, Democracy, the United States, Progress, Civilization, or whatever other abstraction you want, and the bargain still holds): "I will do marvels, such as have not been done in all the earth, nor in any nation. . . . Behold, I drive out before thee the Amorite, and the Canaanite, and the Hittite, and the Perizzite, and the Hivite, and the Jebusite."[56] Our God having long since dispatched these peoples, we can make the list more current by substituting Khoikhoi, Arawak, Pequot, U'wa, or Aborigine. In order to benefit from these marvels, the Chosen People had to promise never to "make a covenant with the inhabitants of the land whither thou goest, lest it be for a snare in the midst of thee: But ye shall destroy their altars, break their images, and cut down their groves."[57] The Israelites had to cut down the groves, just as today we have to deforest the planet, because otherwise it would be too tempting to enter into relationships with other gods, other humans, or the land where we live. And it simply won't do to form those other relationships, because "the LORD, whose name is Jealous, is a jealous God," and to enter into relationship with another is, as the book of Exodus so indelicately puts it, "whoring." To make sure the Chosen People deeply internalized this message, it was drilled into them. We read, again and again, "I will deliver the inhabitants of the land into your hand; and thou shalt drive them out before thee. Thou shalt make no covenant with them, nor with their gods."[58] The reason? Always the same: If these others live, it might be too tempting to gain their ways: "They shall not dwell in thy land, lest they make thee sin against me."

The message is repeated in Deuteronomy, Joshua, indeed, the entire Old Testament.[59] The message is acted out to this day. The message is an extension of the lesson of Noah, letting us know where we dare not look, of R. D. Laing, with his three rules of a dysfunctional family or society. Don't. Don't look. Don't listen. Don't love. Don't let the other be. Don't. The best way to guarantee you won't be in a relationship with something is to not see it. The best way to make certain you won't see something is to destroy it. And, completing this awful circle, it is easiest to destroy something you refuse to see. This, in a nutshell, is the key to our civilization's ability to work its will on the world and on other cultures: Our power (individually and socially) derives from our steadfast refusal to enter into meaningful and mutual relationships.

This refusal—this key to power—was carried forward and used by slavers, Columbus, Pilgrims, the Founding Fathers, Hitler. It is put forward today by politicians who send soldiers to kill at a distance, and by soldiers who do the killing. It is pushed by CEOs and others who wish to reap the benefits of our economic system, and by purveyors of porn who tell us it's okay to represent women as objects to be "fucked in every hole" (or, judging by my Alta Vista search and the prevalence statistics, to be raped) but fail to mention any form of relationship at all. It is okay, we are told incessantly (for incessant repetition is necessary to make this painful and eventually numbing lesson stick) to utilize resources, whether the resources are trees, fish, gold, diamonds, land, labor, warm, wet vaginas, or oil. But one must never enter into relationship with this other who owns or is a resource. To do so would be to break the covenant with your God, whose name is Jealous, whose name is Power, because your power comes directly from your unwillingness (or, perhaps, in time, inability) to maintain relationship: It is much easier to exploit someone you do not consider a living being—a You, as Buber would have put it—much less a friend, a lover, a member of your family. This is the key to understanding the difference between indigenous and civilized warfare: Even in warfare the indigenous maintain relationships with their honored enemy. This is the key to understanding the difference between indigenous and civilized ways of living. This is only one of many things those we enslave could tell us, if only we asked: They, too, are alive, and present another way of living, a way of living that is not-in contradistinction to our God and our Science and our Capitalism and everything else in our lives-jealous. It is an inclusive way of living. They could tell us that things don't have to be the way they are.

⊛　⊛　⊛

Although corporations will undoubtedly determine whether humans survive—or, rather, whether we humans can stop them will determine whether humans survive—we don't talk about them much. We especially don't talk about what they are.

Corporations are entities that we pretend are real, that have been defined (and, institutionally, define themselves) to person status by claiming, in essence, to be a body, a living body (remember that *incorporate* comes from the Latin *in,* in, and *corporare,* to form into a body, from *corpus,* a body). Corporations are the "embodiment," the reification, of a single idea—that of amassing wealth. To that end, they have been "granted perpetual life and diversified ownership, each part of which has limited liability for the debts and other liabilities of the firm." This limited liability means that each owner is not liable for the actions of the corporation. Investors can only lose the amount of money invested, and are not held in any way accountable if the company commits genocide, ecocide, murder, or any other crime. Who, then, is held accountable? The officers? Even when a corporate officer is held liable, the corporation and its stockholders, once again, are not. WMX (which used to be called Waste Management, until that name acquired too foul a reputation), for example, is the largest garbage collection and recycling company in the world. It is responsible for the largest toxic dump in the United States, and has located toxic dumps in minority neighborhoods across the country. It was cited six hundred times by the Environmental Protection Agency in the 1980s, and, between 1983 and 1988, faced eighteen grand juries. Yet, even though the corporation has been sued, convicted, and fined for numerous environmental, price-fixing, bribery, and antitrust violations, paying $46 million in fines for bribery and illegal waste handling alone between 1980 and 1988, and even though some executive officers have gone to jail, the company still has contracts with thousands of cities in forty-eight states and nineteen countries. Through the 1980s, WMX's sales increased from $773 million to $6 billion.[60] The point is that on the extremely rare occasion that an executive goes to jail, nothing happens to the company-to the machine which has as its sole function the amassing of wealth.

Limited liability means more than profits, however, and it means more than toxic waste dumps in minority neighborhoods. It is more than the mere institutionalization of irresponsibility. It is an explicit acknowledgment that it's impossible to amass great wealth without externalizing costs. If costs were not being externalized, there would be no need to limit liability.

Corporations are a legal device that came into use during the eighteenth and nineteenth centuries to deal with the myriad limits exceeded by our culture's social and economic system: The railroads and other early corporations were too big and too technological to be built or insured by the incorporators' investments alone; when corporations failed, or caused gross public damage, as they often did, the incorporators did not have the wealth to cover the damage. No one did. Thus, a limit was placed on the investors' liability, on the amount of damage for which they could be held liable. Because of limited liability, corporations have allowed several generations of owners to economically, psychologically, and legally ignore the limits of toxics, fisheries depletion, debt, and so on that have been transgressed by the workings of our economic system.

By now, we should have learned. To expect corporations to function differently than they do is to engage in magical thinking. We may as well expect a clock to cook, a car to give birth, or a gun to plant flowers. The specific and explicit function of for-profit corporations is to amass wealth. The function is not to guarantee that children are raised in environments free of toxic chemicals, nor to respect the autonomy or existence of indigenous peoples, nor to protect the vocational or personal integrity of workers, nor to design safe modes of transportation, nor to support life on this planet. Nor is the function to serve communities. It never has been and never will be. To expect corporations to do anything other than amass wealth is to ignore our culture's system of rewards, to ignore everything we know about behavior modification: We reward those investing in or running corporations for what they do, and so we can expect them to do it again. To expect corporations to do otherwise is delusional. Corporations are institutions created explicitly to separate humans from the effects of their actions, making them, by definition, inhuman and inhumane. To the degree that we desire to live in a human and humane world-and, really, to the degree that we wish to survive-corporations need to be eliminated.

It would be easy to blame corporations for most of the world's ills. This, however, would not be helpful, because corporations are mere tools for governance and for the transfer of wealth from communities to the governors, the latest in a series of tools running back six thousand years to when civilization originated "in conquest abroad and repression at home."[61]

To provide a clarifying example, although the world's forests might receive a brief reprieve were Weyerhaeuser's corporate charter revoked, we must remember that our culture was deforesting the world long before Weyerhaeuser—either the corporation or its founder, Frederick—was conceived. Corporations don't cause destruction; they are tools to facilitate it, legalize it, rationalize it, make it respectable. Another word for "the externalization of costs," for "limited liability," is theft. But this is a special breed of theft, where even the victim is left feeling that a legitimate and just transaction has taken place; the victim may be frustrated, but is more likely to be jealous than outraged. As we have seen repeatedly in this country's elections, the victim will defend the thief's property rights, and will also spend the rest of his or her life trying to earn back the stolen goods. Political rhetoritician Edmund Burke laid out, presumably with a straight face, the responsibilities of the properly enculturated, the mental and emotional states in which the poor must be maintained if they are to keep themselves at labor and not rebel against the rich: "They must respect that property of which they cannot partake. They must labor to obtain what by labor can be obtained; and when they find, as they commonly do, the success disproportioned to the endeavor, they must be taught their consolation in the final proportions of eternal justice."[62]

But perhaps this is going too far. Perhaps by changing the language, by moving away from the academic—"the privatization of profits and the externalization of costs"—to the vulgar—"theft"—I run the risk of offending. I imperil my credibility. That is precisely the point, and precisely the strength of the corporation as a tool for privatizing profits and externalizing costs, for theft, and murder. The transaction is legitimate. The crime complete. It is acceptable. It is legal. And, of course, it is still theft. And it is still murder. But labels aren't so important; no matter what we call it, poison is still poison, death is still death, and industrial civilization is still causing the greatest mass extinction in the history of the planet.

⊛ ⊛ ⊛

The relationship between economics and hatred is far deeper and more formative than what I've said earlier, that any hatred felt long enough and deeply enough feels like economics, tradition, religion, what have you. There's more to it than that. First, because our economics (and our society) is based on competition, it breeds hatred, insecurity, and fear. In *A Language Older Than Words,* I discussed how the anthropologist Ruth Benedict tried to figure out why some cultures are fundamentally peaceful and others are not, why women and children are treated well in some cultures and in others they are not, and why some cultures are cooperative and others are competitive. She found one simple rule that covered all of these. It has to do with our need as social creatures for esteem. In what she termed good, or synergistic, cultures, selfishness and altruism are merged by granting esteem to those who are generous. Cultures that reward behavior benefiting the group as a whole (and specifically that siphon wealth constantly from rich to poor) while not allowing behavior that harms the group as a whole are peaceful, respectful of women and children, and cooperative. Individual members are secure. If, on the other hand, your culture grants esteem to those who are acquisitive, that is, if your culture rewards behavior that benefits the individual at the expense of the whole (and if your culture funnels wealth from poor to rich), your culture will be warlike, abusive toward women and children, and competitive. Individuals will be insecure. She also found that members of the cultures with the former characteristics are, unsurprisingly, for the most part, happy. Members of the cultures with the latter characteristics are, just as unsurprisingly, not.[63] Valium, anyone?

It's worse than this, actually. Because our economics (and our society) is so based on abstraction, that is, the rewards of our economic system are in dollars, which are nothing but numbers that we as a society all tend to agree are worth something, and not on tangible goods, the acquisitiveness that our culture rewards can never be sated. Like hungry ghosts, we eat the world, but it's never enough. We can accumulate more than we need, more, even, than we can ever use, but because there's no limit to how high numbers can go in a bank account, those who accumulate continue to be rewarded for accumulating ever more.

Because competition is so central to our culture, because acquisition is so deeply rewarded, because this cultural urge to acquire is insatiable, and because this acquisition is inevitably based on the exploitation of others, there can be no limit to how thoroughly our culture will exploit others, both human and nonhuman. And because increasing competition leads so easily and obviously, when our lives are at stake, to increasing hatred of our competitors (as well as hatred of those who resist our exploitation), there can be no limit as to the depth and breadth of our culturally induced hatred, both of our direct peers and of those from whom we wish to steal.

But it is even worse than this. As discussed earlier in this book, another of the central movements of our culture—along with movement toward monolithic control—has been toward increasing abstraction, that is, away from the particular, away from Buber's joining of will and grace, and toward perceiving others as Its, objects, numbers, resources to be used, or, as Kevin Bales said of modern slaves, to be used hard, used up, and thrown away. Thus there can be no limit, then, also to the abstraction of our hate, that is, to the increasing emotional and physical distance over which we can and do destroy, to the veils we place between ourselves and those others we may no longer consider as existing.

⊛   ⊛   ⊛

Our culture's deep foundation of competition creates waves of rage and hatred. Not only does this anger get misdirected because it's easier to express it against the powerless, and not only because we are routinely pitted against others of the powerless, but, most especially, because, if we were to focus on the real sources of that rage and hatred, we would soon find ourselves questioning our very identities. Because so many of us have identified ourselves so deeply as civilized, as producers, consumers, workers, engineers, bakers, writers, soldiers, policemen, teachers, we have forgotten that first we are human beings. And what is that? We have no idea. To identify so deeply with the system of production that permeates the deepest recesses of our bodies, just like dioxin from manufacturing, like radiation from fallout, like heavy metals from mining, is to identify with the founding processes of civilization: conquest and repression. To recognize that our lives are based on these processes would—if we reject instead of embrace them—set us adrift in unknown territory. Who would I be and how would I live if I were not a part of this system?

If we were to truly turn our rage and hatred—not envy, where we wish to take the place of those in power, but rage and hatred, where we wish to destroy their base of power—toward the right targets, we would find ourselves questioning the basis upon which anyone holds power over anyone else. We would find ourselves questioning the basis for our own privilege. We would find ourselves suddenly no longer on the inside, no longer White, but now—light-skinned or dark, deepest ebony to subtlest russet to the most translucent pink—hated by those who were still White, and, far more importantly, searching for a new worldview to replace that into which we were formerly indoctrinated. Our identity would be shaken, then shattered. That's all scary as hell. Which is why we don't do it. We cannot see that which is threatening to us. Truly, if we identify with the culture, to hate the thing causing us pain—the culture—would be to hate ourselves. This is too much.

This is why Sitting Bull could make the speech he did at the Golden Spike ceremony for Northern Pacific: "I hate you. I hate you. I hate all the white people. You are thieves and liars. You have taken away our land and made us outcasts, so I hate you."[64] Because Sitting Bull's identity was not based on being White—civilized—he could, with no major psychological distress, recognize the real source of his misery: whites, or, rather, Whites.

To do so did not alter his perception of who and what he was in relationship to his family, his community, the larger social nexus of which he was a part, and, ultimately, the land. It was a simple statement of fact. Because he did not identify himself as White, he did not have to *not* see the source of his misery.

Conversely, this is why the soldier could not translate Sitting Bull's speech accurately, but instead had to scuttle backward to cover up his father's nakedness. Of course. To do otherwise would be to question his identity, and, in the end, to blow apart his world.

The story of Sitting Bull happens also inside of each of us every day, or, at least, inside those of us who feel rage and hate, which I'd wager is most of us, in that we all feel the loss—at some level even more deeply than our identification with the system of production, even more deeply than the dioxin, radiation, heavy metals, as deep inside of us as it is possible to be—of having had our humanity sacrificed on the altar of production, of having been forced into competition with all others with whom we should be cooperating, of having been coerced into believing that coercion is natural or inevitable, of having signed a Faustian contract—the Faustian contract—giving up happiness and connection for power. Inside of us—once again, deep, deep inside of us—we daily give Sitting Bull's speech, hating the system, hating what it does to others, hating what it does to ourselves. And then, somewhere on the way to the surface, the speech gets translated from our native tongue into English, into the "friendly, courteous speech" that has been prepared for us, and that we now prepare for ourselves. And so we give that courteous speech, and we laugh, and we smile, never perhaps noticing how closely a smile sometimes resembles a grimace. And what about the rage? It gets shunted, turned anywhere but at the source. And when it emerges, we smile. We smile as we lynch the black men, and we smile as we shoot them down. We smile as we shoot the photographs of women legs spread, and we smile as we look at them on our computers. We smile as we deforest the land and vacuum the oceans. We smile, we smile, we smile, never able to see the sources of our hatred, never seeing even that the hate exists.

Our current system of production cannot survive creativity. It cannot survive life being life. It cannot survive humans being humans. It cannot survive each of us simply being who we are. In order for civilization to continue, we must each be tweaked, torn, our psyches twisted to conform to a social reality based on exploitation, or, failing that, our bodies broken, burned, hanged from trees as warnings to others who may otherwise be tempted to refuse to become one of those living dead who value property and production over life. The system cannot survive without each of us sacrificing our humanity and our lives to the goal and the god of production, wasting our lives in quiet desperation, or for those beyond the frontier—the non-White, or even those like the Irish immigrants who wished to be White even to their premature dying gasp—slaving away lives to a grave that may sometimes be a welcome rest after a time of too much toil. The system of production cannot survive if we so much as perceive the diversity that surrounds us, much less experience it, and much less if we shake off our identification with our Whiteness and identify instead with living individuals. For our system of production is, despite its awful momentum, extraordinarily fragile. All it would take to bring it to a halt is creativity, *persistent* creativity.

By *creativity* I do not mean the sort of feeble and febrile cleverness rewarded by our culture, where people beat a path to the door of someone who builds a better mouse- or mantrap—a more efficient way to convert the living to the dead—although, within our mind-set, that sort of cleverness is the kind of mock creativity most often rewarded. But, instead, by creativity, I mean a remembering and realizing—making real—the full range of human possibilities in the service of creation and life. What I'm suggesting is a refusal to have one's own particularity dulled or denied, or, rather, a refusal to hand it over in exchange for mass-produced commodities—the so-called comforts and elegancies—for which we sell our birthrights as living, feeling, empathic human beings.

Forget the rulers. The system of production could not long survive if those of us who are White—light-skinned or no—renounced our Whiteness and reclaimed our ability to perceive ourselves and others as individuals worthy of respect and consideration.

Of course, it's not so easy. If enough of us renounce our Whiteness so that the ability of the remaining Whites to exploit us is threatened, some of us—many of us—will be killed, often spectacularly, in an attempt to terrorize the rest of us back into docility.

⊕ ⊕ ⊕

I think often of a conversation I had several years ago with a friend who's a longtime environmental activist. She's Jewish. She'd been down to visit her parents in Florida, and, while there, had gone to a theater to see *Schindler's List*. It had been a disturbing experience for her, she'd said, because, afterward, "All these blue-haired old Jewish ladies came out shaking their fists and saying, 'Never again!'" When she'd said this to me, my friend had made a quick, almost reflexive motion with her head, as though trying to clear her brain from some grave confusion, or clear her mouth from some bitter taste. She'd said, "Don't they realize it's ongoing?"

"They've probably never heard of the U'Wa," I said.

"The U'Wa?" she replied. "Except for the football team they've probably never even heard of the Seminoles, and they're living on Seminole land."

By pretending that the Holocaust (capital *H* to distinguish it from all the others) is Unique (capital *U,* because, of course, every holocaust is unique) we get to isolate it, pretend it was an aberration, a single incomprehensible act of unparalleled evil committed by a nation inexplicably in the thrall of a monstrously and, somehow, charismatically insane individual. We get to pretend that it was not an inevitable consequence of a way of perceiving others and of being in the world.

I used to wonder at the vehemence—even violence—with which some people insist that the Holocaust is Unique. There's an obvious answer, which is that Jews—as a whole—are White. They're civilized. They believe in Production. They even have flesh-colored skin. All of which is to say that they're people, as opposed to the others whom we kill—the indigenous of Africa, Europe, Asia, the Americas, Oceania, and the islands, as well as the poor the world over (and, most especially, the nonhumans)—who do not worship Production, and who therefore are not people in the fullest sense, and who, therefore, under the rules of civilized intercourse, can be eradicated with impunity. So, of course, there would be a tremendous sense of betrayal. We are, in a sense, killing our own. Brother killing brother. Brother killing sister.

There's more to it than this, though. It's crucial to the perpetuation of our culture that we perceive the Holocaust as Unique. This accomplishes much the same thing as pretending the KKK was not a true grassroots organization, and as pretending the essential cessation of lynchings in the United States manifested a collective racial epiphany. It allows meaningful analysis to stop, or, better, to lead us away from ourselves, so we can spend endless years dissecting some distant *other*—*What if Hitler had been successful at painting? Did he contract syphilis from a Jewish woman? —we* can blame for this otherwise unfathomable series of atrocities. This shunting of attention is necessary if we're going to continue to live the way we do. *Don't look. Don't see. Don't tell.* To pretend the Holocaust is Unique is to allow us to not question our own way of living, to not question our own innocence. It allows us to pretend we share no motivations with the perpetrators, pretend we share no cultural imperatives, no code. It allows us, one and all, to be good Germans in the larger ongoing holocaust of the planet and of its inhabitants. It allows us to continue to be decent White men. It allows us to remain civilized. It allows us to avoid being cursed by Noah, our patriarch. It allows us not to have to walk alone and afraid into the wilderness.

❧   ❧   ❧

Of course, there is a world of difference between them and us. The Nazis were killing human beings, for crying out loud, while we're merely killing the planet (and, of course, killing humans along the way, but even these people are only, to use the U.S. military's term, collateral damage). And the Nazis set out to *eradicate* the Jews, Romani, and so on, while all the deaths we cause are just accidental (time and again) by-products of our economic system, right?

A few months ago I got into a rather sharp conversation with an old friend. We went to a restaurant that had on the walls many photographs of lumberjacks standing smiling in front of the stumps of ancient redwoods they had just felled, trees half as old, themselves, as all of civilization. The trees were big enough to beggar belief, and only 150 years ago they ran from here all down the coast. I said, "Someday, soon, I hope, people will look at these photos with the same disgust with which we now look at photos of grinning soldiers celebrating the Holocaust."

She took offense. "I don't think you want to compare killing trees to killing humans."

"It's all the same imperative," I said.

Her voice grew tense, and I could tell my original assertion had made her angry. "It's not the same at all."

I wanted to tell her that the key word in the mass murder of the trees and the mass murder of the humans is *mass*. In both cases the killing is made inevitable by a utilitarian worldview that blinds us to relationship and makes it impossible to perceive the other as an individual, whether the other is a Jew, Indian, woman, or tree. In both cases—the Holocaust and industrial forestry—the killing was or is demanded not for any personal, that is, human, reason, but rather by abstract ideologies, and the valuing of those ideologies over life. In both cases the killing has been made feasible, rational, necessary by a bigotry that holds us separate from and superior to these others, whomever these others may be, and thus entitled to take these others' lives for no good personal animal reasons. You could argue that at least by killing trees—whom we certainly consider to have less rights to live on this land than we—we're able to harvest wood that can be converted to cash, and also turn unproductive—from an economic perspective—native growth into productive tree farms. I would respond that by killing those they deemed subhumans, Germans were able to harvest gold, glasses, shoes, soap, and hair that could be converted to cash, and were able to gain *lebensraum* that Germans would soon put to good use. The biggest, and, really, the only fundamental, difference between the nightmares is that, insofar as the one killing the planet, we're still on the inside. We still consider ourselves the master race, or species.

⊛　⊛　⊛

This book began as an exploration of hate in the Western world, and it ends, really, with the end of life on the planet. The problem, as near as I can see it, is the valuing of the abstract over the particular: of production over life; of economic (and other) systems over living beings, be they humans, or rivers, or polar bears; of our preconceptions of what niggers or Chinese or Irish dogs are supposed to be, instead of this black man, this Chinese woman, this Irish man, complete with his or her own cultural and personal histories, with desires and hopes and fears; of photographs of women over the women themselves, of the bodies of women over their whole beings, bodies and minds and hearts and sorrows and joys; of truncated conceptions of our own capacities, based on what we have been allowed to express over who we really are. The problem is simply that of seeing ourselves and others as instruments to be used, instead of people to be enjoyed in relationship.

What I propose as a "solution" to this problem of the ascendancy of abstraction is a return to the particular. I support an antisystem to promote a falling in love with the particular. To love this particular tree, that particular person, this glint of sunlight off this dragonfly wing, and, insofar as is possible, to perceive each of those around us as subjects. This is not a simple plea for us to all just get along. I'm not suggesting we replace abstract hate with a love just as abstract. That's pointless, absurd, meaningless, and, in the end, impossible. I am not an abstract being. I have fingers, flesh, bones. I love this person. I do not love that person. Nor am I suggesting we simply step away from violence. I'm suggesting that there is a difference—all the difference in the world, really, between real fights between real people-even when real blood is spilled—and killings based on preconceptions. What I'm suggesting is a return to our humanity.

If we are to do that, the first thing we must do is to see the inhumanity of our current system for what it is, and we must speak about it. If the first rule of a dysfunctional family or society is *Don't*, the first rule of a functioning society is *Do*. Talk about it. Speak out, like Ham, Noah's curses be damned.

Of course, it's not so easy. It's all very fine for me to say how much joy it brings me to listen to birdsong, but my enjoyment, or anyone else's, is irrelevant to the suffering of others, to the degree that it does not compel me to shut down the source of the other's misery. Having fallen in love with our own lives, and the lives of those around us—even our honored enemies (though not McNamara and his likes who, by their actions, show themselves to be willing to exploit)—the next step is to get rid of our whole inhumane system, to quit valuing production over life, and to physically stop those who do. The next step is to bring down that which originated in conquest abroad and repression at home. The next step is a planet liberated from the destruction; the next step is the end of civilization. *Get rid of civilization?* I can hear you say. *That's your solution?* The hatred that char-acterizes so much of our system—the hatred I've described and analyzed in this book—is not a product of biology. People are not fundamentally hateful. Our hate is not a result of several billion years of natural selection. It's a result of the framing conditions under which each of us are raised. It's a result of the unquestioned assumptions that inform us. If we want to stop the hate, we have to get rid of the framing conditions. Until we do that, we're bound to fail. So, yes, that is precisely my solution, we need to get rid of civilization.

Maybe that seems absurd to you. It doesn't to me. It just seems like a lot of work, done by a lot of people in a lot of places in a lot of different ways. But I'll tell you something that does seem absurd to me: the possibility of allowing this inhumane system to continue.

Strangely Like War
The Global Assault on Forests

with George Draffan

Green Books Ltd., 2004

Strangely Like War was something of a homecoming for me, because I began my activism with forest protection. It was an honor for me to be able to write a book defending forests, giving something back to them. Forests have done so much for me. Prairies do just as much for us as forests do, and oceans do just as much, and rivers do just as much, but I happen to live in a forest, so it felt desperately important for me to do at least this act of protecting them. This book and Welcome to the Machine were written with the extraordinary George Draffan.

The very day we wrote the final words of this book, scientists declared that yet another subspecies of tiger has gone extinct in the wild (with only captives remaining, so discouraged they're dosed with Viagra to try to make them breed). Gone extinct. Such a passive way to put it, as though we know no cause, can assign no responsibility. It's almost as though we were to say that victims of murder passed away, or that victims of arson decided to move.

The South China tiger joins its cousins the Caspian tiger, Bali tiger, and Javan tiger, all victims of logging, road building, and the leveling of forests under this excuse or that.[1] The other tigers will almost undoubtedly join them soon.

It doesn't matter much to the tigers whether the forests are cut because Mao decided that "Man Must Conquer Nature," or whether the World Bank decided that "Man Must Develop Natural Resources." The forests are cut, the tigers dead.

                   ⊛   ⊛   ⊛

The forests of the world are in bad shape. About three-quarters of the world's original forests have been cut, most of that in the past century. Much of what remains is in three nations: Russia, Canada, and Brazil. Ninety-five percent of the original forests of the United States are gone.

We don't know how fast the surviving forests are disappearing. We don't know how many acres are cut each year in the United States, nor how much of that is old growth. We have estimates, and we'll give them through the book, but the paucity of information even on present levels of cutting reveals more than it hides: it reveals how desperately out of control is the whole situation.

The United States Forest Service and the Bureau of Land Management sell trees from public forests—meaning they belong to you—to big timber corporations at prices that often do not even cover the administrative costs of preparing the sales, much less at full market value. For example, in the Tongass National Forest in southeastern Alaska, 400-year-old hemlock, spruce, and cedar are sold to huge timber corporations for less than the price of a cheeseburger, and taxpayers have paid for the building of the logging roads as well. The United States Forest Service loses hundreds of millions of dollars a year on its timber sale programs. In other words, if you pay taxes, you pay to deforest your own land.

If you live in the West, Southwest, South, Northeast, Midwest, Alaska, or anywhere else in the United States where there are or were forests, chances are good you've seen or walked clearcuts, sometimes square mile after square mile, cut, scraped, compacted, and herbicided. You've seen lone trees silhouetted on ridgelines, and you've seen once-dense forests reduced to a handful of trees per acre. You've suspected and later learned that these few trees were left so the Forest Service and big timber corporations could say, more or less truthfully, that they did not clearcut this particular piece of ground. And maybe you came back another time and saw that the survivors, too, were gone.

You've probably driven highways lined by trees, then pulled over to look around, only to discover that just like in old westerns, where false fronts hid the absence of real stores, you've been sold a bill of goods: a few yards of trees separate the road from yet more clearcuts. This fringe of trees,

which reveals recognition on the part of timber corporations and government agencies that industrial forestry requires public deception, is ubiquitous enough to have been given a name: the beauty strip.

Do yourself—and the forests—a favor. Next time you fly over a once-
forested region on a clear day, look down. Pay attention to the crazy quilt of
clearcuts you see below, to the red roads linking clearcuts and fragmenting
forests, roads that wash out in heavy rains to scour streambeds and destroy
fisheries.

Only 5 percent of native forest still stands in the continental United
States. 440,000 miles of logging roads run through National Forests
alone.[2] (The Forest Service claims there are "only" 383,000 miles but the
Forest Service routinely lies, keeping double books—a private set showing
actual clearcuts, and a public set showing some of the same acres as old
growth—attempting to mislead the public by labeling clearcuts "temporary meadows," reducing the stated costs of logging roads by amortizing
them over a thousand years, and so on).[3] That's more road than the Interstate Highway System, enough road to drive from Washington, DC, to San
Francisco a hundred and fifty times. Only God and the forests themselves
know how many miles of roads fragment forests altogether.

The forests of this continent have not always been a patchwork of dwindling and increasingly isolated natural communities. Prior to the arrival
of our culture, unbroken forests ran along the eastern seaboard, leading
to the cliché that a squirrel could have leapt tree to tree from the Atlantic
to the Mississippi never having to touch the ground. Today, of course, it
could still do so and never touch ground, but instead walk on pavement.
Polar bears wandered as far south as the Gulf of Delaware, martens were
"innumerable" in New England, wood bison cruised that region, passenger pigeons passed overhead in flocks that darkened the skies for days
at a time, Eskimo curlews did the same, rivers and seas were so full of
fish they could be caught by lowering a basket into the water. American
Chestnuts ran from Maine to Florida so thick on the dry ridgetops of the
central Appalachians that when their crowns filled with creamy-white
flowers the mountains appeared to be covered with snow. Before European
"settlement"—read conquest— of America, there was no such thing as
"old-growth," no such thing as "native forest," no such thing as "ancient
forest," because *all* of the forests were mixed old growth, they were all

native, they were all diverse, ancient communities. Difficult as all of this may be to imagine, living as we do in this time of extraordinary ecological impoverishment, all of these images of fecundity are from near-contemporary accounts easy enough to find, if only we bother to look.

Worldwide, forests are similarly under attack. One estimate says that a hectare (two and a half acres) of forest somewhere in the world is cut every second. That's equivalent to two football fields. One hundred-and-fifty acres cut per minute. That's 214,000 acres per day: an area larger than New York City. Seventy-eight million acres (121,875 square miles) deforested each year: an area larger than Poland.

The reasons for international deforestation are, as we'll explore in this short book, often similar to those for domestic deforestation. Indeed, those doing the deforesting are often the same huge corporations, acting under the same economic imperatives with the same political powers.

Apologists for deforestation routinely argue that because pre-conquest Indians sometimes "managed" forests by setting small fires to improve habitat for deer and other creatures, industrial "management" of forests—deforestation—is acceptable as well. But the argument is as false and unsatisfying as the beauty strips, and really serves the same purpose: sleight of mind to divert our attention from deforestation. The argument—unfortunately as common, also, as beauty strips—is analogous to saying that because someone once clipped a partner's fingernails that it's okay for us to cut those fingers off.

I saw this argument presented again just today in the *San Francisco Chronicle*, in an op-ed piece by William Wade Keye, past chairman of the Northern California Society of American Foresters. He wrote, "Native peoples managed the North American landscape, cutting trees and using fire to perpetuate desirable forest conditions. There is no reason that we cannot equal or better this record of stewardship."[4] Well, there are actually many reasons. Indians lived in place, and considered themselves a part of the land; they did not come in as an occupying force and develop an extractive economy. They did not participate in an economy and culture that values money over life. They were smart enough not to invent chainsaws and fellerbunchers (huge shears on wheels that roll along the ground, severing trees and stacking them into piles). They were smart enough not to invent woodchippers, and not to invent pulp mills. They were smart enough not to invent an economy that ignored everything but cash. They were smart enough not to invent limited liability corporations. They didn't export mountains of timber overseas. They knew trees and other nonhumans as intelligent beings with precious lives worth considering, and not as cash on the stump, nor as resources to be managed, nor even as resources at all. Their spiritual beliefs did not include commands to "subdue the earth," nor was their cosmology based on the absurd notion that one succeeds in life by "outcompeting" one's human and nonhuman neighbors.

And the Indians didn't subdue the earth. There is absolutely nothing in our culture's history to suggest that we can "equal or better this record of stewardship." There is everything in our culture's history and present practices to suggest that the deforestation will continue, no matter the rhetoric of those doing the deforestation, and that ecological collapse will be our downfall, as it has been for earlier civilizations.[5]

But believe neither us, nor even contemporary accounts of early explorers who wrote of the extraordinary richness of native forests, nor especially the handsomely paid liars of the timber industry and the government.

For the truth lies not in what they say, nor even in what we say. The truth lies on the ground. Go out and walk the clearcuts for yourself. Rub the dried soil between your fingertips. Walk the dying streams, listen to the silence in the skies (except for the whine of chainsaws and roar of distant logging trucks). Walk among ancient ones still standing, trees sometimes two thousand years old. Put your hands on their bark, on their skin. Taste the difference in the air. Smell it. Reflect on the beauty of what's still there, and on what has been lost—what has been taken from us.

When you've finished crying, and if you want to know more about the current crisis in the forests—where we are, how we got here, and where we're going—then come back and read the rest of this book.

☙ ☙ ☙

When you consider the current landscape of the cradle of civilization— what is now Iraq and environs—what pictures come to mind? If you're like me, the images are of barren plains and even more barren hillsides, goats or sheep grazing on a few scrubby brushes breaking a monotony of light brown dirt. But it was not always so. As John Perlin states in *A Forest*

Journey: The Role of Wood in the Development of Civilization, "That such vast tracts of timber grew near southern Mesopotamia might seem a flight of fancy considering the present barren condition of the land, but before the intrusion of civilizations an almost unbroken forest flourished in the hills and mountains surrounding the Fertile Crescent."[6] The trees were cut to build the first great cities, and for ships that plied the first great empire. Once the ships were built, wood was imported to make the cities even bigger: down went the great cedar forests of what is now southwest Turkey, the great oak forests of the southeastern Arabian peninsula, and the great juniper, fir, and sycamore forests of what is now Syria.[7]

One of humanity's oldest written stories—one of the formative myths of our culture—is that of Gilgamesh, who destroyed southern Mesopotamia's cedar forests to build a city.[8] According to this story, Enlil, the chief Sumerian deity, who must forever watch out for the well-being of the earth, entrusted the demigod Humbaba to defend the forest from invaders, from the first civilized humans—as opposed to people of the land—to enter the region. But the warrior-king Gilgamesh killed Humbaba and leveled the forest. Enlil sent down curses on the deforesters, "May the food you eat be eaten by fire; may the water you drink be drunk by fire." These curses have followed us now for several thousand years.

Let's move a little west. Picture this time the hills of Israel and Lebanon. I recently asked a man from Israel if his country has trees, and he said, "Oh yes, we have lots of little trees which we water by hand." This fits with the images that come to mind: every picture I've ever seen of the Crucifixion, for example, shows a hilltop mainly or entirely devoid of trees. The same is true for most of the pictures I've seen of Palestinian refugee camps, and for Israeli settlements. What happened to the "land of milk and honey" we read about in the Bible? And what about those famous "cedars of Lebanon" we've read so much about? They're long gone, cut to build temples, cities, and ships, cut for fuel, cooking, metalworking, pottery kilns, and all the trinkets of commerce.

Move west again, to Crete, and then up to Greece, and we see the same stories of trees making way for civilization. Knossos was heavily forested, and now is not. Pylos, the capital of Mycenaean Greece, was surrounded by giant pine forests. Once-forested Melos became barren. The same is true for all of Greece.

When you think of Italy, do you think of dense forests? Italy was once forested. These fell with the rise of the Roman empire.

Or how about North Africa? Surely not. This land is as barren as the Middle East. But here, once again quoting Perlin, "Berbers fulfilled their duty by felling the dense forest growth for their Arab masters. Such large quantities of wood were shipped from these mountains that the local port was named 'Port of the Tree.'"[9] All to make Egyptian warships.

We could continue with this journey, through France, Britain, across North and South America, into Asia and Africa, but by now you see the pattern.

The pattern continues today, accelerating as our culture metastasizes across the globe. Worldwide, forests fall.

⊞ ⊞ ⊞

When a forest is cut, not only trees are killed. Whether it's lions in ancient Greece, spotted owls or coho salmon right now in the Pacific Northwest, or gorillas in Africa, the loss of forests means the loss of the creatures who live there.

The list of plants and animals damaged or extirpated by the deaths of once-great forests is long, and getting longer every day. Golden-crowned lemur, orangutan, Siberian tiger (of whom there are only two hundred and fifty left), marbled murrelet, Port Orford cedar (killed by a fungus transported on logging equipment), black forest-wallaby, aye-aye, red cedar, mahogany, ivory-billed woodpecker, Carolina parakeets, golden-capped fruit bat, Hazel's forest frog, smooth-skinned forest frog, Amur tiger, Amur leopard, forest owlet, Nelson's spiny pocket mouse, Saker falcon, red wolf, panda bear, and on and on.

Scientists estimate an average of 130 species are driven extinct every day. That's about 50,000 each year. That is not just by deforestation, but by the larger effects of industrial civilization. Deforestation plays its part, though, in great measure because forests are home to so many creatures. For example, although rainforests presently cover only 3.5 percent of the planet's land surface, they support more than half of all known life forms. The national forests of the United States provide habitat for 3,000 species of fish and wildlife.

Seventy-five percent of the mammals endangered by the activities of industrial civilization are threatened by loss of forest habitat.[10] For birds, the figure is 45 percent. For amphibians it's 55 percent, and for reptiles it's 65 percent.

Even those apologists for industrial forestry who admit other creatures besides humans live on this planet, and who acknowledge that destroying their homes could possibly—remotely possibly, mind you—harm them the tiniest little bit still then argue that logging is a trivial cause of damage compared to mining and agriculture. They especially like to show pictures of poor (brown) people using slash and burn agricultural techniques in the rainforests. But this argument is as much as deflection as most of their others. Worldwide, logging likely accounts for more than two-thirds of the forests destroyed, as opposed to burning and other causes.[11] In Oceania it's "only" 42 percent. Asia, 50 percent. Central America, 54 percent. South America, 69 percent. Africa, 79 percent. Europe, 80 percent. North America, 84 percent. Russia, 86 percent.

Recent studies show, too, that species extinction likely continues for a century after deforestation.[12] Guy Cowlishaw of the Zoological Society of London cautions, "We should not be lulled into a false sense of security when we see that many species have survived habitat loss in the short term. Many are not actually viable in the long term. These might be considered 'living dead.'" By correlating, for example, the number of individuals of different species of primates living in Africa, their habitat size, and the extent of deforestation of their habitat, he has come to the conclusion that deforestation is leaving Africa with a large extinction debt. Even if no additional forest were to be cut, six countries—Benin, Burundi, Cameroon, Cote d'Ivoire, Kenya, and Nigeria—stand to lose more than a third of their primate species in the next thirty or forty years. That presumes, once again, *no further deforestation*. But scientists estimate that within that same time, 70 percent of remaining West African forests and 95 percent of remaining East African forests will be cut.

It's not just primates. Studies on birds show similar trends. Thomas Brooks, a biologist from the University of Arkansas who has studied avian extinction in Kenya's Kakamega Forest, said, "Even a century after a forest has been fragmented, it may still be suffering from bird extinctions."

Brooks continued, "The good news is we have a brief breathing space.

Even after tropical forests are fragmented, there is still some time to adopt conservation measures to prevent the extinction of their species. The flip side of this is bad news, though: There is no room for complacency."

Healthy forests are crucial not only to the creatures who live there. Forests purify water and air. They mitigate global warming by storing carbon. Forests increase local precipitation (half of the rain in rainforests comes from local water evapotranspirated from the forest itself). They prevent flooding and erosion.

It is common when making a plea to halt deforestation to talk about the ways the loss of these forests hurt *us*, using, for example, the fact that rainforests can be considered great medicine chests, if only we will use the medicines instead of destroying the chests. Just tonight I read on a website deploring tropical deforestation, "The rainforest is the earth's natural laboratory, from where one quarter of today's pharmaceuticals are derived. One seemingly insignificant plant, the rosy periwinkle, gave us medicines which revolutionized the treatment of leukemia in children. According to the National Cancer Institute, seventy percent of the plants used in fighting cancer can only be found in the rainforest. But less than one percent of tropical forest species have been thoroughly examined for their medicinal properties."[13]

While it's certainly true that there are many selfish reasons to stop cutting down forests, we don't want to emphasize them, because ultimately— and even in the short run—we don't think that particularly helps. It doesn't challenge the grotesquely narcissistic and inhuman utilitarian perspective that *is* our worldview and underlies our attempts to dominate the world.

A few years ago I was one of the few environmental representatives at a conference of children's health advocates. That in itself was strange, I thought: how can you possibly discuss the health of children without emphasizing the fact that industrial civilization is rendering the planet uninhabitable for them?

One of the advocates there—a high level federal bureaucrat at the Centers for Disease Control—expressed the need to halt tropical deforestation (it often seems to me that more people in the United States want to halt tropical deforestation than want to stop it here at home) by saying, "We need to save those plants because they're our medicines for the future."

"That's precisely the problem," I responded. "The belief that the forests belong to us. They're not *our* medicines, and they're not *our* forests. First, the plants belong to themselves, and they belong to the forest. Second, if they belong to any humans at all, they belong to the indigenous people who live on that land. We have no more right to take their plants for medicines than we do for timber."

Several people looked at me as though I had suddenly stopped speaking English and begun quacking like a duck.

This is what often happens when you cease to speak the language of unbridled exploitation—untethered selfishness—and begin to suggest that forests, and the creatures who live in them (including indigenous humans), have the right to live on their own, regardless of how useful or not they may be to us. What was happening in that room was in many ways what happens moment-by-moment in the forests: a clash of incompatible world-views and value-systems.

⊛ ⊛ ⊛

At every step of the way there have been humans living in the forests that have fallen to the axe, and now the chainsaw, people who do not look at forests with only an eye toward how they may be turned to profit, but instead toward how the people can live there forever. There were the indigenous conquered near the Fertile Crescent, whose sacred groves were cut by Gilgamesh and those who came after. The Canaanites and many others, conquered in the Promised Land, whose sacred groves were cut by the Israelites lest the Israelites be tempted to worship in their shade. The indigenous of northern Greece, whose forests were cut to serve commerce, and who were called *barbarians* because they did not speak the language of civilization, and thus did not speak, but instead made sounds like *barbarbar*. These people were conquered, their forests cut. The indigenous of Italy, France, England, called *savages* because they lived in forests (*savage* etymologically derives from the root of *forest*: *savage*: "not domesticated, untamed, lacking the restraints normal to civilized human beings," from Middle English *sauvage*; from Middle French, from Medieval Latin *salvaticus*; alteration of Latin *silvaticus*, of the woods, wild; from *silva*, wood, forest). These, too, were killed, their lands deforested.

Move across the ocean to the United States. A standard conceit of the settlers was that they faced not *terra incognita* but *terra vacuuis*, an empty land with trees ripe for cutting. But these were not empty lands, and they are not empty lands today. There are those who live there. There are non-humans, whose lives are as meaningful to them as yours is to you and mine is to me. And there are humans, with lives just as precious.

Wilderness is a social construct. My niece recently moved to Louisiana, and sent me a note in which she stated how uncomfortable she is that an alligator lives on her Coast Guard base. "Call me crazy," she wrote, "But I think it's odd to have wild animals so close to where people are." Not always would this have seemed odd. For almost all of human existence, it was simply how things were. And for some humans it still is. For them there is no city in here, no wilderness out there. No split between humans who exploit and a resource base to be exploited.

What all of this means is that so often when we talk about saving forests we too often forget about the people who call them their home. No, we're not talking about those people with more cash than integrity who buy ecologically-sensitive pieces of ground and threaten to construct vacation homes—with the real purpose being to extort money from those who wish to protect the land. Nor are we talking about (mainly successful) attempts by transnational timber corporations to "gain access to" (in other words deforest) wild forests the world over. Nor are we talking about loggers, many of whom truly do love to walk in the forests they're destroying. Nor are we talking about environmentalists living in yurts and composting their feces into humanure. We're not *even* talking about writers and researchers who love to look at salmon and will do anything possible to help stop deforesation.

We're talking about the indigenous, those who live on the land that their ancestors lived and died on, going back so many generations that the distinction is lost between those who live on the land and the land itself. We're talking about those whom we have never gotten to know, and who have never fit our self-serving stereotype that they are "beastly," "savage," "primitive," somehow subhuman, living lives that are "nasty, brutish, and short." This notion is self-serving because it reinforces the conceit that these people would be better off if we civilize them, take them (by force if necessary) out of their childlike ways to live as adults (and not coincidentally take their lands). As Ronald Reagan put it, "Maybe we made a mistake in trying to maintain Indian cultures. Maybe we should not have humored them in that, wanting to stay in that primitive lifestyle. Maybe we should have said: No, come join us. Be citizens along with the rest of us."[14] Conveniently left unsaid is the theft of their land, and its ultimate despoliation.

Nor do the indigenous live romantic lives wandering about picking a few berries now and then. They have serious long-term relationships with the plants and animals with whom they share their landscape. Ray Rafael, who has written extensively on the concept of wilderness, has said, "Native Americans interacted with their environment on many levels. Fortunately, they did so in a sustainable way. They hunted, they gathered, and they fished using methods that would be sustainable over centuries and even millennia. They did not alter their environment beyond what could sustain them indefinitely. They did not farm, but they managed the environment. But it was different from the way that people try to manage it now, because they stayed in relationship with it."[15]

Theft of indigenous land is not ancient history, something that *only* happened a long time ago, something to express our regrets over as we continue to profit from their land. It happens today, all over the planet. Anywhere there are indigenous people living traditionally in forests, they are being threatened, harassed, arrested, dispossessed, killed, and their forests are being cut down.

⊞ ⊞ ⊞

The whole of the damage caused by the timber industry is greater than the sum of its injuries. The harm accumulates. Imagine someone cutting a pound of flesh from your left thigh. That not only hurts like hell, but now your body has to deal with the wound. Just as you begin to recover, however, the same people take a pound of flesh from your right thigh. *You can live with that, can't you? You've got plenty of flesh. One pound won't hurt.* Then they take a pound of flesh from your upper arm. One from your lower back. One from your abdomen. Each time, those doing the cutting point out that you've not yet died from these ever-so-slight losses—*a pound is less than one percent of your body weight, so quit your crying and let me cut on you, you big baby*—so it must not have hurt you. *You look good,* they say. *Better than ever.* And then they take a pound from your left calf. Next your tormenters do not allow you to sleep. They cut at you incessantly. They drain pint after pint of blood (*Mosquitoes drain blood,* they say, *and you survive that natural process, so this process must be natural, too!*) They put strychnine in your food. And then they cut. And they cut. And they cut. Your flesh becomes a mosaic, a patchwork of scars.

What happens next?

You die, of course.

⊗   ⊗   ⊗

The same is true of forests. Building roads causes a certain amount of damage. The use of heavy machinery causes even more. More by cutting trees, more by dragging them out of the forest. More by burning leftover brush and applying poisons to the landscape. Step by step they injure forests. Step by step the forests die.

It's all even worse than this. When timber corporations do replant, the new "forest" is usually a monocrop. It's no longer a forest at all, but a fiber plantation. Most forest creatures cannot live there any better than they could in an Iowa cornfield, but spruce budworms love monocrops of spruce trees, and mountain pine beetles love monocrops of pine trees. Fiber plantations are far more susceptible than forests to fires, diseases, and catastrophic outbreaks of insects such as spruce budworms, mountain pine beetles, and tussock moths.[16] Of course fires, diseases, and insect outbreaks are expected, predictable, and *beneficial* natural responses to monocrops, beneficial because by helping to destroy monocrops they prepare the long slow way for the return of diversity. And of course fires, diseases, and insect outbreaks prompt calls from politicians, Forest Service bureaucrats, and timber industry spokespeople for increased cutting, not only on the affected plantations, but most especially on all of the natural forests they can get their chainsaws on.

It's all crazy. And it's all killing the forests.

⊗   ⊗   ⊗

But as is always the case when attempting to stop our culture from destroying some part of wild nature, all losses are permanent, all victories temporary. Winning a timber-sale appeal doesn't mean stopping a timber sale. It doesn't mean protecting a piece of ground. It means protecting a piece of ground for the year or two it takes the Forest Service to write up another EA, this time trying harder to bamboozle us. The score today is that less than 5 percent of the ancient forest in the United States remains.[17] Despite the fact that the whole system is rigged in favor of deforestation, a bunch of us were able to use the system's own rigged rules to stop, for a while, most of the illegal logging on the national forests of our region. Because nearly all commercial logging on public lands violates environmental protection laws, this means we stopped nearly all commercial logging on these forests. Activists across the country used similar tactics with similar success to try to enforce the law and protect the forests. The response by our local Forest Service was to hire scores of new employees. Were these new hires biologists, botanists, hydrologists, and anthropologists, brought on in an attempt to better understand forests? No. They hired one timber-sale planner, and all the rest were technical writers directed to produce slicker documents.

The response nationally was rather more severe, and is what helped me learn how unwaveringly committed to the destruction of the planet our culture is. The timber industry, politicians, and the corporate media launched a massive propaganda campaign. This in itself is nothing new: It's what they all *do*. But they took the momentum we had gained in our descriptions of devastated forests and inverted it to declare: *The forests are suffering a major health crisis, so we need to move quickly to cut them down.*

Stop. Reread their declaration. Read it again. I've read that or similar lines too many times, and they still make no sense.

But that's one of the advantages of wielding the sort of totalitarian power held by the government/corporate interlock. While it's certainly more convenient for them to carry out their "preferred alternative" without too much public resistance, public assent to their goals is ultimately incidental. Any time the public finds a way to meaningfully participate in decisions concerning its own landbase—as we did with the appeals process—they simply change the rules. Almost any excuse will serve to allow them to sever public participation in the process. As we've seen time and again, on issue after issue, when the corporate press publishes absurdities often enough, the absurdities begin to seem palatable to some, confusing to others, and discouraging to still others. As long as it paralyzes the public, the corporations win. Ninety-five percent of the old growth is gone, and they're getting away with cutting the rest.

So in 1995 Congress passed and President Clinton signed what became known as the Salvage Rider, which stated that the quickly worsening health of the forests demanded immediate action. Therefore, any timber sale that the Forest Service or Bureau of Land Management (BLM) declared necessary to improve "forest health" would be exempt from all environmental laws. The Salvage Rider contained something called "sufficiency language," a magical phrase meaning no appeals or other legal challenges are allowed. Public participation is explicitly prohibited. Of course.

Can you guess what happened? The Forest Service and BLM predictably declared nearly every timber sale to be necessary for forest health. It was a chainsaw massacre. Ancient forests fell everywhere. In my corner of the world, every one of the thousands upon thousands of acres I had worked to save—every goddamned acre, every fucking acre, every beautiful, vibrant, stunning, gracious, wise, living acre of ancient forest—was clearcut over the next two years. I did not have the courage to return to many of those places. I could not have borne to see them destroyed. Others I did go to see, and walked the moonscapes that until recently had been living, vibrant forests. It is not an experience I look forward to repeating, though of course it is an experience shared by all of us who love wild places and who are facing down the deforesters.

This is how our political system works. Choose your own equivalent example; they are myriad. This is why the system must go.

᠊᠊ ᠊᠊ ᠊᠊

The processes of globalization can be loosely summarized by a sequence of motivations and actions. Insatiable consumption leads those in power to invade other nations to steal their trees. This leads to the conversion of forestland to cropland, which leads to former forest dwellers being forced to work as farmers or factory laborers. This leads to poverty, which leads to refugees entering the receding frontier forests for food and fuel. All of this leads to international debt, which leads to agribusiness and raw materials exports, which leads to more debt and greater poverty. This finally leads to collapsing domestic economies, structural adjustment, and acceleration of the whole cycle.

⊛   ⊛   ⊛

Transnational corporations are eating the world. One of their strengths as a tool of the rich is that they kill so effectively at great distances. Those who profit never have to see the devastation they cause. Another of their strengths is that because they do not actually exist, because they are legal fictions, they can never be killed. Worse, because they have no bodies— being instead the "embodiment" of greed, and because they, like the rest of our culture, are an attempt to separate themselves—for whatever stupid, insane, murderous, and suicidal reason—from the world that surrounds them, they effectively never need to stop growing.

So corporations grow huge as forests grow small. Our lives and local economies grow ever more controlled by these ever larger, ever more distant, and ever more imperious fictional entities.

And they are eating the world.

It's all very strange, very sad, very stupid.

⊛   ⊛   ⊛

Canadian and American taxpayers pay for destruction of their forests. Brazilian and Guatemalan peasants suffer the depredations of their landlords. Future generations will pay for their ancestors' gluttonous consumption. Lemurs and tigers pay through their persecution—and extirpation—by humans.

Some of these externalizations of the worldwide timber industry are unintentional, and cause the destruction of forests as an unplanned consequence. Others are deliberate, and are intended to boost or maintain profits of various wood and paper industries by externalizing costs or increasing consumption of wood and paper products to match the industries' overcapacity, making human "demand" keep up with a supply created by megamills.

From 1996 to 1998, the U.S. wood and paper products industry took $3.6 billion in profits. They paid $500 million in taxes and received $759 million in tax breaks. Two dozen wood and paper companies paid less than nothing in taxes; they actually received taxpayer money. For example, in 1998, Weyerhaeuser's taxes were a *minus* $9.5 million. In 1994 the state of Ohio granted a package of subsidies to International Paper, including a $7 million loan and a $700,000 grant to buy equipment, a $3.4 million tax credit, $420,000 in sales tax exemptions, and a $90,000 job training grant.[18]

The U.S. State Department's Overseas Private Investment Corporation and the U.S. Export-Import Bank fund deforestation overseas, from Russia to Indonesia to Chile. Japan subsidizes the destruction of forests and their replacement by sterile plantations around the world—in Russia, Okinawa, Indonesia, Brazil, Australia, and Vietnam. European nations fund roads and dams that chip away at what's left of the African rainforests.

Planting trees and environmental monitoring of timber corporations are nearly meaningless when tax monies are used to build logging roads and log export docks, to insure politically risky logging operations in regions torn by civil wars, to exempt paper corporations from environmental regulations, to loan tractors and pulping machines to communities which have managed so far to protect their forests, and to provide police and military to attack or kill those who resist.

These subsidies must be changed. Imagine if the same amount of money was thrown at actually trying to help forests—not through bogus schemes where those in power change the names of destructive actions to make them more palatable (remember *temporary meadows*?) but through programs that serve the forests and all who live there?

What a concept.

⊕ ⊕ ⊕

Eventually the decisions about land and resource use need to be controlled by local peoples who know, love, and depend on the forests.

We pride ourselves on our democratic freedoms, but even our façade of democracy stops at the border. Half the extant human languages are in Papua New Guinea, the Congo Basin, and the Amazon.[19] Democracy would look quite different if those who spoke languages other than English were allowed voice. Heck, democracy would look different if those who spoke languages other than high finance were allowed voice.

It is fashionable to condemn "banana republics" for having unequal distributions of land, but U.S. land distribution is no better. In Brazil, 2 percent of the farms occupy 54 percent of the land.[20] In Honduras, two-thirds of the arable land is owned by 5 percent of the farm owners.[21] In the United States, the top 5 percent of *landowners* (not 5 percent of total population) own 75 percent of the land,[22] worse than the ratio in Honduras. In California in 1970, 58 percent of the farmland was owned by twenty-five owners (*not* 25 percent). And as we've seen, many of the largest landowners in the United States and around the world are timber corporations.

The landless worldwide are surviving on a fraction of the acreage that was once theirs, acreage that is still available, being held to keep people landless—so they will be cheap laborers—and for speculative real estate purposes.

We don't need "public participation, consensus, and collaboration," or "community forestry" programs run by corporate and government elites; we need local control of land and markets. Adam Smith's invisible hand of the market only worked when the market was local, face to face, voluntary, transparent, lowtech, and based on ethical, mutual relationships. It's been a long time since that was the case.

☙ ☙ ☙

We have some declarations of our own. Immediately leave remaining frontier forests alone, and confine industrial forestry to existing plantations. Soon, once we have learned how, restore most and then all of the plantations to natural forests. This work could be done by restoration ecologists, who, like traditional forest-dwellers, are grounded in their specific local natural communities. Restoration ecology will be one step toward recovering indigenous knowledge and techniques, which is always specific to place. The purpose of restoration is not fiber production, even sustainable fiber production, but restoring ecosystems and their humans to their natural local patterns and processes.

Perhaps most important of all, relinquish control of land to those who belong to the land. Satellite data has shown that where indigenous people hold land title, there has been less forest destruction. Give back the land to the humans and nonhumans who live there, and who have lived there for a very long time, who belong to the land. Give it back to those from whom those in power stole it, and from whom they continue to steal it.

⊛   ⊛   ⊛

Deforestation boils down to power. Those who deforest do so because they are supported by the full might of the state. It is ludicrous for anyone to suggest that those who stole these lands by force, and who maintain control by force, and who deforest at the point of a gun, will give the land back to its rightful human and nonhuman owners because it is the right thing to do, the sane thing to do, the human thing to do, the nonsuicidal thing to do. No.

They will not leave the forests, and leave the forests alone, until either the forests are gone, or until those of us who love the land force them out of the forests.

We do not hope to stop deforestation with this book alone, any more than we can stop it by filing timber sales, compiling corporate profiles, voting, doing tree sits, or sending money to good organizations like Amazon Watch or Cultural Survival.

We do not know how to stop deforestation. We do not know how to get deforesters out of the forests. No one else—forest dwellers or civilized—has figured that out either, or surely the deforesters would have been removed by now.

But we do know this. Once people see deforestation for the atrocity that it is, they will then stop those who continue to destroy. It is for this we wrote the book. It is to this we have dedicated our lives.

Walking on Water
Reading, Writing, and Revolution

Chelsea Green Publishing, 2005

After I had published a few books, a lot of people told me that my writing was beautiful, but my topics were too gloomy. They said I would reach a larger audience if I wrote a book that was happy and exuberant. So I tried writing a happy and exuberant book—and the book is actually quite happy and exuberant, and was tremendous fun to write—and it didn't sell. But I really like the book.

The promise of education is quite extraordinary, and what we get instead is a toxic mimic of what education should be. Education comes from the root e-ducere, which means to lead forth or draw out. Originally it was a Greek midwife's term meaning "to be present at the birth of." That is the sacred role of educators: to be present at the birth of their students. Instead, our students are being trained how to be slaves.

As is true for most people I know, I've always loved learning. Also as is true for most people I know, I always hated school. Why is that?

The answer, obvious to me now, is that I didn't like what I was learning. I don't think my primary problem was with the subjects themselves: I taught myself about numbers before first grade so I could follow baseball statistics, and I was writing plays (albeit short ones) in second grade. It was something deeper.

One of the difficulties we have in thinking or speaking about the problems of our school system is that we presume the primary purpose of school is to help children learn how to read, write, and do arithmetic. This is an understandable mistake, but one we continue to make at our peril. For more is at stake in the process of schooling than mere booklearning or even the development of character. The process of schooling gives children the tools they can—and often must—use to survive after graduating into "the real world," and teaches them what it is to be a member of our culture. Not often enough asked are the questions: What sorts of tools are these? and, What *is it* to be a member of this culture? In other words, we might be well served to ask what sorts of beings we are creating by the process of schooling.

My own primary experience of school was one of tedium. Year after year I sat in the back of the class, watching the second hand move ever-so-slowly. I can't tell you how many times I calculated the seconds till school was over for the day, the week, the year, thus branding into my mind the importance of arithmetic. When bored I laughed, and almost daily I pinched the insides of my thighs until they turned red, or sometimes bit the inside of my cheek until it was raw, all to keep me from bursting into uncontrollable laughter. I shifted from cheek to cheek of my buttocks, trying to keep my legs from falling asleep. I snuck books into classrooms and read them in my lap. I taught myself American Sign Language in an attempt to silently communicate with a friend in another row, even if for no other purpose than to tell him he looked like a booger. I tested to see how long I could hold my breath. I counted the number of times the teacher said *umm* or *okay* in one hour, with the record being a remarkable two hundred and fifteen (I still remember the number as well as the welt I brought up on the inside of my thigh that day). In sum, one of the primary things I learned was how to kill time.

I learned also to wish away my life. I remember a spring day in eighth grade, standing on the football field with a new friend whose name I no longer recollect. I told him I couldn't wait for the next month to be over, when summer vacation would begin. He looked at me, confused, and said to me something he had clearly heard from a parent: "You're wishing away the only thing you've got." I knew immediately he was right, but that didn't alter the fact that I wished it were June instead of May.

What else did I learn? I learned to not talk out of order, and to not question authority—not openly, at least—for fear of losing recess time, or later of losing grade points. I learned to not ask difficult questions of overburdened or impatient teachers, and certainly not to expect thoughtful answers. I learned to mimic the opinions of teachers, and on command to vomit facts and interpretations of those facts gleaned from textbooks, whether I agreed with the facts or interpretations or not. I learned how to read authority figures, give them what they wanted, to fawn and brown-nose when expedient. In short, I learned to give myself away.

I've talked to friends whose development was similarly shunted (or stunted) by school, though for some the emphatic feeling was anxiety instead of tedium. I am not the only person who twenty years later still dreams anxiously—as I did again about a month ago—that it's the last week of class and I'm frantically preparing for a test on a subject I do not know and do not care about, and for which I've no idea even where the classroom is.

It's not possible to talk about schooling without talking about socialization. It's not possible to talk about socialization without talking about society, and what society values. We hear a lot of talk—a lot of meaningless talk, really—about how terrible it is that high school students cannot locate the United States on a map of the world (which should be easy enough: just look in the center), give the century in which the American Civil War was fought, or name any members of the cabinet of the United States. We are told that standardized testing must be imposed to make sure students meet a set of standardized criteria so they will later be able to fit into a world that is itself increasingly standardized. Never are we asked, of course, is whether it's a good thing to standardize children (sorry, I mean students), knowledge, or the larger world. But none of this—not maps, not dates, not names, not tests—are really the point at all, and to believe so is to fall into the fallacy that school is about learning information, not behaviors.

We hear, more or less constantly, that schools are failing in their mandate. Nothing could be more wrong. Schools are succeeding all too well, accomplishing precisely their purpose. And what is their primary purpose? To answer this, ask yourself first what society values most. We don't talk about it much, but the truth is that our society values money above all else, in part because it represents power, and in part because, as is also true of power, it gives us the illusion that we can get what we want. But one of the costs of following money is that in order to acquire it, we so often have to give ourselves away to whomever has money to give in return. Bosses, corporations, men with nice cars, women with power suits. Teachers. Not that teachers have money, but in the classroom they have what money elsewhere represents: power. We live in a culture that is based on the illusion—and schooling is central to the creation and perpetuation of this illusion—that happiness lies outside of us, and specifically in the hands of those who have power.

Throughout our adult lives, most of us are expected to get to work on time, to do our boss's bidding (as she does hers, and he his, all the way up the line), and to not leave till the final bell has rung. It is expected that we will watch the clock, counting seconds till five o'clock, till Friday, till payday, till retirement, when at last our time will again be our own, as it was before we began kindergarten, or preschool, or daycare. Where do we learn to do all of this waiting?

Also expected is that we will be good citizens, good boys and girls all. We won't question country, God, capitalism, science, economics, history, the rule of law, but in all those areas we will defer—and continue to defer—to experts, just as we were taught. And the experts themselves? It is expected that they will be exquisitely sensitive self-censors, knowing always what or whom to question, what questions to leave unexamined, and most of all which asses to kiss. And none of us, if all goes well, will ever question how these areas—religion, capitalism, science, history, law—trick out in our own lives, even as we give our lives away.

Here are some questions I've been asking lately: What are the effects of schooling on creativity? How well does schooling foster the uniqueness of each child who passes through? Does schooling make children happier? For that matter, does our culture as a whole engender happy children? What does each fresh child receive in exchange for the so-many hours for years on end that she or he gives to the school system? How does school help to make each child who he or she is?

A couple of years ago I was at the public library in Spokane, Washington. A sort of counselor led a bunch of reluctant teenagers through the front door and to the computer card catalogs. There he turned them over to what he must have thought was the hippest of the librarians, a pony-tailed young man in a checked flannel shirt. The kids sulked. It was pretty clear they were from a detention center, or rehab, or perhaps had gotten in trouble at school and been sent here as punishment. The librarian pointed to the terminals, then said, "Give me a subject."

No one spoke.

"Anything," he said. "You tell me what you want to read about, and I'll find you a book on it."

From my vantage point at another bank of terminals, I could see that despite himself, one of the guys began to get interested. He was easy to read. *I can look up any subject*, he was thinking, *just like that?* The kid wore baggy jeans. He was Hispanic, with a bandana on his head. He had as much goatee as any sixteen-year-old can muster. He started to say something, then stopped. Still no one else spoke. Finally he raised his hand, and said, "You got a book on guns?"

The man with the pony tail just looked at him, so the kid said again, clearly, as if the other were hard of hearing, "Guns."

Everyone laughed. The kid stared for a moment before looking down, and away. I could tell he was wishing he had a gun right then, to blow a hole in the front of the computer. I was wishing I had one to help him.

I saw a blonde girl on the other side raise her hand, and I heard her say, "Whales."

The librarian said, "Whales," and typed it in.

That's why children hate school.

🔹 🔹 🔹

I have experienced learning, even in a classroom, as liberation. I have taught—isn't the right word, because I've always considered it my role instead to merely create an atmosphere in which students wish to learn—at a university and at a prison. The original intent of this book—with a working title of *How to Not Teach*—was to sketch my experiences at Eastern Washington University and at Pelican Bay State Prison—experiences that are more similar than one might at first expect—in the hope that someone else might gain from this retelling. I soon realized, however, that to describe my experience in a vacuum would be artificial, and less helpful than if it were embedded in a discussion of the social context that creates our usual experience of schooling. Another way to say this is that before asking whether I or anyone else has been successful in the classroom, we need to ask what we want to accomplish. And before we can rely on our answer to this second question, we need to ask what we are already doing and what we are currently creating by the process of schooling, because that understanding will help us understand—all rhetoric aside—what we *really* want, and will also make clear the stakes involved in the formation of students' characters.

❀ ❀ ❀

Pretend you wish to make a nation of slaves. Or to put it another way, you wish to procure for your nation's commercial interests a steady supply of workers, and a population pacified enough to not resist the expropriation of their resources. The crudest and probably most common means of facilitating such production is through direct force. Simply capture the workers and haul them to your factories and fields in chains. A slightly more sophisticated approach is to dispossess them, once again usually at gunpoint, then give them the choice of starvation or wage slavery. Alternatively, you can force them to pay taxes or purchase your products, thereby guaranteeing they'll enter the cash economy, meaning, ultimately, that they've got to work in your factories or fields to gain the cash.

The primary drawback of each of these approaches is that the slaves still know they're enslaved, and the last thing you want is to have to put down a rebellion. Far better for them to believe they're free, because then if they're unhappy the fault lies not with you but with themselves.

It all starts with the children. If you don't start young enough, you'll never be able to enculturate them sufficiently, and they will still believe in alternatives. And if they honestly believe in alternatives—those not delineated by you—they may attempt to actualize them. And then where would you be?

❀ ❀ ❀

Here's something I wish I would have told my students. The word *education* comes from the Latin root *e-ducere*, meaning to *lead forth* or *draw out*. Originally it was a midwife's term meaning *to be present at the birth of*. I would contrast that with the root of the word *seduce*, which is closely related, but with a striking difference. To *educe* is to lead forth; to *seduce* is to lead astray. I wish I had talked about that with those students years ago, and I wish I had suggested that they think about that difference the next weekend. As they approached someone of their preferred gender, perhaps they would have said, "I would like to educe you" (which would have been great if their intended happened to be conversant in Latin, but otherwise might have led the other to say, "Get away from me, you perv"). More to the point, I wish I had suggested that our departments of education be called, if we were honest, departments of seduction, for that is what they do: lead us away from ourselves.

On second thought, maybe it's best I didn't talk about that. I was having enough trouble talking about prepositions, and spelling. Who knows what sort of trouble I could have gotten into had I begun talking about the relationship between classrooms and seduction.

⊛   ⊛   ⊛

When I entered a classroom at Eastern Washington University, I changed the name of the class from *Principles of Thinking and Writing* to *Intellectual, Philosophical, and Spiritual Liberation and Exploration for the Fine, Very Fine, and Extremely Fine Human Being.* We moved the desks out of rows, and into a circle. As I went around the classroom taking roll, I asked each student what he or she loved. They told me stories, of their families, of farming, of their art, of their love for sports. But more even than learning details of their lives I learned that they were natural storytellers. Just as my jumpers hadn't really needed to be taught how to jump, but rather needed to be led forth into becoming the jumpers they already were, so, too, I realized quickly that my writing students didn't so much need to be taught how to write as they needed to be cheer-led into becoming the writers they had inside of them. They knew how and where to start a story, how to include appropriate details, how to make the story lead to a pay-off. All of this was there in the first stories they told, of what or whom they loved. They just had to realize the gifts they already possessed. I could not create gifts for them from nothing, but, surely and easily, I could help with this.

⊛   ⊛   ⊛

It's the second day of class. I come in a couple of minutes late. The students have already moved the chairs out of rows and into a circle. I announce that we have a seating rule. They stare at me, jaws open. After only one day, I'm already used to this response. "The one rule in seating," I say, "is that you can't sit where you sat yesterday. Nor can you sit next to the same people."

"That's two rules," someone says.

"So it is," I respond.

"You said there was only one seating rule."

It's my turn to stare.

They grumble a little as they move to different seats.

The first reason for making this rule is obvious: I want for them to try to see things from a different perspective each day they come in. The second is sneakier, and something I wish my teachers had done for me when I was in school: I want to give the shyer members of the class an excuse to sit next to someone they might be interested in seducing, or at the very least, talking to, or at the very, very least, admiring from close-up, rather than afar.

"Okay," I say. "Take out paper and pencil. We're going to talk about the rules of writing."

A look of resignation and recognition passes over their faces as they realize that my re-titling of the class and my question the day before about what they love were both merely ways to seduce them into thinking this class was something different. The revolution now safely over, they're ready to meet the new boss, who will be, of course, the same as the old boss. They assume student mode, ready to write down what I say so they can regurgitate it back to me later.

"The first rule of writing is: don't bore the reader."

They write it down.

I continue, "It doesn't matter how important a writer's message is if the book or movie doesn't keep your interest. If you read a book and it's boring, what do you do? If you watch a movie and it's dull, what do you do? Anytime you pick up a book or watch a movie, you could be doing anything else in the world. You could be taking a walk. You could be eating. You could be having a wonderful discussion about what it will take to dismantle civilization."

They nod dutifully. Some take notes.

"You could," I continue, "be having sex."

Pencils stop. I have their attention.

"The same is true for me with the papers you write. I could be doing anything else instead of reading them. So, I have only one requirement for your papers. I don't care what you write about. It can be fiction or nonfiction. I don't care whether I agree or disagree with your opinions."

Blank faces, and I know they don't believe me. Pencils moving.

"But the important thing is this: the papers have to be good enough—interesting enough—that I would rather read them than make love. Is that clear?"

Pencils stop again. Confusion. A woman in her late twenties, sitting in the far corner of the room, issues a single staccato laugh, and, having been given implicit permission by her response, the rest of the class begins to laugh, too. They think I'm kidding.

⊛   ⊛   ⊛

It should surprise us less than it does that the educational system destroys students' souls. From the beginning, that has been the purpose. Don't take my word on this: take it from the people who set up the system. In 1888 (and I'm indebted to the great website *The Memory Hole* and the great educator and writer John Taylor Gatto for collecting these quotes), the Senate Committee on Education, nervous about the high quality of education provided by non-standardized, localized schools (where—the horror! the horror!—teachers actually taught students to think for themselves!), reported, "We believe that education is one of the principal causes of discontent of late years manifesting itself among the laboring classes."

Industrial educators set out to rectify this problem. How? As industrial educator and philosopher John Dewey (inventor of the Dewey Decimal System) said, "Every teacher should realize he is a social servant set apart for the maintenance of the proper social order and the securing of the right social growth."

Next questions: What are the proper social order and the right social growth? In 1906, Elwood Cubberly, who later became dean of education at Stanford, gave his answers: Schools should be factories "in which raw products, children, are to be shaped and formed into finished products . . . manufactured like nails, and the specifications for manufacturing will come from government and industry."

Then in 1906, the Rockefeller Education Board, major backer of the movement for compulsory public schooling, gave its reasons for putting its money into that movement: "In our dreams . . . people yield themselves with perfect docility to our molding hands. The present educational conventions [i.e., the development of children's intellects and characters in homes and local schools] fade from our minds, and unhampered by tradition we work our own good will upon a grateful and responsive folk. We shall not try to make these people or any of their children into philosophers or men of learning or men of science. We have not to raise up from among them authors, educators, poets or men of letters. We shall not search for embryo great artists, painters, musicians, nor lawyers, doctors, preachers, politicians, statesmen, of whom we have ample supply. The task we set before ourselves is very simple . . . we will organize children . . . and teach them to do in a perfect way the things their fathers and mothers are doing in an imperfect way."

Those in charge could not have been more clear. William Torrey Harris, US Commissioner of Education from 1889 to 1906, wrote: "Ninety-nine [students] out of a hundred are automata, careful to walk in prescribed paths, careful to follow the prescribed custom. This is not an accident but the result of substantial education, which, scientifically defined, is the subsumption of the individual."

Finally, bringing this around not only to students' relationships to themselves but to the land, Harris also stated, "The great purpose of school can be realized better in dark, airless, ugly places. . . . It is to master the physical self, to transcend the beauty of nature. School should develop the power to withdraw from the external world."

No wonder we all hate school.

And the fact that we do hate school is a very good thing. It means we're still alive.

⊕ ⊕ ⊕

There's really only one question in life, and only one lesson. This question is whispered endlessly to us from all directions. The moon asks it each night, as do the stars. It's asked by drops of rain that cling to the soft ends of cedar branches, and by teardrops that cluster at the fold of your nose or the edge of your mouth. Frogs, flowers, stones, pieces of broken plastic, all ask this of each other, of themselves, and of you. The question: Who are you? The lesson: We're born or sprouted or hatched or congealed or we fall from the sky, we live, and then we die or are worn away or broken or disperse into a river, lake, or sea, ripples flowing outward to bounce back from the far shore. And in the meantime, in that middle, what are you going to do? How are you going to find, and *be*, who you are? Who are you, and what are you going to do about it?

If modern industrial education—and more broadly industrial civilization—requires "the subsumption of the individual," that is, the conversion of vibrant human beings into "automata," i.e., into a pliant workforce, then the most revolutionary thing we can do is follow our hearts, to manifest who we really are. And we are in desperate need of revolution, on all scales and in all ways, from the most personal to the most global, from the most serene to the most wrenching. We're killing the planet, we're killing each other, and we're killing ourselves.

And still our neighbors—hummingbirds, craneflies, huckleberries, the sharp cracking report of the earthquake that shakes you awake in your bed—ask us, who are you, who are you in relation to each of us, and to yourself?

Our current system divorces us from our hearts and bodies and neighbors, from humanity and animality and embeddedness in the world we inhabit, from decency and even the most rudimentary intelligence (how smart is it to destroy your own habitat: who was the genius who came up with the idea of poisoning our own food, water, and air?). I've heard defenders of this system say that following one's heart is not a good enough moral compass, that Hitler was following his heart when he tried to conquer the world, tried to rid the world of those he deemed unworthy. But Hitler was no more following his heart than are any of the rest of us who blindly contribute to a culture that is accomplishing what Hitler desired but could not himself bring to completion. The truth is—as I have shown elsewhere, exhaustively and exhaustingly—that it is only through the most outrageous violations of our hearts and minds and bodies that we are inculcated into a system where it can be made to make sense to some part of our twisted and torn psyches to perpetuate a way of being based on the exploitation, immiseration, and elimination of everyone and everything we can get our hands on.

Within this context, the question the whole world asks at every moment cannot help but also be the most dangerous: Who are you? Who are you, really? Beneath the trappings and traumas that clutter and characterize our lives, who are you, and what do you want to do with the so-short life you've been given? We could not live the way we do unless we avoided that question, trained ourselves and others to avoid that question, forced others to avoid placing that question in front of us, and in fact attempted to destroy those who do.

As we see.

❀ ❀ ❀

It's the next class period. I begin by saying, "I'd like for you now to do the most important writing exercise there is. It's a finger exercise. Writing is hard work, and just like in track or any other sport you have to stretch before you work out or you could pull a muscle. But even before you stretch you have to warm up a little bit. So, everybody shake out your hands."

They stare.

"For real. Shake 'em up."

They put their hands in front of them and shake.

"Now, hold your hands up, palms facing you."

They do it.

"The first thing is to reach with your thumb all the way over to the pad of your pinky. Stretch, stretch, stretch. Now, fold your pinky to cover your thumbnail. With me? Next, stretch down your first finger to cover the base knuckle of your thumb. That's a hard one. Finally, take your ring finger and cover your thumb's middle knuckle."

It takes a moment for them to catch up.

I say, "That's the most important writing exercise you can ever do. Do it often, at all authority figures, and especially at your internal critic."

They laugh.

They still don't know I'm serious.

❀ ❀ ❀

I've been told by my supervisor that the official policy for attendance to my classes at Eastern Washington University should be that any student who misses more than two (without a note from a doctor) must flunk. That seems crazy to me, yet it also seems there needs to be some consequence for those students who blow off classes not for good reasons but because they're irresponsible. The solution came from a student who, after being absent one day, brought in a long note from her doctor (illegibly hand-written for authenticity), whose name happened to be Frankenstein, describing how her services were needed for some work he was doing and telling me that if I happened to go to the biology department and happened to look at the collection of human brains they happened to have there, and happened to notice that one of them is missing, not to worry about it; the brain is being put to good use.

One of the only ways I got myself through school was by never reading any of the assigned texts, and still attempting to contribute to class discussions. Otherwise it was just too dull. My proudest achievement that way was the half-hour one-on-one conference with a high school English instructor about the thirty tragedies I was supposed to have read: I'd read three. I have to say, however, now having taught for several years, that I suspect my English teacher knew exactly what was going on, but, like me later, chose to reward my creativity (I actually worked much harder to learn enough about the tragedies to be able to fake it than I would have by simply reading them) instead of punish my disobedience. Likewise, I had a friend who routinely wrote book reports for books that didn't exist. I had another, as much a nerd as me and the rest of my friends, whose history papers consisted almost entirely of detailed descriptions of "obscure" battles that never took place and intricate biographies of nonexistent generals.

Of course I want to encourage that sort of thing.

What we—my students and I—come to is that every time a student is absent, one checkmark will be deducted. This checkmark could then be made up with another paper, preferably one as creative as the Dr. Frankenstein excuse.

☟ ☟ ☟

I still had too much control of the class. My best writing comes when I give up control, let the piece lead me where it will. No, *lead* is too solid a word, as though the muse is walking sedately a step ahead of me, holding my hand and gently pulling. Instead the act of writing, when I allow it to be its best, reminds me of nothing so much as something I loved as a child and teenager, which was to run down rocky mountains as fast as I could.

I'd start at the top, then begin trotting, picking up speed with gravity until my body was very nearly moving faster than my legs could churn, and certainly faster than I was able to pick out safe spaces for my feet to land. I'd hurdle downed logs and large rocks, trusting in my eyes and in my feet and in the ground itself that I'd be able to negotiate whatever was on the far side. Of course I tripped often, but that taught me how to fall without hurting myself too much, to roll and regain my feet and start over in one swift motion. Going back further, I can remember as a small child—five, six, seven years old—running down great sand dunes hundreds of feet high in southern Colorado. I'd fly off the top of a ridge and run faster and faster until my feet could no longer keep up, and then I'd tumble and tumble all the way down (frankly just thinking about that part makes me nauseated at this point: my stomach must have been much stronger as a child). I've been told that when I cleared the top of a ridge, the biggest things on me were my eyes, opening wider and wider from simultaneous fear and delight. That (apart from the nausea) is what I want in my writing, that's what I want in my life, and that's what I want in my classroom.

How would I do that?

Finally it occurred to me to break the students into groups, and ask each group to run the class for one two-hour period (students in day classes had only one hour to work with). They could do almost anything they wanted. I insert the word *almost* because every quarter I had to dissuade at least one group from having the entire class play nude Twister. But almost anything else was fair game. One group wanted to play Capture the Flag. I thought, "What does *this* have to do with writing?" But we did it, then wrote about it, and I felt closer to that class after our group's physical activity than I had even after intense emotional discussions. During the next class period we talked about the relationship between shared physical activities and feelings of intimacy. Another group had us eat popsicles and watch cartoons, then draw pictures from our childhood with our opposite hands (it broke my heart when one fellow shared his picture with the class: "This is my father taking me out in the woods to smoke my first vial of crack"). In the same group we played Duck Duck Goose and Hide-and-Go-Seek in the basement of the near-empty building. Many of the people were continuing students, and thus were older. Looking back, I don't know how anyone could possibly say that he or she has successfully run a writing class without having played Hide-and-Go-Seek with overweight old men, twenty-year-olds, middle-aged mothers of five, and a half-dozen men and women whose native language is not English, all of them dead serious about finding or not being found.

Another group brought a bag of questions and handed one to each of us. Each person would answer the proffered question, and then we'd go around the room all giving our answers. Then to the next person's question, and so on. The questions were excellent. Mine was, "How do you want to die?"

One group included a Vietnam veteran. On their night, he and one other student wrote on one chalkboard the words patriotism, heroism, war, bomb, national defense, national interest, missiles, tanks, guns, helicopters, soldiers, generals. Simultaneously ,two other group members wrote on another chalkboard, right next to the first, the words fuck, prick, cunt, sex, come, tit. We went around the room giving our reactions to the words on the two lists. The point soon became clear: why, they were asking, are the words on the second list considered obscene, while the first are not?

In another group, we went around the circle saying the thing we did in our lives we were the most ashamed of. It would have been easy for people to share only superficial acts had the group members not gone first. One had as a child playfully tied a string around a kitten's neck, forgotten about it, and come back a half hour later to find that the string had caught on the edge of the sofa when the kitten jumped off, and the kitten was dead. A man told us of infidelity to his wife. After that, there was no holding back. We were most of us in tears before we got around the circle.

One group taught us how to do a Country and Western dance, the Tush Push. This was especially difficult for me, a confirmed non-dancer. Because the room was too small, we did this in the building's central courtyard. Midway through one of our times pushing our respective tushes, a couple of the department's most humorless administrators walked by, evidently having worked into the evening. I smiled and waved. Even this class taught me much. I had been working on letting go in my writing for years by this point, and I sometimes became frustrated at the baby steps many students were taking toward manifesting their passion in words. But when it came to me attempting to let go in dancing, I suddenly comprehended their inhibitions: I would push my tush only three or four inches, while many who were too shy to open up in words were wildly swinging their hips (including a fifty-year-old sheriff's deputy I never would have pegged for a tush-pusher). In another class we made marsh-mallow figures representing our hopes and dreams. One fellow, a bow hunter, made a big marshmallow buck with toothpick antlers, and a huge toothpick arrow jutting from its chest; mine was a broken marshmallow dam with marshmallow salmon swimming in a river of marshmallow. We played blindfolded soccer in the classroom, with four people at a time blindfolded, being told where to move by sighted partners ("Left, left," my partner shouted as I ran into the wall. "Oh, sorry, wrong way."). We broke into groups, each group picking out of a hat the rough plot for a screenplay (our group was to come down from a mountain to find that everyone else in the world had disappeared), and then each person in the group picked from a different hat a character to be played in the drama (I was to play the actress Sharon Stone), after which we had an hour to write our scripts and to perform what we later dubbed "An Exercise in Embarrassment." For Halloween, we plopped sleeping bags on the floor, sat around a flashlight surrounded by small pieces of wood (simulating a campfire), ate s'mores, and told ghost stories. For Valentine's Day, we wrote stories about first loves, and memories of hearts broken or overflowing. Mainly we had fun.

▣ ▣ ▣

I haven't yet told you why I no longer have to force myself to write. It's because I fell in love. Another way to say this is that I fell into something far larger than myself. Yet another way to say this is, well, I'll just tell you the story.

It's the fall of 1987. I'm living in Spirit Lake, Idaho. I'm still counting words, making myself write. For cash, I work with a partner in a small bee-keeping woodenware business. It doesn't take much time (though it also doesn't provide much cash), so I read and write a lot. My reading is gener-ally purposeful, as I try to figure out how to write: what works and what doesn't. It's also eclectic, everything from Thomas Mann to Kilgore Trout, Aristotle to Edward Gibbon to Albert Camus to the pulpiest of pulp science fiction, mystery, suspense, and romance to *The 29 Most Common Writing Mistakes and How to Avoid Them.* The stories I write are often modeled on the stories I read, as I try to capture the authors' techniques. For example, I read a description by Raymond Chandler of a woman's face, and as I describe a different woman's face using his form I realize his description moves not randomly but directionally, from hair to cheeks to lip to chin to neck to breasts.

I read a book by James Herriot, one of the *All Things Bright and Beautiful* series about the extraordinarily sappy adventures of a Yorkshire veterinarian. One of the stories is about a man whose only friend is his dog. The dog goes with him daily to a bar. Somehow I know from the first line how the story will end: the dog will die and the man will kill himself. I know also that Herriot is going to use every trick he knows to make me cry. I vow he's not going to get me, and I think I can make it stick since the ending will be no surprise. I get to the end of the story, and when I finally quit crying I get mad at being so easily and obviously manipulated. Once I'm done being mad I marvel at his skill: how did he make me cry with ink on paper? When I'm done marveling I decide that's a skill I want to have.

I need a plot. Well, I think, if it worked for James Herriot, maybe it will work for me. The story will be about a man whose only friend is his dog. The dog dies, and, because I was a wimp, instead of killing himself the man moves out of town. Since I've never been to Yorkshire, I set it in Nevada, where I lived a few years prior.

For several months I try to write the story, but I can't, mainly because my idea stinks. This makes me unhappy for several reasons, not the least of which is that my words-per-day average is plummeting. At this rate I'll never make it to a million words.

Meanwhile, my friend's daughter comes up from California. She'd visited the summer before and we'd quickly become close friends. This time, however, isn't a visit. She's seeking refuge from her abusive husband, who's in jail for raping her. We talk a lot about her experience. I share with her the abuse I suffered as a child. But then, and this is a story we've all seen too many times, she drops the charges and starts to think about going back to him. *The beatings weren't that bad*, she says. *Except for this last one, he never hit me in the face. He going to change; I know that. The problem really isn't him, it's the drugs. My children need a father. How will I support us?*

I talk to her. It does no good. Her father talks to her. Her mother. Nothing. One night I talk to her mother and a family friend until three or four in the morning about what we can do to keep her from going back. We get nowhere. I go home. I sleep.

I awaken at nine that morning with the plot—and much of the early language—for a short story laid out before me. It will be from the perspective of a woman married to an abuser. She becomes good friends with a loner whose only other friend is his dog. Their conversations help her realize she deserves to be treated better than she is. But her husband learns of her friendship, and kills her friend's dog. The friend leaves town. Through that trauma the woman finally learns to love herself enough to give her husband the boot. I think it will take about ten pages—a long story for me at the time—and I vow I won't sleep until it's done. I start to write, and the words come easily, as if I'm dreaming them. But when I get to page six, I realize the story may total fifteen, and I quit for the day. At page twelve, I think it might be twenty. At eighteen, thirty. The dream continues for six months, till I finish a three-hundred-page manuscript.

The book was never published. I sent it to one hundred and twelve publishers, and received one hundred and twelve rejections. I know the number because I charted my submissions (I soon came to recognize the appropriateness of the word *submission* in this context) on a paper taped to the wall, then tracked when I got them back and sent them out again: after about forty rejections I moved the papers to the back of a door where I wouldn't have to see them so often. After about sixty, it got so going to the mailbox traumatized me. I wished I had an agent as a buffer, but I got thirty-five rejections from them, too.

The important thing, though, is that I found the door to where the muse lives, the place where words bubble like water from cold mountain springs. And I was in love. With the words. With the story. With the process. With the feeling of absolute engagement in a struggle that meant something to me.

How did that happen? Part of it, of course, is magic. It's still magic every time it happens, and it happens almost every day now. Part of it is practice, becoming comfortable enough with words and ideas and emotions to be able to shape them with some semblance of clarity and beauty. But there's more. It wasn't simply proficiency. There's a qualitative difference in the experience of writing—and in the words that end up on the paper—between work written without that engagement and love, and that written with. There's a click. An opened door. The difference between liking someone and falling in love. When I wrote earlier that the one thing I've learned about writing was how to tell when I'm writing crap, and to stop writing it (I hope the same is true, more broadly, for living as well), I really meant that I can tell when that door is shut, and so when the writing is coming from this side of it. I don't write then. And I can tell when it's open, and then I follow through to the far side, no matter the other circumstances. It could be late, and I have to get up early, but if the door opens, I go through it. I could be driving, in which case I pull over. I could be having sex (I meant what I said about the best writing being better than sex, but this statement is an oversimplification: the truth is that sex and creativity are tightly tied. Not much is more erotic to me than creativity, and not much is more creative than the erotic. But you knew that already: I'm sure the muse told you, too.)

The question becomes, how/why did the door open, and, most definitely segueing away from sex, how can I help my students to find and open their own doors?

Here's what I know. I cared deeply. I did not want my dear friend to return to her husband. I did not want for her to be hurt. I had a reason to write. I was trying to communicate. I wasn't writing to raise my word count. Nor was I writing precisely to get published (which is a good thing, since it never was). Nor was I writing as practice. I was desperately trying to convey a message—an experience—to my friend. It was like Charles Johnson said: this is what I would write if someone were going to kill me after I wrote the last word. The same holds true for every book I've written.

This ties back to everything I've been talking about. How do I help my students uncover their passion? What infuriates or terrifies or enraptures them? What does all three, and more? What messages do my students desperately want and need to convey, and to whom must they convey them? How can I help them trigger these passions strong enough that they lose self-consciousness and fall completely into the feelings, the words, the messages?

It goes back to the same old questions, but with a new one at the end. Who are you? What do you love? And the new one: What do you want?

If you tell me who you are, tell me what you love, tell me what you want, I'll tell you what you should probably be writing. Or maybe I won't have to. You will already have started.

Welcome to the Machine
Science, Surveillance, and the Culture of Control

with George Draffan

Chelsea Green Publishing, 2004

A young person sent me a 'zine describing a lot of technologies of the surveillance state. It was terrifying, and I thought it was science fiction. But the writer had carefully cited his sources, and when I followed his footnotes I found that all this extraordinarily terrifying surveillance technology was not actually science fiction, but was readily available and was being used. I realized I had to write a book about the surveillance state, and more broadly, about this culture's drive for control. It remains in some ways one of my scariest books, and unfortunately reality has gotten far more terrifying since the book came out.

When I was a child, I was taught—as a fundamentalist Christian—that while the devil could not read my mind, he watched everything I did, scanning for the slightest shift of my body or expression that would reveal my thoughts. He did this, I was told, because he wanted to know me. And he wanted to know me not because he loved me—as God did, who watched me also and who knew in addition what went on in my head and in my heart—but because he wanted to tempt and even control me.

My response as a child was to attempt to control myself, to let neither my face nor body, nor especially my actions, reveal my thoughts. I'd fool him! But I knew even at age five that this was a waste of time. I knew—though of course, I could not have used this language—that if the devil, or for that matter anyone, could assemble a large enough body of data about my external habits, he could in time effectively read my mind. I knew also that the capacity to read my mind, whether by God, man, or devil, would lead necessarily to the capacity to control me: Surveillance controls, and absolute surveillance controls absolutely.

What I didn't realize at the time was that by attempting to control myself I was effectively surrendering my freedom. I was allowing my fear—of the devil, and in retrospect even more so of God—to determine my actions, my expressions, my thoughts, and most damning of all, what I did not think.

I no longer believe in a devil, nor in a God, at least not the sort about which I was taught as a child. I do, however, carry with me the lessons I learned about the relationship between information held by a distant authority and control by that authority. This relationship has always been understood by those in power. It is a relationship we all need to remember.

⊛ ⊛ ⊛

By now, most of us can see the central movement of our culture: for the last several thousand years it has relentlessly expanded its region of control from its original base in ancient Mesopotamia—the "cradle of civilization"—through the Middle East and Levant, around the Mediterranean, into Europe, then Africa, the Americas, Asia, Oceania. In exerting this control, the culture has deforested more than 90 percent of the world, depleted more than 90 percent of the world's fisheries, similarly destroyed the great flocks of birds, the great herds of ungulates. It has destroyed, subsumed, or forcibly assimilated nearly all the cultures in its path, until most of these other ways of perceiving and being in the world have been forgotten. This much is clear. These are simply facts. They are beyond dispute.[1]

It would be a mistake, however, to think that this movement toward the attempt at absolute control extends only into the external world. It extends as surely into our inner worlds, into what we think and who we are, with the attempted control as complete, the devastation as severe, as that in the outer world.

⊛ ⊛ ⊛

One of the pioneers of modern surveillance was the eighteenth-century utilitarian philosopher Jeremy Bentham, designer of the Panopticon. The Panopticon is a blueprint for a prison designed as a cylinder, with cells radiating from the central guard station. There are no nooks or crannies where prisoners can hide. The cells are always lit, while the guard station is dark. Because prisoners can never tell whether or when they are being watched, they have no choice but to presume that at every moment they are under surveillance.

Here is what, with the Panopticon, Bentham proposed to accomplish: "Morals reformed—health preserved—industry invigorated instruction diffused—public burthens lightened—Economy seated, as it were, upon a rock—the gordian knot of the Poor-Laws are not cut, but untied—all by a simple idea in Architecture!"[2] Perhaps more to the point, whoever ran the Panopticon would gain a "new mode of obtaining power of mind over mind, in a quantity hitherto without example."[3]

Bentham was ambitious. This power was to be used widely, for *"punishing the incorrigible, guarding the insane, reforming the vicious, confining the suspected, employing the idle, maintaining the helpless, curing the sick, instructing the willing* in any branch of industry, or *training the rising race* in the path of *education*: in a word, whether it be applied to the purposes of *perpetual prisons* in the room of death, or *prisons for confinement* before trial, or *penitentiary-houses, or houses of correction, or work-houses, or manufactories, or mad-houses, or hospitals, or schools."*[4]

Here's how it works: "It is obvious that, in all these instances, the more constantly the persons to be inspected are under the eyes of the persons who should inspect them, the more perfectly will the purpose X of the establishment have been attained. Ideal perfection, if that were the object, would require that each person should actually be in that predicament, during every instant of time. This being impossible, the next thing to be wished for is, that, at every instant, seeing reason to believe as much, and not being able to satisfy himself to the contrary, he should *conceive* himself to be so."[5]

Bentham's ideas have been influential. For example, the Panopticon serves as the model for modern supermaximum security prisons such as Pelican Bay State Prison, here in Crescent City, California.

Indeed, as Michel Foucault wrote in the 1970s, the Panopticon has become a model for the entire culture. Thus the Panopticon has become not only a "simple idea in Architecture," but also a metaphor for the power relations that undergird modern civilization. Foucault wrote:

"Hence the major effect of the Panopticon: to induce in the inmate a state of conscious and permanent visibility that assures the automatic functioning of power. So to arrange things that the surveillance is permanent in its effects, even if it is discontinuous in its action; that the perfection of power should tend to render its actual exercise unnecessary; that this architectural apparatus should be a machine for creating and sustaining a power relation independent of the person who exercises it; in short, that the inmates should be caught up in a power situation of which they are themselves the bearers. To achieve this, it is at once too much and too little that the prisoner should be constantly observed by an inspector: too little, for what matters is that he knows himself to be observed; too much, because he has no need in fact of being so. In view of this, Bentham laid down the principle that power should be visible and unverifiable. Visible: the inmate will constantly have before his eyes the tall outline of the central tower from which he is spied upon. Unverifiable: the inmate must never know whether he is being looked at at any one moment; but he must be sure that he may always be so. In order to make the presence or absence of the inspector unverifiable, so that the prisoners, in their cells, cannot even see a shadow, Bentham envisaged not only venetian blinds on the windows of the central observation hall, but, on the inside, partitions that intersected the hall at right angles and, in order to pass from one quarter to the other, not doors but zig-zag openings; for the slightest noise, a gleam of light, a brightness in a half-opened door would betray the presence of the guardian. The Panopticon is a machine for dissociating the see/being seen dyad: in the peripheric ring, one is totally seen, without ever seeing; in the central tower, one sees everything without ever being seen."[6]

⊛　⊛　⊛

That is bad enough, but Foucault continues, "It is an important mechanism, for it automizes and disindividualizes power. Power has its principle not so much in a person as in a certain concerted distribution of bodies, surfaces, lights, gazes; in an arrangement whose internal mechanisms produce the relation in which individuals are caught up. . . . There is a machinery that assures dissymmetry, disequilibrium, difference. Consequently, it does not matter who exercises power."[7]

⊞ ⊞ ⊞

Right now, military researchers at MIT and elsewhere are working hard to fabricate technologies that will—and we have to stress that we're not making this up—allow soldiers to leap buildings, deflect bullets, and even become invisible. Shoes containing power packs will store energy when soldiers—or state police, or corporate security guards, insofar as there's a difference—walk, then release this energy in bursts to allow them to jump over walls. Soldiers—cops, corporate goons—will be given exoskeletons, like insects, to deflect bullets. These exoskeletons will have the capacity to turn into offensive weapons as well. These exoskeletons will also deflect light so that those wearing them will be as invisible as the man at the center of the Panopticon, as invisible as God. Ned Thomas, director of the Institute for Soldier Nanotechnologies at MIT, explains why he wants to try to create these übersoldiers—and I picture him laughing like all the mad scientists in all the bad science fiction movies as he speaks:

"Imagine the psychological impact upon a foe when encountering squads of seemingly invincible warriors, protected by armour and endowed with superhuman capabilities, such as the ability to leap over 20-foot walls."[8]

⊞ ⊞ ⊞

Military scientists long ago figured out how to put electronics into the brains of rats, and to cause them to move forward, backward, left, right by pushing buttons on computer keyboards. Imagine the fun these scientists will have if they figure out how to do this to women's hips.

Recent research has been aimed at co-opting the rats' will. Scientists put an electrode near a pleasure center in the rat's brain, and others to stimulate whiskers on each side of the rat's nose. The scientists then trigger, for example, implants near the left whiskers, and follow that by triggering the pleasure center. This convinces the rat to move left. After only ten days of this, rats can be trained to climb trees, walk, and stand in the open, or do many other things rats don't normally like to do, controlled by technicians issuing commands from laptop computers up to 550 yards away. As a reporter for the *Washington Post* put it, not disapprovingly, "The rat thus becomes a living robot, controlled remotely by a human handler but able to go anywhere a rat can go."

"I like the results," said a scientist at Northwestern University, who gave his reason: "This is the first time where you have control of a whole complex animal."

A scientist at New York's Downstate Medical Center put the final word on this, "The rat looks normal and isn't feeling any pain because he's getting rewards for doing the right thing."[9]

The rat is no longer a rat. It is a puppetrat, controlled by "providence," by God, by a man with a laptop.

Imagine putting electrodes near pleasure centers in human brains. Imagine getting humans to feel pleasure for doing things that are against their nature. Imagine getting them to feel pleasure for "doing the right thing," for doing that which is favored by providence, defined, of course, by those at the center. Imagine getting humans—or what used to be humans—to feel pleasure working for Wal-Mart (attaching RFID chips). Imagine getting them to feel pleasure purchasing items (containing RFID chips) from Wal-Mart. Imagine getting them to feel pleasure watching propaganda for the corporate state. Imagine getting them to feel pleasure voting in meaningless elections to put in power people who do not represent them. Imagine getting them to feel pleasure in following laws laid down seemingly not by those in power but by providence. Imagine getting them to feel pleasure as they narc out those who do not have implants or who otherwise do not choose to do "the right thing." Imagine getting them to feel pleasure in hunting down and killing those miscreants.

Imagine the fun these scientists will have when they put electrodes into the pleasure centers of women, to get them to feel pleasure—whether they feel pleasure or not—for "doing the right thing." They already do this: scientists have long since discovered that if they implant electrodes in women's brains—they use their patients in mental hospitals—they can bring the women, even women in what they describe as "a low mood," to have "repetitive orgasms."[10]

They may want to order a set of electrodes for use around the house.

Remote-controlled rats may be the least of our worries.

⊛　⊛　⊛

Tomorrow's warfare, according to experts at a conference on the future of weaponry, will be "revolutionised by computing, robotics and biotechnology to create 'killer insects' that can hunt down their prey in bunkers and caves and eat humans alive." Paul Hirst, professor of social theory at Birkbeck College, London University, gives some details: "micro aircraft that fly by their own sensors and carry many deadly sub-munitions; intelligent jumping mines that shower selected targets with small guided bomblets. . . . The result would be really effective substitutes for chemical and biological weapons: deadly bio-machines of finite life that could be released by sub-munitions, showering opponents in millions of nanobots . . . that could literally eat humans alive."[11]

And how will those in power find those they wish to have eaten? First, in addition to the RFID chips that can identify the location of someone who has bought any tagged consumer items, those in power will, according to Charles Heyman, editor of *Jane's World Armies,* be able to drop thousands of minimicrophones, cameras, and vibration sensors at crucial sites to relay information back to the center of the Panopticon.[12]

In case all of that doesn't suffice, military researchers are currently working hard to fabricate radar devices that will identify people by how they walk. It seems that our gait is as distinctive as our fingerprints, and scientists at Georgia Tech have been able to gain identification success rates of 80 to 95 percent.

A reporter asked Gene Greneker, head of gait research at Georgia Tech, whether he was concerned about the ends to which his work would be put. His response could have been spoken by the creators of mobile killing vans used by Nazis, creators of nuclear bombs, creators of electrodes to be put into the brains of rats (or women), creators of suits to turn the servants of those in power into übersoldiers. He said, "We are research and development people. We think about what's possible, not what the government will do with it. That's somebody else's job."[13]

The article did not report whether Greneker felt pleasure—electronically induced or otherwise—as he said this.

⊛ ⊛ ⊛

A couple of years ago, the United States government began bringing together information-gathering programs under a vast surveillance network called Total Information Awareness (TIA). TIA was a program of the Information Awareness Office, which in turn is part of the Defense Advanced Research Projects Agency (DARPA), run by the Pentagon.

Those in charge would like to be able to provide their agents with instantaneous access to records from around the world. A lot of records. In its advice to corporations that may contract to provide some of this information, DARPA states, "The amount of data that will need to be stored and accessed will be unprecedented, measured in petabytes." One byte is the amount of memory it takes to store one letter. One petabyte is one quadrillion bytes. That's one with fifteen zeros after it. This means that those in power want to maintain a database that would be more than fifty times larger than all of the books in the Library of Congress, or somewhere on the order of a billion books.[14]

This information could include financial, health, shopping, telephone, employment, and library records, fingerprints, DNA samples, gait analyses, brain scans, surveillance photographs, information on whom and how you love (including audio and video recordings of your most intimate moments), recordings of phone conversations, copies of emails, maps of Internet activities, information on addictions or other exploitable weaknesses, and all sorts of other information no sane person could even dream of collecting. Even if the project were to use only one petabyte of storage, that would still be enough to amass forty pages of text for each person on the planet.

In response to criticism, the United States government changed the name of Total Information Awareness—though not, of course, its function—to the less accurate Terrorism Information Awareness. Presumably it also began dossiers on everybody who complained about the program.

The Information Awareness Office logo consists of the name of the organization surrounding a blue background against which we have the truncated pyramid and the by-now-familiar all-seeing eye. This eye, of providence, of God, of the police, of the military, of representatives of major corporations, emits a ray of golden light to illuminate and overlook the globe. In the upper right are the initials DARPA, and in the lower left is *Scientia est Potentia,* a Latin phrase they translate as *Knowledge is Power.*

⊕   ⊕   ⊕

Knowledge is not always power. There are other ways to be and perceive in the world. Knowledge can be love. It can be relationship. It can be connection. It can be neighborliness or familiarity. Knowledge can simply be knowledge.

Last week I had one of the most exciting and wonderful mornings of my life. I live near a pond. I often sit at its edge. I love to watch tadpoles swim, watch them over time grow legs, slowly lose tails, take their first hops onto land, make their first awkward flips of the tongue (sometimes before they learn how to use their tongues, they wildly miss their targets and their whole bodies tumble till they land on their noses!). I also love to watch whirligig beetles who skate in incomprehensibly complex patterns—or maybe in no patterns at all—over the surface of the water, and backswimmers who hang motionless then glide quickly toward potential prey. Newts who swim to the top for great gulps of air, then back down again too deep for me to see. I watch mating dragonflies, the male joining his genitals to the female's near the base of the female's head, leaving her back end free to dip into the water and drop eggs even as they mate.

That morning a large brown insect crawled from the pond, covered with mud. I'd seen insects like this, and I'd also seen their skins hanging empty from blades of grass. I didn't know who they became. So I watched.

I watched as the creature made its way slowly across spaces of bare ground and through patches of grass until it found the blade it wanted to climb. It made its way to near the top, then grabbed on tight.

I waited. I looked away to water skippers and willows and rushes. When I looked back a furry hump had formed on the creature's back, between where the shoulder blades would be on you or me. The hump got larger.

Again I waited. The wind played with the tips of redwood branches. Wrentits sang, as did sparrows and thrushes, and some other bird I could not name but whose trilling song made me smile. A jay cocked its head and looked at me.

The hump became a head, and over time first one, another, then a third pair of legs became visible. They were all the palest yellow, nearly white. They unfolded slowly.

I had no idea who this creature was. The sun rode the sky. It grew warm on my back. More of the creature emerged, and more. It began to hang from the shell that used to be its skin. Sometimes it would move vigorously, sometimes it would slowly expand, and sometimes it would rest. I wondered if it would keep pushing itself from its former skin until it fell to the ground. Then suddenly it thrust itself upward to grasp the grass with its legs. It pulled hard, and pulled again. Finally it was free.

I still had no idea who it was. It was pale and stubby, with ruffles on its back.

I wanted to take a picture to show my friends, to post on my Web site. But I knew, because the creature told me, that this would be wrong.

The ruffles on its back began to expand. Slowly. Everything was slow. I'd been sitting by then for probably two hours, but it seemed much less because each moment I wanted to know what would happen the next.

The ruffles unfolded, the abdomen expanded. Longer, longer. The ruffles became wings, four of them. The eyes clarified. Colors came alive.

It was a dragonfly. No longer pale pink but very bright blue. "Now," it said. "Now get the camera." I did. It spread its wings. I took pictures. It waited.

I was hungry. I walked the path—three-eighths of a mile through dense forest—to my mom's. As I walked I pondered how many times I've walked this path these past three years. Easily three to four thousand. For the first year or so I used to carry a lantern at night, but then I quit because I got to know the path well enough to walk it at a normal pace even on the darkest nights (hint: look up to see the slight break in the forest canopy that signals the path). This time, of course, it was early afternoon. I got to my mom's. I ate there. I often do. I made my own meal, but she often cooks for me. She likes to cook and knows how to do it well. She also knows what foods I like, or don't. Afterward I helped her in her garden. She tells me what chores she would like me to do, and I (eventually) get them done. It works. We each know what helps the other, and want to help the other how we can.

I walked back home, expecting the dragonfly to be gone. But it remained through the afternoon, and into the night.

I awoke around 9:30 the next morning. The first thing I did was go outside, expecting, again, to see only the husk of the dragonfly, clinging to the grass. But the dragonfly remained. I stopped a few feet away. It did not move. I looked down to my feet for just a moment—to make sure I wouldn't step on any baby frogs if I shifted my weight—and when I looked back up it was gone.

There was only one large dragonfly on the pond. It was bright blue. It circled, then rose up to fly around the meadow, then back down to the pond. Then back up, in wider and higher spirals till it felt it knew the landscape. Higher and higher it spiraled, until it flew over the top of the redwoods and into the world.

Knowledge, whether it is of a dragonfly, a path, my mother, me, a landscape, is not always power. There are other ways to be and perceive in the world. Knowledge can be love. It can be relationship. It can be connection. It can be neighborliness or familiarity. Knowledge can simply be knowledge.

☙ ☙ ☙

Surveillance, and this is true for science as well—indeed, this is true for the entire culture, of which surveillance and science are just two holographic parts—is based on unequal relationships. Surveillance—and science—requires a watcher and a watched, a controller and a controlled, one who has the right to surveil or observe—with knowledge, truth, providence, and most of all might on his side—and one who is there for the other to gain knowledge—as power—about.

These unequal relationships require a split, a separation. There can be no real mixing of categories, of participants. The lines between watcher and watched, controller and controlled, must be sharp and inviolable. Humans on one side, nonhumans on the other. Men on one side, women on the other. Those in power on one side, the rest of us on the other. Guards on one side, prisoners on the other. At Pelican Bay State Prison, where I taught creative writing for several years, I once received a chiding letter from my supervisor after I innocently answered an inmate's friendly question as to what I was doing for Thanksgiving: to even let him know I was spending it with my mom was to make myself too known—too visible—to this other who must always be kept at a distance.

If this sounds a lot like the pornographic relationship, that's because it is. Pornography—cousin to surveillance, and bastard child of science—requires the same dynamic of watcher and watched, the same dyad of unchanged subject gazing at an object to be explored at an emotional distance, the same relationship of powerful viewer looking at powerless object. (This may explain at least some of the popularity of pornography: people who are powerless in every other aspect of their lives get to feel some power as they look at these pictures and read the attached text.) When I read that we must not "make scruple of entering and penetrating into these holes and corners," I wonder whether I am reading a letter by the father of science Sir Francis Bacon to King James I (describing how the methods of interrogating witches—that is, restraint and torture—must be applied to the natural world), or whether I'm reading a description at www.perfectlypussy.com. When I read about using the "mechanical arts" (that is, once again, restraint and torture) so that she "betrays her secrets more fully . . . than when in enjoyment of her natural liberty," am I still reading Bacon's words on science, or have I landed at www.fetishhotel.com?[15]

These unequal relationships—insofar as we can even call them relationships—must be oppositional. Predator and prey must not be working together for the benefit of both of their communities, and for the benefit of the land. Instead, from this perspective—this perspective based on selves being separate, and knowledge being gained through splitting off—predator and prey (and this applies to humans as well) must be locked in an eternal battle, good against evil, a battle that ends in Armageddon.

As civilization plays out its grim endgame, and as those in power move ever closer to their ultimately unattainable goal of absolute control (through absolute surveillance), converting in their efforts the wild both inside and out to devastated psyches and landscapes, it might be well past time to reconsider the premises that underlie much of this destructive way of being (or not being) and perceiving (or not perceiving). For in many ways, perception shores up the whole bloody farce.

※　※　※

What does science do? It calls for everything to be measured. It calls for everything that cannot be measured to be ignored or destroyed, and everything that can be measured to be analyzed (according to the rules of science). It calls for calculations to be made as to how everything that can be measured and analyzed can best be used. It calls for those doing the measurements, calculations, and analyses (and most especially their masters) to rule over everything that can be measured. We are describing the

Welcome to the Machine   **165**

methods and effects of science, not the conscious motivations of every scientist.

What is science for? To analyze. Why? To predict. Why? To reduce risk for those doing the calculations (and their masters) and to control those about whom (or, to use their lingo, which) these predictions are made. Why do they do this? So those performing these analyses and predictions can rule over everything they can analyze (and destroy everything they cannot).

Under this rubric, what is power? It is the ability to control outcomes. What, then, is a bureaucracy? It is administration by rules, efficiency, and quantification. It is the administration of control.

What, then, is a culture administered by a bureaucracy?

It is a machine.

What are the necessary preconditions for the conversion of a living human community into a machine? Members of this community must begin to perceive themselves not as fluid threads in a complex and ever-changing web of relationships—where they may play this or that role as appropriate, necessary, and desired (by them and by others)—but as gears within cogs within gears in what they now perceive as a giant machine over which they have no fundamental agency, no loving stake. They must perceive their value no longer as inherent but as strictly utilitarian: they must be converted from human beings to workers. They must be made to perceive relationships as strictly hierarchical, where those closer to the outside of the Panopticon serve those closer to the inside, where rewards run from outside to inside, and only if there are any rewards left do some trickle back out. Everything must be perceived in terms of its short-term utility. Nothing must be given back.

Why would nanotech plants outcompete and overwhelm real live plants? Why would nanotech bacteria outcompete and overwhelm real live bacteria? Why does our machine culture outcompete and overwhelm real live cultures? Because machines are more efficient than living beings. Why are machines more efficient than living beings? Because machines do not give back. All living beings understand that they must give back to their surroundings as much as they take. If they do not, they will destroy their surroundings. By definition, machines—and people and cultures that have turned themselves into machines—do not give back. They use.

And they use up. This gives them short-term advantages in power over the ability to determine outcomes. They outcompete. They overwhelm. They destroy.

Once people have been converted into cogs in their machine culture, the division of labor is increased, the skills of those in the outermost rings of the Panopticon are diminished, all heads are separated from hands (and hearts). Those in the innermost rings refuse to pay attention to anything that cannot be measured, and they convince everyone else to do so as well, if necessary at the point of a gun. They produce and convince everyone else to do so as well, once again at the point of a gun if those to be yoked to the machine have not been inculcated sufficiently to smile as they draw on their traces. Productivity is strictly defined in action (though it's best to not speak of this directly except when necessary) as the conversion of the living to the dead: living forests to two-by-fours; living rivers to hydroelectricity and from there to smelted aluminum (and from there to beer cans); living human beings to human resources. This conversion first takes place perceptually—subjects must stop perceiving others as subjects themselves and instead as objects—and then in the physical world. Efficiency is simply the rate and completeness with which the conversion takes place.

If people—or cogs that used to be people—are to be integrated into production, they must be recruited to be efficient. In practice, this means that nothing must stand in the way of production. Not leisure, not love, not a living landbase, not life on earth. That nothing human or animal is allowed to stand in the way of production can be surprising until we remember that production is, once again, the conversion of the living to the dead. That people are efficient simply means that life is not allowed to get in the way of its own killing.

Central to all of this is that it is more difficult by far to control diverse beings than it is to control objects that are all alike. Diversity must be destroyed. All cultures serving gods other than production—death—must be destroyed. All languages that do not serve this end must be forgotten. All creatures we can't use must be eliminated. All people must be standardized as well. (What do you think schooling is for?) One religion. One way of knowing the world. One economic system. One way of living on the land. If this language seems too strong to you, look around and ask what is happening to cultural diversity, to diversity of languages, to biodiversity, to all forms of diversity. They're disappearing. If you cannot perceive that, there is no hope for you. You will, however, always be welcome in the Panopticon.

☙ ☙ ☙

We should never be deceived that technologies are neutral. They are controlled by those in power, which means those in power have the ability to gain access to information about those under their power, whether or not those under their power desire this information known. This information is then used to reward those the powerful choose to reward, and to harm those they wish to harm. At the same time, those not in power do not have access to the same sort of information about those in power. Within this rubric, information, like power, is a one-way street, and a dead-end one, at that.

☙ ☙ ☙

The smoke and mirrors of high technology appear to be, Wizard-of-Oz-like, our masters. But the fearsome wizard behind the curtain turns out to be everyman and everywoman, fussing with levers, no longer in control or quite sure what the purpose of it all is. We've become caretakers of the machine, janitors in the machine's warehouse, sweeping and tidying up and oiling the levers and attaching the new gizmos that are supposed to control this or that section of the machine. (How come those gizmos are called *governors?*) We aren't sure anymore what it's all for, but someone somewhere must be keeping track, we tell ourselves, and those engineers sure are clever. And we really can't imagine how the machine could be replaced with anything else. It's big and it's more complicated than anyone can imagine, and it's all humming on into the future . . .

Gee-whiz, science is transforming our world. Isn't it amazing, this technology that can be used for good and ill? One of the world's "most advanced humanoid robots" is feted for "walking, turning and even dancing with children."[16] But it's not just for kids! When the Japanese prime minister paid an official visit to the Czech prime minister in the summer of 2003, he brought along a four-foot-tall robot that could tell jokes and make a toast. The robot offered to dance, but the Czech prime minister is not into dancing, and declined.[17] Dancing isn't the only thing robots are good for! The Pentagon's DARPA is offering a million bucks to see if someone can create a machine capable of going from Los Angeles to Las Vegas without human intervention.[18] The purpose of DARPA's "autonomous ground vehicle" race in March 2004 was to "leverage American ingenuity to accelerate the development of autonomous vehicle technologies that can be applied to military requirements."[19]

Big-tech apologist Witold Rybcynski begins his book *Taming the Tiger* with the declaration that "we must live with the machine; we have little choice"—marshaling what has become, pathetically enough, one of the primary arguments in favor of the current deathly system: we're committed to it, so if you don't like the death of the planet, tough luck—and ends it with the suggestion that we can control technology "by directing its evolution, by choosing whether and how to use it, or by deciding what significance it should have in our lives." But he conveniently forgets that civilization is predicated on power differentials. How do I direct the evolution of the unmanned predator drone? How do we choose whether to use the forty million closed circuit television cameras that capture our movements down the street? How do I decide what significance the recording of my e-mail messages and the revoking of my biometric-encoded passport shall have? Rybcynski rightly reminds us that "the struggle to control technology has all along been a struggle to control ourselves."[20] We say rightly, but we mean partly rightly, since we aren't the ones creating predator drones. We aren't the ones at the center of the Panopticon. This is why it's not a matter of controlling technology, but of changing the power relations in society. Our obsession with comfort makes us addicts to technology, and our attachment to security makes us servants of authority. As addicts and servants we neither control technology nor change the nature of power.

Jean-Jacques Rousseau's *Social Contract* declares that governance is by consent, not by mandate. He wrote, "Man is born free; and everywhere he is in chains." He also wrote, "The brain may become paralyzed and the individual still live. A man may remain an imbecile and live. . . ."[21] We see both of these around us each day. And he wrote further, as we also see each day, "Slaves lose everything in their chains, even the desire of escaping them. . . . If then there are slaves by nature, it is because there have been slaves against nature. Force made the first slaves, and their cowardice perpetuated the condition."[22]

Rousseau was perhaps following the analysis of Éttiene de La Boétie, who in 1564 wrote that it is, "the inhabitants themselves who permit, or, rather, bring about, their own subjection, since by ceasing to submit they would put an end to their servitude. A people enslaves itself, cuts its own throat, when, having a choice between being vassals and being free men, it deserts its liberties and takes on the yoke, gives consent to its own misery, or, rather, apparently welcomes it."[23]

A few of the layers of modern society maintain some semblance of consent, but amidst all the high-tech communication and the waves of data, is it informed consent? Is it the consent of consumers who purchase the latest piece of technology wrapped in layers of disposable ancient forest paper? Is it the consent of employees who are utterly dependent upon their paychecks for food, clothing, and shelter? Is it the consent of the wards of hightech hospitals, unprepared to die, who consent to millions of dollars worth of invasive medicine to prolong life for another month? Is it the consent of voters who choose between corporatefunded and party-chosen candidates?

⊛ ⊛ ⊛

We know that traditional functioning human communities have from the beginning of human existence been based on the exchanges of gifts, exchanges where relationships are more important than goods. We know that modern culture is organized around markets, where goods are more important than relationships.

We know that many traditional human communities have no privacy. People know each other. The quest for privacy develops as a response to oppressive relationships. Within real communities, where relationships are not oppressive, the central panoptic statement "So long as we do what they tell us, we have nothing to hide, and nothing to fear" can be inverted to "Because we have nothing to fear, we are free to do as they tell us (or not), and we need hide nothing." In fact, there's no "they" telling us what to do, and nothing to hide.

Think about it. What need would you have for privacy if you had no fear of anyone using what you do against you? That's not to say you won't have modesty or secrecy, or things that you keep to yourself for reasons known only to yourself, for you very well may. But there is a world of difference between not sharing from modesty or secrecy or simply because you don't want to, and not sharing from privacy. Privacy is based on fear, and it is based on unequal power in relationships. The need for privacy is a product of living in the Panopticon. The word itself arose in the fifteenth century (think enclosure) to mean "the quality or state of being apart from observation; freedom from unauthorized intrusion."

We know that traditional human communities have been based on social relations, rather than rules. Power was distributed, and fluid, based on circumstance and experience. Within modern culture, power is concentrated, and static. The rule of law is used to try to control what used to be shared in relationships. Power relations of course lead to irresponsibility: if relations are more important than rules, I am responsible to the relationships, but if relationships are not primary and are based on rules and control anyway, I will try to use or avoid those rules, to twist them to my advantage. Responsibility disappears, to be replaced by moral strictures. Relationships disappear, to be replaced by lawyers.

☖ ☖ ☖

The most dangerous, foolish, and false thing any of us can possibly say is the mantra of, "So long as we do what they tell us, we have nothing to hide, and nothing to fear." The degree to which we internalize the laws and rules of the machine is the degree to which we have no hope of survival, no hope of escape, and certainly no hope of smashing the machine (which means, once again, no hope of survival). The degree to which we internalize the laws and rules of the machine is the degree to which its control is complete, and self-discipline and self-responsibility become moot.

Toward the end of Orwell's *1984*, a secret policeman says to one of his victims, "The command of the old despotisms was 'Thou shalt not.' The command of the totalitarians was 'Thou shalt.' Our command is 'Thou art.'"[24]

The control the powerful wish to exert (and the powerless wish to obey) extends not only over the wilderness, but into the most intimate areas of our minds and hearts. If knowledge is power, as the Information Age cliché has it, and as DARPA contends, then if they know us, they can control us.

Technology separates us from nature, and social indoctrination—our training from birth on, our molding into machineshape—separates us from ourselves. When an "external" entity (either hard-wired, like the mass media or the government, or soft-wired, like my obsession with reading the news, or obeying Officer Friendly) controls the information I receive, it controls my experience of the world. And because my experience of the world controls my actions, whosoever controls my experience of the world controls me.

Those in power have known that all along. By now, we should know it, too.

⊛ ⊛ ⊛

I often think about the religion of my youth and wonder what it would have been like to grow up knowing another sort of God, one who did not sit at the center of a universal Panopticon and sort us all, to heaven with this one and to hell with this other. One who did not wield power from a distance. One who did not live in the sky. One without an oversight that is always possible and (conveniently) never verifiable.

How would our culture be different if the sacred resided instead in our own bodies and in the bodies of those humans and nonhumans who are our neighbors and companions in the world? What sort of society would evolve if we understood that truth is multivaried and relative and just as manifest in dragonflies, phoebes, and drops of rain as in humans? How would our society look—and how would the world look—if our gods and our goddesses valued life over power? How would *we* look if *we* valued life over power? What would happen if we rejected the myth of the machine, and the machine itself?

⊛ ⊛ ⊛

I'll tell you a secret. We may be spending our days and nights in the center of the machine, working for this thing that does not exist to get things that don't make us happy. We may be living in the outer ring of this Panopticon that does not really exist. But unlike those in the SHU, we can walk away. We can deliberately refuse the bargain offered by the machine, refuse to give up the richness of our own lives and the lives of those we love for shining things that please our eyes, for protection from their weapons that are more effective than our own, and above all for the spirits of the machine that make us forget for a time old age, weakness, and sorrow. We can go on total strike against the terms offered by the machine.

Here is the secret: We can say *no* to the machine.

And we can say *yes* to our own lives, and to the lives of those we love.

My students at the prison told me that plastered all over the walls of the L.A. County Jail, and across the walls of jails everywhere, is the phrase "Make a Deal." Many of my students told me that one of the easiest, most direct ways to shut down the whole court system would be for prisoners to stop making it easier on their captors, to not make deals.

We are not deceived that it is always easy to say *no* to the machine. Although it does not exist, there are those who believe in it enough that they may imprison you, they may torture you, they may kill you. This is true. There are few of us who are naïve enough to believe otherwise.

But here's one more secret. Until those in power figure out a way to completely replace humans with machines, there will always be more of us than there are of them. All it will take for this whole rotten system to collapse is for enough of us to learn to say *no*. And to say *no* again. And again. And again. And again.

And for enough of us to learn to say *yes*.

Endgame, Volume One
The Problem of Civilization

Seven Stories Press, 2006

Endgame *is the third work in an informal trilogy.* A Language Older than Words *is about how this culture is irredeemable on a psychological level.* The Culture of Make Believe *is about how it's irredeemable sociologically, on the level of social rewards. But you can still argue with those. With* Endgame, *I wanted to show how this culture is unsustainable and unjust on the level of resource movement, because you cannot argue with resource movement. Cities require a countryside which they must exploit.*

Of course, I also wanted to make it the best book ever written about the irredeemability of this entire culture. I wanted to address the sociopathological nature of the culture and of those who run it. I wanted to address men's hatred of women. And in Volume Two, *I wanted to begin to address the importance of resistance, of stopping this culture from killing the planet.*

Premises

PREMISE ONE: Civilization is not and can never be sustainable. This is especially true for industrial civilization.

PREMISE TWO: Traditional communities do not often voluntarily give up or sell the resources on which their communities are based until their communities have been destroyed. They also do not willingly allow their landbases to be damaged so that other resources—gold, oil, and so on—can be extracted. It follows that those who want the resources will do what they can to destroy traditional communities.

PREMISE THREE: Our way of living—industrial civilization—is based on, requires, and would collapse very quickly without persistent and widespread violence.

PREMISE FOUR: Civilization is based on a clearly defined and widely accepted yet often unarticulated hierarchy. Violence done by those higher on the hierarchy to those lower is nearly always invisible, that is, unnoticed. When it is noticed, it is fully rationalized. Violence done by those lower on the hierarchy to those higher is unthinkable, and when it does occur is regarded with shock, horror, and the fetishization of the victims.

PREMISE FIVE: The property of those higher on the hierarchy is more valuable than the lives of those below. It is acceptable for those above to increase the amount of property they control—in everyday language, to make money—by destroying or taking the lives of those below. This is called *production*. If those below damage the property of those above, those above may kill or otherwise destroy the lives of those below. This is called *justice*.

PREMISE SIX: Civilization is not redeemable. This culture will not undergo any sort of voluntary transformation to a sane and sustainable way of living. If we do not put a halt to it, civilization will continue to immiserate the vast majority of humans and to degrade the planet until it (civilization, and probably the planet) collapses. The effects of this degradation will continue to harm humans and nonhumans for a very long time.

PREMISE SEVEN: The longer we wait for civilization to crash—or the longer we wait before we ourselves bring it down—the messier the crash will be, and the worse things will be for those humans and nonhumans who live during it, and for those who come after.

PREMISE EIGHT: The needs of the natural world are more important than the needs of the economic system.

Another way to put Premise Eight: Any economic or social system that does not benefit the natural communities on which it is based is unsustainable, immoral, and stupid. Sustainability, morality, and intelligence (as well as justice) require the dismantling of any such economic or social system, or at the very least disallowing it from damaging your landbase.

PREMISE NINE: Although there will clearly someday be far fewer humans than there are at present, there are many ways this reduction in population may occur (or be achieved, depending on the passivity or activity with which we choose to approach this transformation). Some will be characterized by extreme violence and privation: nuclear Armageddon, for example, would reduce both population and consumption, yet do so horrifically; the same would be true for a continuation of overshoot, followed by a crash. Other ways could be characterized by less violence. Given the current levels of violence by this culture against both humans and the natural world, however, it's not possible to speak of reductions in population and consumption that do not involve violence and privation, not because the reductions themselves would necessarily involve violence, but because violence and privation have become the default of our culture. Yet some ways of reducing population and consumption, while still violent, would *consist* of decreasing the current levels of violence—required and caused by the (often forced) movement of resources from the poor to the rich—and would of course be marked by a reduction in current violence against the natural world. Personally and collectively we may be able to both reduce the amount and soften the character of violence that occurs during this ongoing and perhaps longterm shift. Or we may not. But this much is certain: if we do not approach it actively—if we do not talk about our predicament and what we are going to do about it—the violence will almost undoubtedly be far more severe, the privation more extreme.

PREMISE TEN: The culture as a whole and most of its members are insane. The culture is driven by a death urge, an urge to destroy life.

PREMISE ELEVEN: From the beginning, this culture—civilization—has been a culture of occupation.

PREMISE TWELVE: There are no rich people in the world, and there are no poor people. There are just people. The rich may have lots of pieces of green paper that many pretend are worth something—or their presumed riches may be even more abstract: numbers on hard drives at banks—and the poor may not. These "rich" claim they own land, and the "poor" are often denied the right to make that same claim. A primary purpose of the police is to enforce the delusions of those with lots of pieces of green paper. Those without the green papers generally buy into these delusions almost as quickly and completely as those with. These delusions carry with them extreme consequences in the real world.

PREMISE THIRTEEN: Those in power rule by force, and the sooner we break ourselves of illusions to the contrary, the sooner we can at least begin to make reasonable decisions about whether, when, and how we are going to resist.

PREMISE FOURTEEN: From birth on—and probably from conception, but I'm not sure how I'd make the case—we are individually and collectively encultured to hate life, hate the natural world, hate the wild, hate wild animals, hate women, hate children, hate our bodies, hate and fear our emotions, hate ourselves. If we did not hate the world, we could not allow it to be destroyed before our eyes. If we did not hate ourselves, we could not allow our homes—and our bodies—to be poisoned.

PREMISE FIFTEEN: Love does not imply pacifism.

PREMISE SIXTEEN: The material world is primary. This does not mean that the spirit does not exist, nor that the material world is all there is. It means that spirit mixes with flesh. It means also that real world actions have real world consequences. It means we cannot rely on Jesus, Santa Claus, the Great Mother, or even the Easter Bunny to get us out of this mess. It means this mess really is a mess, and not just the movement of God's eyebrows. It means we have to face this mess ourselves. It means that for the time we are here on Earth—whether or not we end up somewhere else after we die, and whether we are condemned or privileged to live here—the Earth is the point. It is primary. It is our home. It is everything. It is silly to think or act or be as though this world is not real and primary. It is silly and pathetic to not live our lives as though our lives are real.

PREMISE SEVENTEEN: It is a mistake (or more likely, denial) to base our decisions on whether actions arising from them will or won't frighten fence-sitters, or the mass of Americans.

PREMISE EIGHTEEN: Our current sense of self is no more sustainable than our current use of energy or technology.

PREMISE NINETEEN: The culture's problem lies above all in the belief that controlling and abusing the natural world is justifiable.

PREMISE TWENTY: Within this culture, economics—not community wellbeing, not morals, not ethics, not justice, not life itself—drives social decisions.

Modification of Premise Twenty: Social decisions are determined primarily (and often exclusively) on the basis of whether these decisions will increase the monetary fortunes of the decision-makers and those they serve.

Re-modification of Premise Twenty: Social decisions are determined primarily (and often exclusively) on the basis of whether these decisions will increase the power of the decision-makers and those they serve.

Re-modification of Premise Twenty: Social decisions are founded primarily (and often exclusively) on the almost entirely unexamined belief that the decision-makers and those they serve are entitled to magnify their power and/or financial fortunes at the expense of those below.

Re-modification of Premise Twenty: If you dig to the heart of it—if there is any heart left—you will find that social decisions are determined primarily on the basis of how well these decisions serve the ends of controlling or destroying wild nature.

⊛   ⊛   ⊛

As a longtime grassroots environmental activist, and as a creature living in the thrashing endgame of civilization, I am intimately acquainted with the landscape of loss, and have grown accustomed to carrying the daily weight of despair. I have walked clearcuts that wrap around mountains, drop into valleys, then climb ridges to fragment watershed after watershed, and I've sat silent near empty streams that two generations ago were "lashed into whiteness" by uncountable salmon coming home to spawn and die.

A few years ago I began to feel pretty apocalyptic. But I hesitated to use that word, in part because of those drawings I've seen of crazy penitents carrying "The End is Near" signs, and in part because of the power of the word itself. Apocalypse. I didn't want to use it lightly.

But then a friend and fellow activist said, "What will it take for you to finally call it an apocalypse? The death of the salmon? Global warming? The ozone hole? The reduction of krill populations off Antarctica by 90 percent, the turning of the sea off San Diego into a dead zone, the same for the Gulf of Mexico? How about the end of the great coral reefs? The extirpation of two hundred species per day? Four hundred? Six hundred? Give me a specific threshold, Derrick, a specific point at which you'll finally use that word."

❦ ❦ ❦

Do you believe that our culture will undergo a voluntary transformation to a sane and sustainable way of living?

For the last several years I've taken to asking people this question, at talks and rallies, in libraries, on buses, in airplanes, at the grocery store, the hardware store. Everywhere. The answers range from emphatic *no*s to laughter. No one answers in the affirmative. One fellow at one talk did raise his hand, and when everyone looked at him, he dropped his hand, then said, sheepishly, "Oh, voluntary? No, of course not." My next question: how will this understanding—that this culture will not voluntarily stop destroying the natural world, eliminating indigenous cultures, exploiting the poor, and killing those who resist—shift our strategy and tactics? The answer? Nobody knows, because we never talk about it: we're too busy pretending the culture will undergo a magical transformation.

This book is about that shift in strategy, and in tactics.

❦ ❦ ❦

It should be clear to everyone by now—even those with a vested interest in ignorance—that industrial civilization is killing the planet. It's causing unprecedented human privation and suffering. Unless it's stopped, or somehow stops itself, or most likely collapses under the weight of its inherent ecological and human destructiveness, it will kill every living being on earth. It should be equally clear that the efforts of those of us working to stop or slow the destruction are insufficient. We file our lawsuits; write our books; send letters to editors, representatives, CEOs; carry signs and placards; restore natural communities; and not only do we not stop or slow the destruction, but it actually continues to accelerate. Rates of deforestation continue to rise, rates of extinction do the same, global warming proceeds apace, the rich get richer, the poor starve to death, and the world burns.

At the same time that we so often find ourselves seemingly helpless in facing down civilization's speeding train of destruction, we find that there's a huge gap in our discourse. We speak much of the tactics of civil disobedience, much of the spiritual politics of cultural transformation, much of the sciences of biotechnology, toxicology, biology, and psychology. We talk of law. We also talk often of despair, frustration, and sorrow.

Yet our discourse remains firmly embedded in that which is sanctioned by the very overarching structures that govern the destruction in the first place. We do not often speak of the tactics of sabotage, and even less do we speak of violence. We avoid them, or pretend they should not be allowed to enter even the realm of possibility, or that they simply do not exist, like disinherited relatives who show up at a family reunion.

Several years ago I interviewed a long-term and well-respected Gandhian activist. I asked him, "What if those in power are murderous? What if they're not willing to listen to reason at all? Should we continue to approach them nonviolently?"

He responded, reasonably enough, "When a house is on fire, and has gone far beyond the point where you can do anything about it, all you can do is bring lots of water to try to stop its spread. But you can't save the house. Nonviolence is a precautionary principle. Before the house is on fire you have to make sure you have a fire hydrant, clearly marked escape routes, emergency exits. The same is true in society. You educate your children in nonviolence. You educate your media in nonviolence. And when someone has a grievance, you don't ignore or suppress it, but you listen to that person, and ask, 'What is your concern?' You say, 'Let's sit down and solve it.'"

I agreed with what he said, so far as it went, but that didn't stop me from understanding that he'd sidestepped the question.

Before I could bring him back, he continued, "Say a father beats his children. Once he has already reached that stage, you have to say, 'What kind of a childhood did he have? How did he not learn the skills of coping with adverse situations in a calm, compassionate, composed way?'"

This Gandhian's compassion, I thought, was entirely misplaced. Where was his compassion for the children being beaten? I responded that I believed the first question we need to ask is how we can get the children to a safe place. Once safety has been established, by any means possible, I said, and once the emotional needs of the children are being met, only then do we have the luxury of asking about the father's emotional needs, and his history.

What happened next is really the point of this story. I asked this devoted adherent of nonviolence if in his mind it would ever be acceptable to commit an act of violence were it determined to be the only way to save the children. His answer was revealing, and symbolizes the hole in our discourse: he changed the subject.

After I transcribed and edited the interview, I sent it to him with a new question inserted, attempting once again to pin him down. What did he do this time? He deleted my question.

Too often this is the response of all of us when faced with this most difficult of questions: when is violence an appropriate means to stop injustice? But with the world dying—or rather being killed—we no longer have the luxury to change the subject or delete the question. It's a question that won't go away.

❀ ❀ ❀

If I'm going to contemplate the collapse of civilization, I need to define what it is. I looked in some dictionaries. *Webster's* calls civilization "a high stage of social and cultural development."[1] *The Oxford English Dictionary* describes it as "a developed or advanced state of human society."[2] All the other dictionaries I checked were similarly laudatory. These definitions, no matter how broadly shared, helped me not in the slightest. They seemed to me hopelessly sloppy. After reading them, I still had no idea what the hell a civilization is: define *high*, *developed*, or *advanced*, please. The definitions, it struck me, are also extremely self-serving: can you imagine writers of dictionaries willingly classifying themselves as members of "a low, undeveloped, or backward state of human society"?

I suddenly remembered that all writers, including writers of dictionaries, are propagandists, and I realized that these definitions are, in fact, bite-sized chunks of propaganda, concise articulations of the arrogance that has led those who believe they are living in the most advanced—and best—culture to attempt to impose by force this way of being on all others.

I would define a civilization much more precisely, and I believe more usefully, as a culture—that is, a complex of stories, institutions, and artifacts—that both leads to and emerges from the growth of cities (*civilization*, see *civil* : from *civis*, meaning *citizen*, from Latin *civitatis*, meaning *city-state*),with cities being defined—so as to distinguish them from camps, villages, and so on—as people living more or less permanently in one place in densities high enough to require the routine importation of food and other necessities of life. Thus a Tolowa village five hundred years ago where I live in Tu'nes (*meadow long* in the Tolowa tongue), now called Crescent City, California, would not have been a city, since the Tolowa ate native salmon, clams, deer, huckleberries, and so on, and had no need to bring in food from outside. Thus, under my definition, the Tolowa, because their way of living was not characterized by the growth of city-states, would not have been civilized. On the other hand, the Aztecs were. Their social structure led inevitably to great city-states like Iztapalapa and Tenochtitlán, the latter of which was, when Europeans first encountered it, far larger than any city in Europe, with a population five times that of London or Seville.[3] Shortly before razing Tenochtitlán and slaughtering or enslaving its inhabitants, the explorer and conquistador Hernando Cortés remarked that it was easily the most beautiful city on earth.[4] Beautiful or not, Tenochtitlán required, as do all cities, the (often forced) importation of food and other resources. The story of any civilization is the story of the rise of city-states, which means it is the story of the funneling of resources toward these centers (in order to sustain them and cause them to grow), which means it is the story of an increasing region of unsustainability surrounded by an increasingly exploited countryside.

German Reichskanzler Paul von Hindenburg described the relationship perfectly: "Without colonies no security regarding the acquisition of raw materials, without raw materials no industry, without industry no adequate standard of living and wealth. Therefore, Germans, do we need colonies."[5] Of course someone already *lives* in the colonies, although that is evidently not of any importance.

But there's more. Cities don't arise in political, social, and ecological vacuums. Lewis Mumford, in the second book of his extraordinary two-volume *Myth of the Machine*, uses the term *civilization* "to denote the group of institutions that first took form under kingship. Its chief features, constant in varying proportions throughout history, are the centralization of political power, the separation of classes, the lifetime division of labor, the mechanization of production, the magnification of military power, the economic exploitation of the weak, and the universal introduction of slavery and forced labor for both industrial and military purposes."[6] (The anthropologist and philosopher Stanley Diamond put this a bit more succinctly when he noted, "Civilization originates in conquest abroad and repression at home."[7]) These attributes, which inhere not just in this culture but in all civilizations, make civilization sound pretty bad. But, according to Mumford, civilization has another, more benign face as well. He continues, "These institutions would have completely discredited both the primal myth of divine kingship and the derivative myth of the machine had they not been accompanied by another set of collective traits that deservedly claim admiration: the invention and keeping of the written record, the growth of visual and musical arts, the effort to widen the circle of communication and economic intercourse far beyond the range of any local community: ultimately the purpose to make available to all men [sic] the discoveries and inventions and creations, the works of art and thought, the values and purposes that any single group has discovered."[8]

Much as I admire and have been influenced by Mumford's work, I fear that when he began discussing civilization's admirable face he fell under the spell of the same propaganda promulgated by the lexicographers whose work I consulted: that this culture really is "advanced," or "higher." But if we dig beneath this second, smiling mask of civilization—the belief that civilization's visual or musical arts, for example, are more developed than those of noncivilized peoples—we find a mirror image of civilization's other face, that of power. For example, it wouldn't be the whole truth to say that visual and musical arts have simply *grown* or become more highly advanced under this system; it's more true that they have long ago succumbed to the same division of labor that characterizes this culture's economics and politics. Where among traditional indigenous people—the "uncivilized"—songs are sung by everyone as a means to bond members of the community and celebrate each other and their landbase, within civilization songs are written and performed by experts, those with "talent," those whose lives are devoted to the production of these arts. There's no reason for me to listen to my neighbor sing (probably off-key) some amateurish song of her own invention when I can pop in a CD of Beethoven, Mozart, or Lou Reed (okay, so Lou Reed sings off-key, too, but I like it). I'm not certain I'd characterize the conversion of human beings from participants in the ongoing creation of communal arts to more passive consumers of artistic products manufactured by distant experts—even if these distant experts are *really* talented—as a good thing.

I could make a similar argument about writing, but Stanley Diamond beat me to it: "Writing was one of the original mysteries of civilization, and it reduced the complexities of experience to the written word. Moreover, writing provides the ruling classes with an ideological instrument of incalculable power. The word of God becomes an invincible law, mediated by priests; therefore, respond the Iroquois, confronting the European: 'Scripture was written by the Devil.' With the advent of writing, symbols became explicit; they lost a certain richness. Man's word was no longer an endless exploration of reality, but a sign that could be used against him. . . . For writing splits consciousness in two ways—it becomes more authoritative than talking, thus degrading the meaning of speech and eroding oral tradition; and it makes it possible to use words for the political manipulation and control of others. Written signs supplant memory; an official, fixed and permanent version of events can be made. If it is written, in early civilizations [and I would suggest, now], it is bound to be true."[9]

I have two problems, also, with Mumford's claim that the widening of communication and economic intercourse under civilization benefits people as a whole. The first is that it presumes that uncivilized people do not communicate civilization or participate in economic transactions beyond their local communities. Many do. Shells from the Northwest Coast found their way into the hands of Plains Indians, and buffalo robes often ended up on the coast. (And let's not even mention noncivilized people communicating with their nonhuman neighbors, something rarely practiced by the civilized: talk about restricting yourself to your own community!) In any case, I'm not certain that the ability to send emails back and forth to Spain or to watch television programs beamed out of Los Angeles makes my life particularly richer. It's far more important, useful, and enriching, I think, to get to know my neighbors. I'm frequently amazed to find myself sitting in a room full of fellow human beings, all of us staring at a box watching and listening to a story concocted and enacted by people far away. I have friends who know Seinfeld's neighbors better than their own. I, too, can get lost in valuing the unreality of the distant over that which surrounds me every day. I have to confess I can navigate the mazes of the computer game *Doom 2: Hell on Earth* far better than I can find my way along the labyrinthine game trails beneath the trees outside my window, and I understand the intricacies of Microsoft Word far better than I do the complex dance of rain, sun, predators, prey, scavengers, plants, and soil in the creek a hundred yards away. The other night, I wrote till late, and finally turned off my computer to step outside and say goodnight to the dogs. I realized, then, that the wind was blowing hard through the tops of the redwood trees, and the trees were sighing and whispering. Branches were clashing, and in the distance I heard them cracking. Until that moment I had not realized such a symphony was taking place so near, much less had I gone out to participate in it, to feel the wind blow my hair and to feel the tossed rain hit me in the face. All of the sounds of the night had been drowned out by the monotone whine of my computer's fan. Just yesterday I saw a pair of hooded mergansers playing on the pond outside my bedroom. Then last night I saw a television program in which yet another lion chased yet another zebra.

Which of those two scenes makes me richer? This perceived widening of communication is just another replication of the problem of the visual and musical arts, because given the impulse for centralized control that motivates civilization, widening communication in this case really means reducing us from active participants in our own lives and in the lives of those around us to consumers sucking words and images from some distant sugar tit.

I have another problem with Mumford's statement. In claiming that the widening of communication and economic intercourse are admirable, he seems to have forgotten—and this is strange, considering the sophistication of the rest of his analysis—that this widening can only be universally beneficial when all parties act voluntarily and under circumstances of relatively equivalent power. I'd hate to have to make the case, for example, that the people of Africa—perhaps 100 million of whom died because of the slave trade, and many more of whom find themselves dispossessed and/or impoverished today—have benefited from their "economic intercourse" with Europeans. The same can be said for Aborigines, Indians, the people of pre-colonial India, and so on.

I want to re-examine one other thing Mumford wrote, in part because he makes an argument for civilization I've seen replicated so many times elsewhere, and that actually leads, I think, to some of the very serious problems we face today. He concluded the section I quoted above, and I reproduce it here just so you don't have to flip back a couple of pages: "ultimately the purpose [is] to make available to all men [*sic*] the discoveries and inventions and creations, the works of art and thought, the values and purposes that any single group has discovered." But just as a widening of economic intercourse is only beneficial to everyone when all exchanges are voluntary, so, too, the imposition of one group's values and purposes onto another, or its appropriation of the other's discoveries, can lead only to the exploitation and diminution of the latter in favor of the former. That this "exchange" helps all was commonly argued by early Europeans in America, as when Captain John Chester wrote that the Indians were to gain "the knowledge of our faith," while the Europeans would harvest "such ritches as the country hath."[10] It was argued as well by American slave owners in the nineteenth century: philosopher George Fitzhugh stated that "slavery educates, refines, and moralizes the masses by bringing them into continual intercourse with masters of superior minds, information, and morality."[11] And it's just as commonly argued today by those who would teach the virtues of blue jeans, Big Macs™, Coca-Cola™, Capitalism™, and Jesus Christ™ to the world's poor in exchange for dispossessing them of their landbases and forcing them to work in sweatshops.

Another problem is that Mumford's statement reinforces a mindset that leads inevitably to unsustainability, because it presumes that discoveries, inventions, creations, works of art and thought, and values and purposes are transposable over space, that is, that they are separable from both the human context and landbase that created them. Mumford's statement unintentionally reveals perhaps more than anything else the power of the stories that hold us in thrall to the machine, as he put it, that is civilization: even in brilliantly dissecting the myth of this machine, Mumford fell back into that very same myth by seeming to implicitly accept the notion that ideas or works of art or discoveries are like tools in a toolbox, and can be meaningfully and without negative consequence used out of their original context: thoughts, ideas, and art as tools rather than as tapestries inextricably woven from and into a community of human and nonhuman neighbors. But discoveries, works of thought, and purposes that may work very well in the Great Plains may be harmful in the Pacific Northwest, and even moreso in Hawai'i. To believe that this potential transposition is positive is the same old substitution of what is distant for what is near: if I really want to know how to live in Tu'nes, I should pay attention to Tu'nes.

There's another problem, though, that trumps all of these others. It has to do with a characteristic of this civilization unshared even by other civilizations. It is the deeply and most-often-invisibly held beliefs that there is really only one way to live, and that we are the one-and-only possessors of that way. It becomes our job then to propagate this way, by force when necessary, until there are no other ways to be. Far from being a loss, the eradication of these other ways to be, these other cultures, is instead an actual gain, since Western Civilization is the only way worth being anyway: we're doing ourselves a favor by getting rid of not only obstacles blocking our access to resources but reminders that other ways to be exist, allowing our fantasy to sidle that much closer to reality; and we're doing the heathens a favor when we raise them from their degraded state to join the highest, most advanced, most developed state of society. If they don't want to join us, simple: we kill them. Another way to say all of this is that something grimly alchemical happens when we combine the arrogance of the dictionary definition, which holds this civilization superior to all other cultural forms; hypermilitarism, which allows civilization to expand and exploit essentially at will; and a belief, held even by such powerful and relentless critics of civilization as Lewis Mumford, in the desirability of cosmopolitanism, that is, the transposability of discoveries, values, modes of thought, and so on over time and space. The twentieth-century name for that grimly alchemical transmutation is genocide: the eradication of cultural difference, its sacrifice on the altar of the one true way, on the altar of the centralization of perception, the conversion of a multiplicity of moralities all dependent on location and circumstance to one morality based on the precepts of the ever-expanding machine, the surrender of individual perception (as through writing and through the conversion of that and other arts to consumables) to predigested perceptions, ideas, and values imposed by external authorities who with all their hearts—or what's left of them—believe in, and who benefit by, the centralization of power. Ultimately, then, the story of this civilization is the story of the reduction of the world's tapestry of stories to only one story, the best story, the real story, the most advanced story, the most developed story, the story of the power and the glory that is Western Civilization.

⊞ ⊞ ⊞

Our capacity and propensity for self-delusion—indeed the *necessity* of self-delusion if we're to continue to propagate this culture—means I need to be explicit. The first premise is: *Civilization is not and can never be sustainable. This is especially true for industrial civilization.*

Years ago I was riding in a car with friend and fellow activist George Draffan. He has influenced my thinking as much as any other one person. It was a hot day in Spokane. Traffic was slow. A long line waited at a stoplight. I asked, "If you could live at any level of technology, what would it be?"

As well as being a friend and an activist, George can be a curmudgeon. He was in one of those moods. He said, "That's a stupid question. We can fantasize about living however we want, but the only sustainable level of technology is the Stone Age. What we have now is the merest blip—we're one of only six or seven generations who ever have to hear the awful sound of internal combustion engines (especially two-cycle)—and in time we'll return to the way humans have lived for most of their existence. Within a few hundred years at most. The only question will be what's left of the world when we get there."

He's right, of course. It doesn't take a rocket scientist to figure out that any social system based on the use of nonrenewable resources is by definition unsustainable: in fact it probably takes anyone *but* a rocket scientist to figure this one out. The hope of those who wish to perpetuate this culture is something called "resource substitution," whereby as one resource is depleted another is substituted for it (I suppose there is at least one hope more prevalent than this, which is that if we ignore the consequences of these actions they will not exist). Of course on a finite planet this merely puts off the inevitable, ignores the damage caused in the meantime, and begs the question of what will be left of life when the last substitution has been made. Question: When oil runs out, what resource will be substituted in order to keep the industrial economy running? Unstated premises: a) equally effective substitutes exist; b) we want to keep the industrial economy running; and c) keeping it running is worth more to us (or rather to those who make the decisions) than the human and nonhuman lives destroyed by the extraction, processing, and utilization of this resource.

Similarly, any culture based on the nonrenewable use of renewable resources is just as unsustainable: if fewer salmon return each year than the year before, sooner or later none will return. If fewer ancient forests stand each year than the year before, sooner or later none will stand. Once again, the substitution of other resources for depleted ones will, some say, save civilization for another day. But at most this merely holds off the inevitable while it further damages the planet. This is what we see, for example, in the collapse of fishery after fishery worldwide: having long-since fished out the more economically valuable fish, now even so-called trash fish are being extirpated, disappearing into civilization's literally insatiable maw.

Another way to put all of this is that any group of beings (human or nonhuman, plant or animal) who take more from their surroundings than they give back will, obviously, deplete their surroundings, after which they will either have to move, or their population will crash (which, by the way, is a one sentence disproof of the notion that competition drives natural selection: if you hyper-exploit your surroundings you will deplete them and die; the only way to survive in the long run is to give back more than you take. Duh). This culture—Western Civilization—has been depleting its surroundings for six thousand years, beginning in the Middle East and expanding now to deplete the entire planet. Why else do you think this culture has to continually expand? And why else, coincident with this, do you think it has developed a rhetoric—a series of stories that teach us how to live—making plain not only the necessity but desirability and even morality of continual expansion—causing us to boldly go where no man has gone before—as a premise so fundamental as to become invisible? Cities, the defining feature of civilization, have always relied on taking resources from the surrounding countryside, meaning, first, that no city has ever been or ever will be sustainable on its own, and second, that in order to continue their ceaseless expansion cities must ceaselessly expand the areas they must ceaselessly hyperexploit. I'm sure you can see the problems this presents and the end point it must reach on a finite planet. If you cannot or will not see these problems, then I wish you the best of luck in your career in politics or business. Our collective studied-to-the-point-of-obsessive avoidance of acknowledging and acting on the surety of this end point is, especially given the consequences, more than passing strange.

Yet another way to say that this way of living is unsustainable is to point out that because ultimately the only real source of energy for the planet is the sun (the energy locked in oil, for example, having come from the sun long ago; and I'm excluding nuclear power from consideration here because only a fool would intentionally fabricate and/or refine materials that are deadly poisonous for tens or hundreds of thousands of years, especially to serve the frivolous, banal, and anti-life uses to which electricity is put: think retractable stadium roofs, supercolliders, and aluminum beer cans), any way of being that uses more energy than that currently coming from the sun will not last, because the noncurrent energy—stored in oil that could be burned, stored in trees that could be burned (stored, for that matter, in human bodies that could be burned)—will in time be used up. As we see.

I am more or less constantly amazed at the number of intelligent and wellmeaning people who consistently conjure up magical means to maintain this current disconnected way of living. Just last night I received an email from a very smart woman who wrote, "I don't think we can go backward. I don't think Hunter/Gatherer is going to be it. But is it possible to go forward in a way that will bring us around the circle back to sustainability?"

It's a measure of the dysfunction of civilization that no longer do very many people of integrity believe we can or should go forward with it because it serves us well, but rather the most common argument in its favor (and this is true also for many of its particular manifestations, such as the global economy and high technology) seems to be that we're stuck with it, so we may as well make the best of a very bad situation. "We're here," the argument goes, "We've lost sustainability and sanity, so now we have no choice but to continue on this self- and other-destructive path." It's as though we've already boarded the train to Treblinka, so we might as well stay on for the ride. Perhaps by chance or by choice (someone else's) we'll somehow end up somewhere besides the gas chambers.

The good news, however, is that we don't need to go "backward" to anything, because humans and their immediate evolutionary predecessors lived sustainably for at least a million years (cut off the word *immediate* and we can go back billions). It is not "human nature" to destroy one's habitat. If it were, we would have done so long before now, and long-since disappeared. Nor is it the case that stupidity kept (and keeps) noncivilized peoples from ordering their lives in such a manner as to destroy their habitat, nor from developing technologies (for example, oil refineries, electrical grids, and factories) that facilitate this process. Indeed, were we to attempt a cross-cultural comparison of intelligence, maintenance of one's habitat would seem to me a first-rate measure with which to begin. In any case, when civilized people arrived in North America, the continent was rich with humans and nonhumans alike, living in relative equilibrium and sustainability. I've shown this elsewhere, as have many others,[12] most especially the Indians themselves.

Because we as a species haven't fundamentally changed in the last several thousand years, since well before the dawn of civilization, each new child is still a human being, with the potential to become the sort of adult who can live sustainably on a particular piece of ground, if only the child is allowed to grow up within a culture that values sustainability, that lives by sustainability, that rewards sustainability, that tells itself stories reinforcing sustainability, and strictly disallows the sort of exploitation that would lead to unsustainability. This is natural. This is who we are.

In order to continue moving "forward," each child must be made to forget what it means to be human and to learn instead what it means to be civilized. As psychiatrist and philosopher R. D. Laing put it, "From the moment of birth, when the Stone Age baby confronts the twentieth-century mother, the baby is subject to these forces of violence . . . as its mother and father, and their parents and their parents before them, have been. These forces are mainly concerned with destroying most of its potentialities, and on the whole this enterprise is successful. By the time the new human being is fifteen or so, we are left with a being like ourselves, a half-crazed creature more or less adjusted to a mad world. This is normality in our present age."[13]

▣ ▣ ▣

What do we do with this information: Phoenix, Arizona, could sustain a human population of maybe one hundred and fifty. What about the rest of them, living right now on stolen resources? The land under New York City could probably sustain several thousand, or at least it could have if there were still passenger pigeons, bison, salmon, eel, and Eskimo curlews. What happens to the rest? I'm a bit luckier here in Tu'nes. The population might be remotely sustainable at a hunter-gatherer level, if salmon, steelhead, elk, and lamprey were still here in significant numbers.

To reverse the effects of civilization would destroy the dreams of a lot of people. There's no way around it. We can talk all we want about sustainability, but there's a sense in which it doesn't matter that these people's dreams are based on, embedded in, intertwined with, and formed by an inherently destructive economic and social system. Their dreams are still their dreams. What right do I—or does anyone else—have to destroy them?

At the same time, what right do they have to destroy the world?

▣ ▣ ▣

If Nazis or other fascists took over North America, what would we all do? What would we all do if they implemented Mussolini's definition of fascism: "Fascism should more appropriately be called Corporatism because it is a merger of State and corporate power"? And what would we do if they then instituted laws allowing them to put a significant portion—say one-third—of all Jewish males between the ages of eighteen and thirty-five into concentration camps? What if this occupied country called itself a democracy, but most everyone understood elections to be shams, with citizens allowed to choose between different wings of the same Fascist (or, following Mussolini, Corporate) party? What if anti-government activity was opposed by storm troopers and secret police? Would you fight back? If there already existed a resistance movement, would you join it? Substitute the word *African-American* for *Jewish* and ask yourself the same questions.

Now, would you resist if the fascists irradiated the countryside, poisoned food supplies, made rivers unfit for swimming (and so filthy you wouldn't even *dream* of drinking from them anymore)? What if they did this because . . . Hell, I can't finish that sentence because no matter how I try I can't come up with a motivation good enough even for fascists to irradiate and toxify the landscape and water supplies. If fascists systematically deforested the continent, would you join an underground army of resistance, head to the forests, and from there to boardrooms and to the halls of the Reichstag to pick off the occupying deforesters and most especially those who give them their marching orders?

Okay, so maybe your sense of kin, and your sense of skin, doesn't extend to the natural world. Maybe you don't yet love the land where you live enough that you will fight for it. But what if the fascists toxify not only the landscape but the bodies of those you love? What if their actions put dioxin—one of the most toxic substances known—and dozens of other carcinogens into the flesh of your lover, children, mother, brother, sister, father? Would you then fight back? What if the fascists toxify your own body? Would you still cling to the illusion that their edicts carry more weight than that brought to bear by their secret and not-sosecret police? Would you work for this regime? Would you teach others its virtues? Or would you fight back? If you will not fight back when they toxify your own body (and toxify your mind with propaganda leading you to believe their edicts carry moral weight), when, precisely, will you fight back? Give me— and more importantly yourself—a specific threshold at which you will finally take a stand. If you can't or won't give that threshold, why not?

None of these questions are rhetorical. The questions are real. They are, at this point, some of the most important questions there are.

❦ ❦ ❦

In his extraordinarily important book *The Nazi Doctors,* Robert Jay Lifton explored how it was that men who had taken the Hippocratic oath could participate in prisons where inmates were worked to death or killed in assembly lines. He found that many of the doctors honestly cared for their charges, and did everything within their power—which means pathetically little—to make life better for the inmates. If an inmate got sick they might give the inmate an aspirin to lick. They might put the inmate to bed for a day or two (but not for too long or the inmate might be "selected" for murder). If the patient had a contagious disease, they might kill the patient to keep the disease from spreading. All of this made sense within the confines of Auschwitz. The doctors, once again, did everything they could to help the inmates, except for the most important thing of all: They never questioned the existence of Auschwitz itself. They never questioned working the inmates to death. They never questioned starving them to death. They never questioned imprisoning them. They never questioned torturing them. They never questioned the existence of a culture that would lead to these atrocities. They never questioned the logic that leads inevitably to the electrified fences, the gas chambers, the bullets in the brain.

We as environmentalists do the same. We work as hard as we can to protect the places we love, using the tools of the system the best that we can. Yet we do not do the most important thing of all: We do not question the existence of this death culture. We do not question the existence of an economic and social system that is working the world to death, that is starving it to death, that is imprisoning it, that is torturing it. We never question a culture that leads to these atrocities. We never question the logic that leads inevitably to clearcuts, murdered oceans, loss of topsoil, dammed rivers, poisoned aquifers.

And we certainly don't act to bring it down.

⊗ ⊗ ⊗

If you've gotten this far in this book—or if you're simply anything other than entirely insensate—we probably agree that civilization is going to crash, whether or not we help bring this about. If you don't agree with this, we probably have nothing to say to each other (How 'bout them Cubbies!). We probably also agree that this crash will be messy. We agree further that since industrial civilization is systematically dismantling the ecological infrastructure of the planet, the sooner civilization comes down (whether or not we help it crash) the more life will remain afterwards to support both humans and nonhumans.

If you agree with all this, and *if* you don't want to dirty your spirituality and conscience with the physical work of helping to bring down civilization, and *if* your primary concern really is for the well-being of those (humans) who will be alive during and immediately after the crash (as opposed to simply raising this issue because you're too scared to talk about the crash or to allow anyone else to do so either), then, given (and I repeat this point to emphasize it) that civilization is going to come down anyway, you need to start preparing people for the crash. Instead of attacking me for stating the obvious, go rip up asphalt in vacant parking lots to convert them to neighborhood gardens, go teach people how to identify local edible plants, even in the city (*especially* in the city) so these people won't starve when the proverbial shit hits the fan and they can no longer head off to Albertson's for groceries. Set up committees to eliminate or, if appropriate, channel the (additional) violence that might break out.

We need it all. We need people to take out dams and we need people to knock out electrical infrastructures. We need people to protest and to chain themselves to trees. We *also* need people working to ensure that as many people as possible are equipped to deal with the fallout when the collapse comes. We need people working to teach others what wild plants to eat, what plants are natural antibiotics. We need people teaching others how to purify water, how to build shelters. All of this can look like supporting traditional, local knowledge, it can look like starting rooftop gardens, it can look like planting local varieties of medicinal herbs, and it can look like teaching people how to sing.

The truth is that although I do not believe that designing groovy eco-villages will help bring down civilization, when the crash comes, I'm sure to be first in line knocking on their doors asking for food.

People taking out dams do not have a responsibility to ensure that people in homes previously powered by hydro know how to cook over a fire. They do however have a responsibility to support the people doing that work.

Similarly, those people growing medicinal plants (in preparation for the end of civilization) do not have a responsibility to take out dams. They do however have a responsibility *at the very least* to not condemn those people who have chosen that work. In fact they have a responsibility to support them. They especially have a responsibility to not report them to the cops.

It's the same old story: the good thing about everything being so fucked up is that no matter where you look, there is great work to be done. Do what you love. Do what you can. Do what best serves your landbase. We need it all.

This doesn't mean that everyone taking out dams and everyone working to cultivate medicinal plants are working toward the same goals. It does mean that if they are, each should see the importance of the other's work.

Further, resistance needs to be global. Acts of resistance are more effective when they're large-scale and coordinated. The infrastructure is monolithic and centralized, so common tools and techniques can be used to dismantle it in many different places, simultaneously if possible.

By contrast, the work of renewal must be local. To be truly effective (and to avoid reproducing the industrial infrastructure) acts of survival and livelihood need to grow from particular landbases where they will thrive. People need to enter into conversation with each piece of earth and all its human and nonhuman inhabitants. This doesn't mean of course that we can't share ideas, or that one water purification technique won't be useful in many different locations. It does mean that people in those places need to decide for themselves what will work. Most important of all, the water in each place needs to be asked and allowed to decide for itself.

I've been thinking a lot again about the cell phone tower behind Safeway, and I see now how these different approaches manifest themselves in this one small place. The cell phone tower needs to come down. It is contiguous on two sides with abandoned parking lots. Those lots need to come up. Gardens can bloom in their place. We can even do our work side by side.[14]

● ● ●

I've been bashing hope for many years. Frankly, I don't have much of it, and I think that's a good thing. Hope is partly what keeps us chained to the system. First there is the false hope that suddenly somehow the system may inexplicably change. Or technology will save us. Or the Great Mother. Or beings from Alpha Centauri. Or Jesus Christ. Or Santa Claus. All of these false hopes—all of this rendering of our power—leads to inaction, or at least to ineffectiveness: how, for example, would Philip Berrigan have acted had he not believed—hoped—God would help solve things?

One reason my mother stayed with my father was that there were no battered women's shelters in the fifties and sixties, but another was because of the false hope that he would change. False hopes, as I've written elsewhere, bind us to unlivable situations, and blind us to real possibilities. Does anyone really believe that Weyerhaeuser is going to stop deforesting because we ask nicely? Does anyone really believe that Monsanto will stop Monsantoing because we ask nicely? If only we get a Democrat in the White House, this line of thought runs, things will be okay. If only we pass this or that piece of legislation, things will be okay. If only we *defeat* this or that piece of legislation, things will be okay.[15] Bullshit. Things will not be okay. They are already not okay, and they're getting worse.

One of the smartest things Nazis did to Jews was co-opt rationality, co-opt hope. At every step of the way it was in the Jews' rational best interest to not resist: many Jews had the hope—and this hope was cultivated by the Nazis—that if they played along, followed the rules laid down by those in power, that their lives would get no worse, that they would not be murdered. Would you rather get an ID card, or would you rather resist and possibly get killed? Would you rather go to a ghetto (reserve, reservation, whatever) or would you rather resist and possibly get killed? Would you rather get on a cattle car, or would you rather resist and possibly get killed? Would you rather get in the showers, or would you rather resist and possibly get killed?

But I'll tell you something important: the Jews who participated in the Warsaw Ghetto uprising, including those who went on what they thought were suicide missions, had a higher rate of survival than those who went along peacefully. Never forget that.

卐   卐   卐

The seventeenth premise of this book is that *it's a mistake (or more likely, denial) to base our decisions on whether our actions will or won't frighten fence-sitters or the mass of Americans.* Sure, we can let the potential response of these people be one more piece of information that helps to influence our choices, but we must always remember that we are only responsible for our own actions. Just as we are not responsible for the choices—retributive or otherwise—made by those in power as putative response to any action we may take, so, too, we are not responsible for the response or non-response of the mass of Americans (or Czechs, Liberians, or Indonesians, for that matter).

Here's another way to put the seventeenth premise: The mass of civilized people will never be on our side.[16] I'm not saying by this that we should give up on educating or informing people (I am, after all, a writer: educating and informing is what I *do*). I'm saying, first, that we need to try to be aware of where our identification lies—with whom or what we identify—and we need to ask ourselves: If what the mass of Americans want is in opposition to what your own particular landbase needs, which do you choose to support? If it comes down to stark choices—which of course it already does—on which side will you take your stand (recognizing also that refusing to choose is just another way of choosing the default)?[17]

Second, I'm saying that we need to be aware that we have a finite amount of time each day and a finite amount of time in our lives, so if we actually hope to accomplish something tangible we need to choose wisely how we spend that time. Some people may feel it's the best use of their time to inch fence-sitters closer to falling to the side of the living, and by all means they should do that. I don't think most fence-sitters are effectively reachable, and so I do not write for them. I write for people who already know how horrible civilization is, and who want to do something about it. I want to encourage them to be more radical, more militant, just as others have encouraged me.

Further, we need to recognize that educating people will only go so far toward saving salmon, sturgeon, marlins, prairie dogs, forests, rivers, glaciers, oceans, skies, the planet. At some point we have to actually *do* something.

The problem is not and has never been that the mass of people do not have enough information, such that if we just present them with enough facts they will strive for justice, for sanity, for what is best for their landbase. Think again about rape. Rape is not caused by a lack of information. Similarly, it doesn't take a genius to figure out that dams kill salmon, or that deforestation kills creatures who live in forests. Would it have merely required information to get the whites who slaughtered Indians (or who took their land after the soldiers had done the slaughtering) to stand with these Indians against members of their own culture? Would it require that today, as traditional indigenous people continue to be put in reserves, concentration camps, prisons, and graves, and as their land continues to be stolen? When cancer kills those we love—our grandparents, brothers, sisters, children, friends, lovers—when chemicals cause little girls to develop breasts and pubic hair, when pesticides make children stupid and sickly, the problem is not education. The problem has never been education. To believe that it is, is to buy into yet one more lie that keeps us from acting to protect ourselves.

⊛ ⊛ ⊛

I have to admit it discourages me that at this late date we still have to fend off this argument that we must not tell the truth for fear we will frighten or anger the mass of people. Certainly an examination of history shows a greater willingness of the mass of people to participate in the atrocities of the culture than to oppose them. How is it that a surefire way for a president to increase his standing in the polls is to invade yet another defenseless country? Or, compare how many Germans were in the Wehrmacht in World War II—or how many were just good Germans—to how many were part of the resistance. One of the reasons members of the resistance knew they had to kill Hitler was because he was so magnificently popular among the majority of people: if Hitler were allowed to speak, they knew the people would listen.

⊛ ⊛ ⊛

If we're going to talk about bringing down civilization, we need to talk about fulcrums.[18]

If you recall, Archimedes said something to the effect of "Give me a long enough lever and a place to stand and I can move the world." Well, he was being concise; by emphasizing the length of the lever and the place to stand he left off the lever's other crucial component: the fulcrum. Archimedes could have the longest and strongest board in the universe, and the most solid place to stand, and he still wouldn't have been able to leverage his strength without that pivot point.

The purpose of a lever is to transmit or modify (often magnify) power or motion. I can bend metal with a crowbar I couldn't budge using muscles alone. I can crack nuts easily with a nutcracker, and moving heavy weights is a piece of cake with a wheelbarrow.

What does this have to do with taking down civilization?

Everything.

So long as the dominant culture is still dominant—by which I mean so long as its exploitative mindset holds sway over what's left of the hearts and minds of the people who run this culture—there will always be a disproportionate number of people willing to kill to perpetuate it (to gain or maintain the power, or the promise of power, associated with being an exploiter[19]) compared to the number who are willing to fight to protect life. It's Jefferson's line all over again: "In war they shall kill some of us; we shall destroy all of them." And those who are willing, ready, and oftentimes eager to destroy those who threaten the hegemony of those in power often include their hired guns: Those in power worldwide have about 20 million soldiers and 5 million cops at their command. In the U.S. alone, these numbers are about 1.4 million soldiers and 1.4 million cops (one-third of whom are prison guards), the primary function of whom is to use violence or its threat to serve those in power. Far worse, nearly all of us have allowed ourselves to become convinced of the righteousness of Premise Four of this book: that violence flows only one direction, that it is right and just for servants of power to kill in that service (yet it is proper for their leaders to inevitably declaim on the regrettability of these inevitable murders), and it is blasphemy for the rest of us to fight back.[20] This latter is as true for mountain lions who fight back against those who wish to destroy their habitat as it is for humans who fight back against those who wish to destroy their habitat.

All of this is a roundabout way of saying that those in power have the luxury of using that power inelegantly. They can and often do simply overwhelm us with sheer force. ("Shock and awe" is one of the currently preferred terms.) Those of us fighting for life, on the other hand, need to learn how and where to find appropriate fulcrums to amplify our efforts.

If we're going to talk about fulcrums, we need to also talk about bottle-necks. Anyone who has ever driven on a freeway knows precisely what a bottleneck is. You're driving along fine at 69 miles per hour on a six-lane highway. You top a hill and hit your brakes because the person in front of you hit his brakes, because the person in front of him hit her brakes. Traffic slows to a crawl. People begin frantically changing lanes, trying to find one that will get them through this mess three minutes sooner. Even-tually people realize they need to get into the center lane (you realize this about ten seconds after you got into the left lane, and just as three semis creep by you in the center). At long last you come across the problem: a car broken down in the left lane and a cop parked on the right. Moments later, you're zooming again at precisely four miles over the speed limit, but for that forty-five minutes of traffic jam, you had the full bottleneck experience so beloved of motorists everywhere. Or one more example. Take a hose (or a pipeline). Kink it (or disable a pumping station). It doesn't matter how smoothly the water (or oil) flows through the rest of the hose (or pipeline). If there is a kink (or a disabled pumping station) in even one place, the water (or oil) will not flow. Bottleneck!

Now, how does this apply on a larger scale?

Albert Speer, Minister of Armaments for the Third Reich, later com-mented that the Allied bombing efforts could have been more effective had they more often targeted bottlenecks. One small example of this was that when the Allies bombed tractor factories, the Germans were no longer able to manufacture engines for tanks and airplanes there until the facto-ries had been rebuilt, but when the Allies bombed ball bearing plants, the Germans were hindered from rebuilding factories. You need ball bearings to manufacture manufacturing plants. Ball bearing plants were bottle-necks in the process.

Here's an example of the Allies not hitting bottlenecks: the firebombing of Hamburg, which killed tens of thousands of people and destroyed much of the city, cost less than two months of productivity.[21] As a result of not targeting bottlenecks, Allied bombing reduced total German produc-tion by only 9 percent in 1943, and by building new factories, overworking undamaged factories, and diverting consumer production towards mili-tary ends, the Germans still met their production targets.[22]

But it ends up that ball bearing plants were trivial bottlenecks compared to others. Transportation networks, for example, were an even larger bottleneck. Eventually the Allies were able to destroy about two-thirds of the German rolling stock.[23] A United States military analysis later determined that the difficulties this caused the Germans in moving raw materials and finished goods made the attacks on railroads "the most important single cause of Germany's ultimate economic collapse."[24]

We all know (and Hitler knew this too) that oil was another bottleneck. You can have the most powerful tanks in the world, and without oil they're just big hunks of steel. Without oil you have no modern army. Heck, without oil, you have no modern civilization. Keep that in mind. Hitler's understanding of these basic facts was one reason for his ultimately fatal choice to try to take the oil fields of the Caucasus instead of just pushing toward Stalingrad. Further, once the Allies started pounding the German synthetic oil industry—hitting the selected targets again and again and again—they were able to reduce monthly oil production from 316,000 to 17,000 tons.[25] These shortages obviously crippled the German war economy.

Just so we're clear that there are lots of bottlenecks, and that a little creativity can discover them, here's another bottleneck from World War II: industrial diamonds. Industrial grinding and drilling is almost impossible without diamonds. Both the Nazis, who had on hand only an eight-month supply, and DeBeers, which controlled the world's diamond supply, knew this. The Nazis smuggled several million carats into Germany. DeBeers could have acted to stop them—and thus effectively stopped wartime production,which means effectively stopped the war—but did not.

The new questions become: What are some of civilization's bottlenecks? What are some of civilization's limiting factors? Like transportation networks, oil, and industrial diamonds for the Nazis, what are some of the objects or processes that, if interdicted, could cause civilization to grind to a halt?

Similarly, where can we find fulcrums, pivot points, to magnify our efforts? Where do we put the levers, what do we use for fulcrums, how and when and how hard to we push to help topple this culture of death?

Are these fulcrums psychological? I hear all the time that it would do no good to take out dams, for example, because that would leave intact the mindset that leads to their erection in the first place. We need to change hearts and minds, I am told, and once these hearts and minds have been changed everything else will fall into place. Civilization will disappear because people are no longer insane enough to want it.

But maybe that question is too vague. Whose hearts and minds are we trying to reach? Where do we place our efforts in changing hearts and minds to achieve the most effect? Is it among the politically and economically powerful? Is it among the "mass of Americans"? Is it among the disaffected? Is it among the poor? Is it among the so-called criminal classes? Is it among the cops and the military? These latter, after all, have a lot of guns.[26] Where will we achieve the most good?

Are the fulcrums spiritual? People value what they consider sacred. They sacralize what they value. Perhaps we should attempt to desacralize power for power's sake. Perhaps we should attempt to break down the divine right of science, the divine right of corporations, the divine right of production, the divine right of nation-states. Perhaps we should attempt to help people to remember that spiders who live in their bathrooms are sacred, as are salmon who spawn in rivers outside their homes, plants who push up through sidewalks, salamanders who live high in the hollows of ancient redwoods, their own bodies, their own experiences, their own sexuality, their own flesh free from industrial carcinogens. Where do we place the levers, the fulcrums, to help people remember that they are humans living in a landbase, that they are animals?

Are these fulcrums personal, such that, like Hitler, the "removal" of this or that person will make a tangible difference? Would it help the redwoods and workers of northern California to make sure Charles Hurwitz, CEO of MAXXAM, does not damage them from his high-rise home in Houston, Texas? If so, where and how and when do we act in this way?

In cases where it's not the individual CEO, but the position—where social framing conditions make it so that most people who would take up that position share the same deadened worldview that would cause them to commit the same atrocities—where then do we place the levers and fulcrums? Do we go CEO to CEO, "removing" them one by one? We always hear that the machinelike characteristics of corporations mean that CEOs are simply cogs—albeit large ones—in these community-destroying institutions, and so it would do no good to remove them. It's an odd argument to make, even when I make it myself (as I did a few pages ago).[27] There are few who suggest that simply because arresting or killing one rapist does not stop other men from raping, that this means we should not stop whatever rapists we can through any means necessary. Yet when it comes to CEOs the argument seems to hold: Someone else will just take this one's place, so we must not stop this one personally. In fact, we must allow him to continue to be rewarded with millions of dollars per year in salaries and stock options.Where are the fulcrums to stop these people, these institutions? Where are the bottlenecks?

Or perhaps the fulcrums are social. Perhaps instead of (or in addition to) removing individual CEOs,we need to change the social institutions that themselves amplify the destructive efforts of these individuals. Charles Hurwitz does not kill redwoods by cutting them down. He kills them by ordering them cut down, or even more abstractly, by ordering someone to maximize profits. Are there counterlevers we can use to pry away his levers of power? Are there social means by which we can do that?

Or perhaps, as was also true of the Nazis, some of the fulcrums are infrastructural. John Muir is famously noted as saying, "God has cared for these trees, saved them from drought, disease, avalanches, and a thousand tempests and floods. But he cannot save them from fools." The thing is, a fool couldn't cut down trees by him or herself. I used to think that we were fighting an incredibly difficult battle in part because it takes a thousand years of living to make an ancient tree, while any fool can come along with a chainsaw and cut it down in an hour or two. I've since realized that's all wrong. The truth is that thriving on a living planet is easy—the whole forest, for example, conspires to grow that tree and every other, and *we* don't have to do anything special except leave it alone—while cutting down a tree is actually a very difficult process involving the entire global economy. I wouldn't care how many ancient redwoods Charles Hurwitz cut down, if he did it all by himself, scratching pathetically with bloodied nails at bark, gnawing with bloody teeth at heartwood, sometimes picking up rocks to make stone axes. To cut down a big tree you need the entire mining infrastructure for the metals necessary for chainsaws (or a hundred years ago, whipsaws); the entire oil infrastructure for gas to run the chainsaws, and for trucks to transport the dead trees to market where they will be sold and shipped to some distant place (once Charles had downed the tree by himself, I would wish him luck transporting it without the assistance of the global economy); and so on. It takes a whole lot of fools to cut down a tree, and if we break the infrastructural chain at any point, they won't be able to do it.

The same is true, of course, for the rest of this culture's destructive activities, from vivisection to factory farming to vacuuming the oceans to paving the grasslands to irradiating the planet: every one of this economy's destructive activities requires immense amounts of energy and worldwide economic, infrastructural, military, and police support to accomplish. If any one of these supports fail—I want to emphasize, if *any one of these supports fail*—the destructive activities will be curtailed. Where do we place our levers?

Or maybe the fulcrums are all of the above. Maybe changing people's hearts' fulcrums has a place. Maybe so do all the others, and maybe we should pursue them all, according to our gifts, proclivities, and opportunities. The bottom line so far as fulcrums and bottlenecks: What will it take to stop this culture of death before it kills the planet?[28]

🙢 🙢 🙢

I don't have a lot of patience for those who blame "all of us consumers" for damage caused by the economic and social system, those who say, "We're all in this together,"[29] and who point out, "If we didn't buy tickets, the airline industry would go broke." Well, first, if we didn't buy airline tickets, the feds would bail them out. All major industries rely on massive subsidies of public moneys to stay afloat. Second, if we're going to throw out a fantasy about the mass of Americans rising up to not buy airline tickets, why dream so low? Why not dream big and have this same fantastic mass of people start taking out dams? Why don't we have them storm vivisection labs and factory farms to liberate tormented animals? Why not have them dismantle the entire infrastructure? (Oh, because that might lead to real change, and we don't even want to *dream* about that.) The same people who tell me I can make a difference by not buying an airline ticket quite often tell me I shouldn't try to take out a dam because taking out one lone dam wouldn't accomplish anything. And not buying one lone airline ticket will?

The point, once again and as always, is leverage.[29] Sure, I support individuals and sometimes even industries I believe are headed the right direction through spending my hard-earned dollars in places and ways that are less destructive,[30] and similarly, insofar as possible, I don't support through my spending individuals and industries that are especially destructive, but I also recognize that far more needs to be done than this. I am not *merely* a consumer, much as those in power would like for me to define myself as such. The tools of consumerism are but one set available to me. The trick is to know when and how to use that set, and when and how to use others. The trick, to put it another way, is to leverage my efforts, to make my own small force have larger effects. The questions: What do I want to move?[31] What do I use for levers? Where do I place the fulcrums? How hard and when do I push?

⊛   ⊛   ⊛

There are other problems with attempting to spend or boycott our way to sustainability. The first is that it simply won't work. Spending won't work because within an industrial economy nearly all economic transactions are destructive. Because the industrial economy—indeed a civilized economy—is systematically, inherently, functionally, and inescapably destructive, even buying "good things" isn't really doing something good for the planet so much as it is doing something not quite so bad. Let's say I purchase organic lettuce at the grocery store. That's a good thing, right? Well, not particularly. The problem is that the mass cultivation of lettuce—organic or not—still destroys soils, and spending our way to sustainability its transportation to market still requires the use of oil. I suppose if I purchased lettuce grown in small-scale permaculture beds from my next door neighbor, I'd be doing something even less bad, but this is rare enough to be the exception that makes the rule crystal clear.[32] For an act to be sustainable, it must benefit the landbase, which means the soil, the critters who live in the soil, the plants who live on the soil, the animals who eat the plants, the animals who eat the animals, the insects and others who turn the dead back into soil. Producing, marketing, or purchasing organic lettuce doesn't do that. Rare indeed within our culture is the economic activity that improves the landbase (and that doesn't pay taxes, to boot, since more than 50 percent of the discretionary federal budget goes to pay for war). And don't throw up your hands in despair and give me the old saw about how *all* human activities damage landbases: noncivilized people have lived on landbases for a very long time without destroying them, in fact enhancing their landbases according to the needs of the landbases.

The problem is not our humanity. The problem is this culture—this *entire* culture—and slight changes in spending habits won't significantly stop the destruction.

That's not to say we shouldn't enact whatever changes we can to make whatever difference we can—remember, we do need it all—and buying organic lettuce is better than buying pesticide lettuce, on any number of levels. It's just to say that when I spoke earlier of this culture being a culture of occupation, of the government being a government of occupation, of the economy being an economy of occupation, I wasn't speaking metaphorically or hyperbolically. I was speaking sincerely, literally, physically, in all seriousness and truth. If we were Russians living under the German occupation in 1943, would we believe we could stop the Nazis by buying products made by German companies we like a little more and not buying them from I.G. Farben and other companies we don't like?

The same is true for boycotts. We can't boycott our way to sustainability any more than we can spend our way to it. The industrial economy, as is true for any economy of occupation (which means any civilized economy), is fundamentally a command economy (defined as "an economy that is planned and controlled by a central administration"). I know, I know, we've all been fed the line that "our" economy is based on some mythical thing called the free market, and that whatever it produces is by definition what we want. But I don't want depleted uranium any more than I want depleted oceans. Do you? So how did we get them? If the economy really were free, why are armed military and police necessary to secure producers' access to resources? And even if it were a "free market," that wouldn't help our landbases, since these markets do not value those parts of our landbases not perceived as productive (in other words, not obviously amenable to exploitation). And as mentioned before, in a global economy, free market or not, any wild thing that is vulnerable to exploitation (in other words, is valuable) will either be domesticated—enslaved—or exploited to extinction. But it's worse than this. It's not a free market anyway. Remember the words of Dwayne Andreas: "There's not one grain of anything in the world that is sold in the free market. The only place you see a free market is in the speeches of politicians."[33] Economist Brad DeLong puts this another way: "As producers and employees many of us live in an economy that is better thought of as a *corporate* economy: an economy in which patterns of economic activity are organized by the hands of bosses and managers, rather than one in which the pattern of activity emerges unplanned by any other than the market's invisible hand."[34] Yet another way to say all this is to note that, as alluded to above, all sectors of the economy, in fact the economy as a whole, would collapse almost immediately without huge subsidies. If every person in the country suddenly decided to somehow boycott, for example, the oil industry—which of course won't happen, for any number of obvious reasons—the U.S. and other governments would merely increase the subsidies to that sector of the economy, and probably for good measure arrest the boycott organizers on racketeering charges.[35]

Another reason we can't spend our way to sustainability is that we will *always* be outspent by those who are actively destroying the world. Destroying the world is how they make their money. It is always how they have made their money: through production, through the conversion of the living to the dead, through forcing others (the natural world, human communities) to pay the price for their activities. If you don't produce—that is, destroy—you won't make money. That still isn't to say that there aren't degrees of destructiveness: the damage caused by a permaculture farmer hand-delivering his lettuce leaves to his neighbors would be trivial compared to the damage caused by a full-on industrio-chemical lettuce agricorporation, but, and this is the point, so would his profits. That's why those who profit from this destructiveness will *always* have more money than we do, and will always be able to outspend us. An example should make this clear. Let's say I make a boatload of money writing and selling books. Oops, scratch that, since the manufacture of books—even on recycled paper using soy-based inks—requires lots of water, energy (ghost slaves), and raw materials. In other words, it's very destructive. Okay, so let's say instead I make a boatload of money making a boatload of money (in other words, I haul out my trusty printing press, and I just *make* the damn stuff). Oops, I can't do that, since the counterfeiting of money requires high-quality papers and lots of presumably toxic inks, lots of energy, and so forth. In other words, that's very destructive too. So okay, darnitall, let's say instead I just walk to a bank (wearing only used clothing taken from the dumpster behind Goodwill), and I *take* a boatload of money. I do this at night, because I don't want to threaten or scare any of the tellers, or perform any other action that might be construed as violent. Even better, I don't go to a bank, but go at night to Wal-Mart, and sneak in through an open door. I don't want to break a window, because there are those who would consider this an act of violence. I don't blow the safe because there are those who would consider *this* an act of violence. But let's say the safe is open. I take a boatload of money. Or if the safe isn't open, I take a bunch of consumer items, fabricate some receipts (okay, so this takes paper, but we'll just ignore that) and return them over the next days and weeks and months for a boatload of money. Wal-Mart, with its $258.6 billion in revenues, isn't going to miss it.[36] The point is that I somehow find a way to acquire a boatload of money that a) didn't cause me to "produce"—in other words, destroy—anything, and b) didn't cause me to pay taxes—in other words, to pay the government so it can destroy things. The question becomes, what am I going to do with this cash? Let's say I do what I actually would do if I acquired a boatload of cash: I buy some land and set it aside. Let's ignore the fact that in so doing I'm reinforcing the extremely damaging idea that land can be bought and sold. I buy an entire small creek drainage, and I set to work to improve habitat in that drainage for salmon, Port Orford cedars, mountain lions, Pacific lampreys, red-legged frogs, and so on. I create a sanctuary, a place where salamanders, newts, tree frogs, towhees, phoebes, and spotted owls can thrive and live as they did before the arrival of our awful culture. I've done a good and great thing, maybe even as good and great as what Elser tried to do. But now I find I want to protect more land, because these creatures need more habitat. What do I have to do? Because I pulled this land out of production, and thus am not "making any money" off of it, I have to write more books, print more money, make more trips to Wal-Mart, and unless I've figured out non-destructive ways to acquire cash—like the nocturnal trips to Wal-Mart—then I'm basically creating sacrifice zones elsewhere that I do not see so that the land I do see can be protected. I have to do this every time I want to protect more land.

Now, let's contrast that with someone who purchases this entire watershed not to create a sanctuary but to cut the trees. That person will "make money" off the land by harming it, and can use that money to purchase more land, where that person can cut more trees and make more money, and use that money to buy more land, and so on until there's nothing left. See, for example, Weyerhaeuser, or any other timber (or other) corporation.

Because the civilized economy is extractive, because it rewards those who exploit humans and nonhumans, that is, because it rewards those who do not give back to the landbase what it needs, that is, because it rewards people for disconnecting themselves from the reciprocal relationships that characterize (human or nonhuman) sustainable economies (and relationships), those who value the accumulation of money or power over life will always have more money or power than those who value life over money or power.

⊛   ⊛   ⊛

I went to see *Star Wars* when I was in high school, which seems about the right time to see it. I liked it a lot. I wasn't one of those people who saw it a hundred times or anything. I wasn't *that* much of a nerd. Besides, I was too busy playing *Dungeons and Dragons*. I saw it again recently. It's not so good as I remember. In fact it's pretty bad. The characters are flat, the dialog hokey, the acting nondescript. But I still loved the ending, where Luke remembers to "use the force" to blow up the Death Star. For those of you who may have forgotten, the Death Star (according to the official *Star Wars* website) "was the code name of an unspeakably powerful and horrific weapon developed by the Empire. The immense space station carried a weapon capable of destroying entire planets. The Death Star was to be an instrument of terror, meant to cow treasonous worlds with the threat of annihilation. While the massive station is evidence of the evil that was the Galactic Empire, it was also proof of the New Order's greatest weakness— the belief that technology and terror were superior to the will of oppressed beings fighting for freedom." That's all pretty interesting stuff, and of course applicable to the discussion at hand: civilization as Death Star.

The website also says, "The Death Star was a battle station the size of a small moon. It had a formidable array of turbolasers and tractor beam projectors, giving it the firepower of greater than half the Imperial Starfleet. Within its cavernous interior were legions of Imperial troops and fighter-craft, as well as all manner of detention blocks and interrogation cells. The

Death Star was spherical, and dark gray in color. Located on the Death Star's northern hemisphere was a concave disk housing the station's main laser weapon. . . . In a brutal display of the Death Star's power, Grand Moff Tarkin targeted its prime weapon at the peaceful world of Alderaan. [Rebel princess] Leia Organa, an Imperial captive at the time, was forced to watch as the searing laser blast split apart her beloved world, turning the planet and its populace into orbital ash and debris." I'm not sure if you feel a stab of recognition at being a captive of the empire, forced to watch your beloved world and its (human and nonhuman) populace turned into orbital ash and debris. I do.

The website continues, "Using . . . stolen technical data, [rebel] Alliance tacticians were able to pinpoint a crucial flaw in the Death Star's design. A small ray-shielded thermal exhaust port led directly from the surface of the station into the heart of its colossal reactor. If the port could be breached by proton torpedoes, then the resulting chain reaction would destroy the station."[37] We all know what happened next: By using the force, and with the help of Han Solo and Chewbacca, as well as the spirit of Obi-Wan Kenobi, Luke Skywalker was able to drop a proton torpedo right down the tiny port, and blow up the Death Star.

One small proton torpedo destroyed the Death Star. This would be a prime example of leveraging your power by using a properly placed fulcrum. In our case, to switch metaphors, where do we place the charges? Where is the correct thermal exhaust port? How do we start a chain reaction that will cause the "Death Star" before us to self-destruct?

⊕ ⊕ ⊕

You know, don't you, that this wasn't the movie's original ending. I have in my hands an extremely rare early draft of the *Star Wars* film script, never before published.[38] It may surprise you to learn that the early drafts were written by environmentalists.[39] In this version, the rebels do not of course blow up the Death Star, but instead prefer to use other tactics to slow the intergalactic march of Empire. For example, they set up programs for people on planets about to be destroyed to produce luxury items like hemp hacky sacks and gourmet coffee for sale to inhabitants of the Death Star. Audience members will also discover that there are plans afoot to encourage loads of troopers and other citizens of the Empire to take eco-tours of doomed planets. The purpose will be to show to one and all that these planets are economically important to the Empire and so should not be destroyed. In a surprise move that will rivet viewers to the edges of their seats, other groups of rebels file lawsuits against the Empire, attempting to show that the Environmental Impact Statement Darth Vader was required to file failed to adequately support its decision that blowing up this planet would cause "no significant impact." Viewers will thrill to learn of plans to boycott items produced by corporations that have Darth Vader on the board of directors, and will leap to their feet in theaters worldwide when they see bags full of letters written directly to Mr. Vader himself asking that he please not blow up anymore planets. (Scribbled in the margin is a note from one of the screenwriters: "For accuracy's sake, when we show examples of these letters, it is *imperative* that all letters to Mr. Vader be respectful and courteous, and that they stress the need to find coopera-tive solutions to the differences between the rebels and the Empire. Under no circumstances should the letters be such that they would alienate or anger Mr. Vader. If the letters upset Mr. Vader, the rebels' letter campaign to the Grand Moff Tarkin would certainly fail as well.") Other plans include sending petitions and filing lawsuits.

Now, you and I both know that all of this should be sufficient not only to bring the Empire to its knees but to make a damn fine and exciting movie. The thing is: there's more. Thousands of renegade rebels, unhappy with what they perceive as toadying on the part of the mainstream rebels, decide, in a scene guaranteed to bring tears to the eyes of even the most cold-hearted theatergoers, to stand on the planets to be destroyed, link arms (or, in some cases, tentacles), and sing "Give Peace a Chance." They send DVDs of this to both Darth Vader and his boss the Grand Moff Tarkin, to whom they also send wave after wave of lovingkindness™. Some few rebels sneak aboard the Death Star and lock themselves down to various pieces of equipment. (Early in this draft of the film, the screenwriters included a long scene showing the extensive training in nonviolent com-munication that is a prerequisite to joining the rebels. Most writers had originally, by the way, called it a rebel army, but several objected to the vio-lence inherent in that word. Next came "rebel force," but nearly as many objected to that word as well. In any case, the nuanced scene of nonviolence training was dropped in later drafts and the infamous [and horribly violent] Cantina scene was, incomprehensibly to some, put in its place.[40]) Stirring debates are held onscreen among these rebels as to whether they should voluntarily surrender on approach of the troopers, or whether they should remain locked down to the end. In a brilliant and brave touch of authenticity, the rebels are never able to come to consensus.

The writers themselves entered into a debate as to whether the troopers should decapitate the locked-down rebels on or off screen, with one writer pleading that instead rebels must be explicitly shown being taken alive to interrogation cells: "Showing," he wrote in the margin, "or even implying that the troopers would ever commit these acts of violence, even in response to such obvious challenges to their authority as rebels *invading* their space and doing *violence* to their machinery by interfering with that machinery's lawful use would send absolutely the wrong message to theatergoers, and would give the wrong impression of Mr. Vader's ultimately peaceful intentions."

Once inside the Death Star, a splinter group breaks off from those about to lock themselves down. They rush down long hallways, somehow avoiding the myriad troopers. They burn a couple of transport ships, and use chemicals to etch "Galaxy Liberation Front" on the walls of the Death Star. This group miraculously escapes back to the planet about to be destroyed, where they're held by the peaceful protesters so they can be immediately and rightly turned over to troopers. That same writer comments in the margin, "Not only is it vital, once again, that the right message be sent to audience members by showing these rebels being put in a position to take responsibility for their actions, but it would also be terribly unrealistic to expect these peaceful rebels to put up with these actions that would simply give Darth Vader the excuse he needs to blow up the planet. The disrespectful hooligans *must* be turned over to the Empire promptly and without question."

Near the end of the movie another debate is held among the rebels. (One problem I had with this environmentalist screenplay was that there was a bit too much debate and not quite enough action.[41]) As the Death Star looms directly overhead, a few of the rebels advocate picking up weapons to fight back. These rebels are generally shouted down by pacifist rebels, who argue that attacking those who run the Death Star is "just another example of the Empire's harmful philosophy coming in by the back door." They state that the rebels who want to fight back are simply being co-opted by the need to control things. If we want to change Darth Vader, they say, we must all first *become* the change. To change Darth Vader's heart, we must first change our own. We must above all else have compassion for Darth Vader, and remember that he, too, was once a child. One writer put in the margins: "Excellent! This will be sure to moisten the cheeks of sensitive people everywhere!" He did not mention whether or not these tears would be of frustration. Finally Leia, Luke, Han, Chewbacca, and a couple of robots show up and tell these others they've found a way to blow up the whole Death Star. The rest of the rebels—even those who'd previously been in favor of surgical strikes aimed at "removing" Darth Vader—are horrified. They point out that blowing up the Death Star will do nothing to change the hearts and minds of those who create Death Stars, and so will accomplish nothing. Han Solo replies, "It will stop this Death Star from destroying this planet." The pacifist rebels are unmoved. They remind the unruly four that the Death Star has a crew of 265,675, plus 52,276 gunners, 607,360 troops, 25,984 stormtroopers, 42,782 ship support staff, and 167,216 pilots and support crew.[42] Each of these people on the Death Star has a family. Do you want to make their children orphans? The pacifists themselves begin to cry. (That same screenwriter comments: "If that doesn't yank the tears out of audience members' tiny ducts, I don't know what will!") They say, voices firm behind the sobs, "You cannot blow up the Death Star. What about the custodial engineers? What about the cooks? What about the people who work the shopping malls? What about those who joined the empire's armed services just so they could go to college? You—Leia, Han, Luke, and Chewbacca—are heartless and cruel."

In the exciting final scene of the environmentalist version, a scuffle breaks out between Leia, Luke, Han, and Chewbacca on one side, and the pacifists on the other. At last the pacifists chase those four from the room and from the film. They're never seen again, which isn't really important since in this version they're minor characters anyway. The Death Star looms closer and closer. Audience members chew their fingernails as they wait to see whether the letters and petitions and lawsuits will work their magic. Viewers see lasers inside the Death Star warming up to destroy the planet. The lasers glow a hellish red. The camera switches to cover the endangered planet. Suddenly a cheer will rise up from the audience as they see a small bright speck emerge from the planet's surface and speed into space. "Yes!" they will roar, as they learn that all of the intrepid environmentalist protesters were able to get off the planet moments before it got blown up!

Coda: The final shot of the movie, revealing what a complete triumph this was for the rebels, will be a still showing an article on the lower-left of page forty-three of the *New Empire Times* devoting a full three sentences to the destruction of the planet. Yes! The protesters got some press![43]

Endgame, Volume Two
Resistance

Seven Stories Press, 2006

Before we talk about how to win, we have to talk about what we mean by winning. And before we talk about what we mean by winning, we have to talk about the difference between strategy and tactics.

Strategic goals are your largest-scale objectives. For the rebels in *Star Wars* this winning might have meant overthrowing the Empire, or at least liberating the rebel planets from its oppression. A strategic goal of the Nazis in World War II was to gain access to necessary resources, such as oil. A strategic goal of the United States right now is to gain access to necessary resources, such as oil.

Strategic goals can also be slightly smaller. While the largest strategic goal of the United States in the Civil War was to defeat the Confederacy—disallow the states that constituted the Confederacy from seceding from the United States—there were certainly many intermediate strategic goals: dismember it by gaining control of the Mississippi, cut it off from the rest of the world through a blockade of its ports, and so on.

Individuals can have strategic goals, too. One person's strategic goal may be to make a lot of money. Another's may be to get married. Still another's may be to bring down civilization.

One's strategies will be the methods or means by which one hopes to accomplish one's strategic goals. Or, as my dictionary puts it on the grandest scale, a strategy is "1(a)1: the science and art of employing the political, economic, psychological, and military forces of a nation or a group of nations to afford the maximum support to adopted policies in peace and war: (2) the science and art of military command exercised to meet the enemy in combat under advantageous conditions." If one has as a strategic goal of gaining and maintaining access to necessary resources, such as oil, one's strategies will be *how* one goes about achieving this goal. One example would be invading Iraq. Another would be maintaining "friendly" governments (which by virtue of their "friendliness" by definition become freedom-loving™ and democratic™) wherever there happens to be oil. Another would be the Nazis invading Russia to get at the oil fields in the Caucasus (and to push the Russians farther from the Ploesti oil fields in German-allied Rumania).

Intermediate between strategic and tactical goals are operational goals. For the rebels in *Star Wars*, an operational level goal was to take out the Death Star. For the Nazis in World War II one operational goal was to attempt to secure Rostov so they could push on toward Baku and Tblisi in the Caucasus. Another operational goal was to suppress partisans so rail lines remained open. An early operational goal for the Allies in the Normandy invasion was to establish enough of a beachhead to allow troops massed there some level of mobility. And so on.

Tactical goals are the smallest scale. If we're going to take Rostov, we may need to take this particular bridge to our left. And that factory to our right has snipers in it who must be killed or driven out. Tactics are *how* you accomplish these tactical goals. How are you going to secure this bridge? How are you going to take out the snipers in that factory? Who will provide covering fire as this unit moves forward? So far as the *Star Wars* example we've been discussing, the rebels have to get a ship close enough to the exposed tube to drop in a proton torpedo. How can Luke fly in without getting killed?[1]

The person who wants to get rich thinks strategically, operationally, and tactically as well. So does the person who wants to get married. We've all heard, for example, of men or women who tactically feign excitement over their potential partner's hobbies (the man goes dancing, the woman goes to football games, both feign an interest in sexual intimacy, and each feigns an interest in getting to know *who* the other person *is*[2]) in order to achieve their strategic goal of marriage. This reveals the importance of precision in defining one's goals, since while these tactics may lead to marriage, nobody said anything about a happy marriage.

I thought about strategic, operational, and tactical goals when I was in college. Strategic goal: Become a writer. To do this I would need the time to find out who I am and what I love, the time to think about what I want to write, the time to practice writing. This meant that part of my strategy had to involve gaining large blocks of uninterrupted time for me to think and to not think, to feel and then to feel some more.

The question became, how do I get that? Operational goal: Find a way to not get a job. (This became in a sense its own strategic goal: Whether or not I ever became a writer, I vowed to never work a job I didn't love: Why would anyone ever do anything so silly as sell one's life for mere money?) Because my first degree was in mineral engineering physics from the Colorado School of Mines, I knew I could get a high paying job right after college. One possible means to achieve my operational goal[3] would be to get a job for an oil company, work hard and live cheaply, then retire in a few years with a nest egg that would, the plan went, last me long enough for me to start making money writing. Another plan I considered, and you may not believe this, was to join the military.[4] The navy has something called the Nuclear Propulsion Officer Candidate (NUPOC) Program where they pay college juniors and seniors signing bonuses of something like 10K, and then pay another $2,500 per month while they go to school. In exchange the kids have to spend five years in the navy and another three years in the reserves after they graduate. In the navy they act as nuclear officers on submarines.[5] I thought I'd have a heck of a nest egg when I got out. I'd save my money through college, and then of course I'd save all my money when I was on the submarine: what could I buy there? I was very excited about this. I'd waste eight years of my life, and then I'd be ready to go. How bad could it be?

I found out. The Navy flew me to Charleston, South Carolina, where recruiters gave me and a bunch of other students tours of submarines. That's when I understood.

Do you ever have those moments where suddenly you make a quantum jump in understanding, where you see the world so differently that you cannot imagine how you could have perceived it any other way before? Do you have those times when this new understanding makes you feel as though up until that moment you must have been deluded or asleep or just plain stupid?

I had a couple of those on the submarine. The first came as we stood cramped in a narrow hallway, and one of the recruiters spoke excitedly about the nuclear missiles this sub could launch at targets thousands of miles away. Suddenly, and remember that I warned you my realization made me feel I must have been stupid, I understood that this program had nothing to do with Jacques Cousteau. I had somehow been under the impression that the United States government would pay me to go to school so that afterwards I could go down in a submarine to research the habitat needs of squid and octopi. I hadn't considered that I might actually be doing my part to blow up the world.

Soon after, I realized that while most of the sailors I talked to hated their jobs and would gladly have blown up the submarine if they could have gotten away with it and if it would have meant they could have gone home (and still received a paycheck), many of the officers actually got off on the power that they held, not only over the sailors but over the technologies at their fingertips. They clearly enjoyed the fact that they—though they didn't use this language—were part of a team that together could more or less end life on the planet. I felt stupid about this as well: Given how much I had read about the military, the banal nature of their arrogance and depravity surprised me far more than it should have.

A question not often enough discussed among those killing the world and too often discussed among those of us trying to stop them has to do with the relationship between strategy and tactics on one hand, and the morality of those strategies and tactics on the other. The first option of those in power is to explicitly delink the latter from the former. Indeed, as George Draffan and I discuss in *Welcome to the Machine*, the civilized generally consider themselves to be highly rational, and a great working definition of rationalization is that it is the deliberate elimination of information unnecessary to achieving an immediate task. This information to be eliminated can certainly include morality. As we say in that book: "If your goal is to maximize profits for a major corporation, all you need to do is ignore all considerations other than that. If your goal is to maximize gross national product (that is, the rate at which the world is converted into products), then all you need to do is ignore everything that might stand in the way of production." The second option of those in power, when delinking morality and strategy (or tactics, what have you) falls too absurd, is, as mentioned earlier, to invoke claims to virtue. Given the insanity of most of the members of this culture, it is almost not possible for claims to virtue to be too absurd to be accepted by many. The point is that by *claiming* the moral high ground on the strategic level those in power then exempt themselves from morality on the tactical level: Under the sign of the Cross (or capitalism, freedom™, democracy™, or civilization) we just this once (time and again) go forth and conquer, and in doing so we may end up committing any number of what would be considered, if someone else did them, atrocities.

Those of us at least pretending to oppose the horror that is this culture play a complementary role in this same game. We de-link morality and strategy as surely as do those in power. They carry strategy and tactics without effectively concerning themselves about morality. We carry morality without concerning ourselves with effective strategies and tactics.[6] As always, this is a convenient game for all who play it. We all get what we want (or have been encultured to want). Those in power get to extend their perceived control. We get to feel moral. All players get to reap the material rewards of playing this game. And of course wild humans and wild nonhumans are still driven extinct.

We members of the too-loyal opposition often spend far more time discussing morality than we do either strategy or tactics (or heaven forbid, actually accomplishing something). More to the point, when we do discuss strategy or tactics, nine times out of ten (or more realistically 999 times out of 1000) we don't *need* to discuss morality because we've so thoroughly internalized premise four of this book that the possibility of violating it under any circumstances is entirely incomprehensible. Our morality has become morality™, defined, designed, and pre-packaged for us by those who are killing the planet. And our strategy and tactics have become strategy™ and tactics™, also defined, designed, and pre-packaged for us by those who are killing us, defined, designed, and preplanned to be ineffective.

It's all a very interesting, bizarre, unfortunate, and self- and other-destructive manifestation of the same old fragmentation that characterizes this culture. Those in power define and carry for us morality on the largest scale, ignoring morality on the smallest scale, and we carry and abide by (their) morality on the smallest scale, ignoring morality on the larger scale.[7] Because those in power come to represent all that is good and great, because they are advancing civilization, because they are developing natural resources, because they are bringing democracy™, freedom™, the free market™, Christianity™, and McDonald's, they may lie and kill to achieve their ends. We, on the other hand, must always act honorably (or rather, honorably™). We must never lie. We must never kill. They do the dirty work. We carry the morality. The world burns. In the meantime, those of us on the inside of this game continue to have our participation purchased through the carrot of the material benefits of exploitation and the stick of repression.

It should come as no surprise that the mass of us abandon responsibility for interpreting large (and small) scale morality to our betters. That's how the system *works*. It cannot work without this. This abandonment is central to *everything* that is wrong with this culture, and it is central to the explorations in this book. From birth we are trained to abandon this responsibility by almost every authority figure, at almost every moment. Priests mediate between us and God, and translate God's words and intents and desires for us (no, not *for* us, but *instead of* us, *against* us) so that we will not be tempted to misunderstand, to come to any conclusions not in line with their own, to interpret our own morality ourselves, or get it directly from God, or get it directly from the land. Scientists, too, stand between us and God—now called knowledge™—and interpret God's (or nature's, or knowledge's) will for us. Instead of us. Against us. Judges mediate between us and God—now called justice™—and interpret God's (or the law's) will for us. Instead of us. Against us. Teachers mediate between us and education, determining for us what and how and when we shall learn. The list goes on.

In each of these cases we give up our own authority, our own responsibility to make choices.

Violence—the defensive birthright of every being, no matter how peaceful—falls prey within this culture to the same fragmentation, the same insane division of labor. Soldiers and police (and rapists and abusers) carry violence for the rest of us. *They* are violent. *We* are not. We are, instead, moral. But the distinction isn't really that distinct; it's all part, as I've been saying, of the same sick game. Civilization—like any other abusive relationship—is based on force, on violence, on theft, on murder, on exploitation, and whether or not the pacifist (or you or I) pulls a trigger matters not a whit for culpability.

Hate is the same. Soldiers hate. They are trained to hate. That's what boot camp is *for*.[8] We, on the other hand, do not hate.[9] Hatred is for killers. We do not kill. Thus we do not hate. The soldiers carry our hatred for us, and they do our killing for us. So do the police. We carry morality for them. And the poor groan under the weight of the soldiers' boots—surrogates for our own—stomping on their faces.

We are cut off from our own violence, and we are cut off from our own nonviolence. We have neither. We are split off shells, partial people pretending to be whole but only completed by our split off and disavowed twins.

It's all about disconnection. This culture is based on disconnection. Man (strong) versus woman (weak), man (good) versus nature (flawed), thought (honest) versus emotion (misleading), spirit (pure) versus flesh (polluted), love (good) versus hate (bad), serenity (good) versus anger (bad), nonattachment (good) versus attachment (bad), nonviolence (righteous) versus violence (evil), and so on *ad nauseum*. So often I've heard pacifists and others say we need to get rid of all dualism, that by speaking of those who are killing the planet as my enemy I am perpetuating the same dualisms that got us here. But striving to eradicate dualism is perpetuating the same dualism! This time it's nondualism (good) versus dualism (bad). It's all nonsense. The problem isn't that there are pairs of opposites. Opposites simply exist. Nor is the problem that there are values assigned to these opposites. We can—and I certainly would—argue against the values chosen by this culture for each of these poles, but the truth is that the different poles do have different values. And that leads to the real problem, which is the word *versus*. Yes, men and women are different. But they are not in opposition; instead they work together. Yes, humans are different than nonhumans (as it would also be true that salmon are different than nonsalmon, and redwoods are different than nonredwoods). But they are not in opposition; instead they work together. Thought is different than emotion. But they are not in opposition; instead they work together. Spirit is different than flesh. But they are not in opposition; instead they work together. Love is different than hate, serenity is different than anger, nonattachment is different than attachment, nonviolence is different than violence. But they're not in opposition; each of these paired opposites works together. Dualism is different than nondualism. But they are not in opposition; instead they work together. Duh.

What happens if you reconnect? What happens if you make choices as to when you should think, and when you should feel? What happens if your thoughts and feelings merge and diverge and flow in and out of each other, with each one taking the fore when appropriate (and sometimes when inappropriate, since perfection does not exist in the real world, and emotions and thoughts each sometimes make mistakes: That's life) and with them working sometimes together and sometimes in opposition? What happens if you make choices as to when you should think and feel dualistically—in opposition to some other—and when you should work with this other? What happens if you sometimes make choices and sometimes you do not? What happens if when appropriate you are violent ("Hear! Hear!" say the wolverine and the shrew), and when not appropriate you are not?

The "answer" is not to try ever more desperately to eradicate hate from our own hearts, to carry more and more of the love that is split off from the rest of the culture—as if it's the case that if we can only carry enough love to make up for everyone else that things will be all right, or even that any love we feel might in any way counter someone else's hate—and split off ourselves the hate we do not allow ourselves to feel. The "solution" is to reintegrate, to feel what we feel, to determine our own moralities (large and small scale) and to act on them.

In any case, I decided that while my larger goal of finding a way to live without a wage job was certainly moral, the proximate strategy of joining the navy would be an immoral way to achieve it. Robbing banks or shoplifting from Wal-Mart would undoubtedly have been more moral. I did neither. As I wrote about in *A Language Older Than Words*, I became a beekeeper, partly because I love bees, and partly because as a beekeeper I would work hard part of the year and I would have time to just be for part of the year.

Now that we've defined what are strategies and tactics, but before we start trying to determine what ours are, we need to talk about what we want to accomplish. What do we want?

By now I'm pretty sure you have a good idea what I want. A world not being killed. A world being renewed. A world with more wild salmon each year than the year before. A world with more wild humans each year than the year before.

What do you want? The question is not rhetorical. Don't just pass it by and move to the next chapter. Stop. Put down the book. Go outside. Take a long walk. Look at the stars. Pet the bark of trees. Smell the soil. Listen to a river. Ask them what they want. Ask your heart what you want. Ask your head. Ask your heart again. Then figure out how you're going to get it.

☙ ☙ ☙

We have been too kind to those who are killing the planet. We have been inexcusably, unforgivably, insanely kind.

I understand now. For years I have been asking whether abusers believe their lies, and I'm finally comfortable with an answer.

This understanding came in great measure because I finally stopped focusing on the lies and their purveyors and I began to focus on the abusers' actions. I realized, following Lundy Bancroft, that to try to answer the question of whether the abusers believe their lies is to remain under the abusers' spell, to "look off in the wrong direction," to allow myself to be distracted so I "won't notice where the real action is." To remain focused on that question is exactly what abusers want.

Bancroft helped me realize some very important things. He writes specifically about abusers, emphasizing perpetrators of domestic violence, but what he says applies as well to this whole culture of abuse, and to perpetrators of the larger scale abuse I've been writing about.

His central thesis seems to be that the primary problem is not that abusers particularly "lose control" or that they are particularly prone to "flying into a rage," but instead that they feel entitled to exploit, will do anything in order to exploit, and will exploit precisely as much as they can get away with.

Bancroft excels at exploding misconceptions. When a woman stated that her abusive partner Michael loses control and breaks things in a rage, only to feel remorse afterwards, Bancroft asked whether the things that were broken were Michael's or hers. She answered, "I'm amazed that I've never thought of this, but he only breaks my stuff. I can't think of one thing he's smashed that belonged to him." Bancroft asked who cleans up. She does. He responded, "Michael's behavior isn't nearly as berserk as it looks. And if he really felt so remorseful, he'd help clean up."[10]

I remember a time my father was berating and beating my teen-aged sister, and her boyfriend showed up an hour early for their date. My father immediately ceased calling her a slut, dropped his hands to his sides, smiled, and walked to greet her boyfriend as if nothing had happened. His rage was not out of control, but something he was able to turn on and off like a light switch.

Or picture this. My father hits my mother. He has hit her many times before. But this time she slips into another room, calls the police. She comes back out. My father hits her again and again. He is interrupted by the doorbell. He points one finger at her, runs his other hand through his hair, walks to the door, opens it. There are two policemen. My father is cool, calm, as though nothing has happened. My mother is frantic, frightened, having just been beaten. The cops sympathize with my father for living with someone so emotional—they also sympathize because their allegiance already runs to the abuser—and they leave. The door closes. My father resumes beating my mother. His rage, once again, could be turned on and off.

My mother can perhaps be forgiven for her naïveté in relying on authorities to assist her. She was, after all, nineteen years old, with two children and pregnant with a third. But at this point, especially on the larger scale, the rest of us should not be so naïve.

Abusers are not out of control. They are very much in control. I never understood that till I read Bancroft's book.

Similarly, I speak of this culture's destructive urge, and how those in power destroy those things they cannot control. I have written of clearcuts, of devastated oceans, of murdered poor and extirpated species. But corporations and those who run them do not flail willy-nilly at everything around them. Like Michael, they do not destroy what belongs to them. And of course they do not clean up their messes, no matter how much remorse they may feign, and no matter how much they may claim to have moved beyond petroleum, or into new forestry, or whatever other words they may wish to throw around.

Bancroft asks the abusers he works with what are the limits of their violence. He might say, "You called her a fucking whore, you grabbed the phone out of her hand and whipped it across the room, and then you gave her a shove and she fell down. There she was at your feet, where it would have been easy to kick her in the head. Now, you have just finished telling me that you didn't kick her. What stopped you?" His point is not so much the question as the answer. He says the abusers *can always give . . . a reason.*"[11] Some of the reasons: "I wouldn't want to cause her a serious injury." "I realized one of the children was watching." "I was afraid someone would call the police." "I could kill her if I did that." "The fight was getting loud, and I was afraid neighbors would hear." The most frequent response is, "Jesus, I wouldn't do *that*. I would never do something like that to her." Only twice in fifteen years has Bancroft heard the answer, "I don't know."[12]

His point is that when abusers are committing their atrocities, they remain acutely aware of the following questions, "Am I doing something that other people could find out about, so it could make me look bad?[13] Am I doing something that could get me in legal trouble? Could I hurt myself? Am I doing anything that I myself consider too cruel, gross, or violent?"[14] These questions are asked word for word in corporate boardrooms. I spoke at length a few years ago with a former corporate lawyer who recovered her conscience, quit, and began working against the corporations. "The people who run these corporations," she said, "know exactly what they're doing. They know they're killing people. They know they're destroying rivers. They know they're lying. And they know they're making a lot of money in the process." Bancroft continues, "A critical insight seeped into me from working with my first few dozen clients. *An abuser almost never does anything that he himself considers morally unacceptable.* He may hide what he does because he thinks *other* people would disagree with it, but he feels justified inside. I can't remember a client who ever said to me: 'There's no way I can defend what I did. It was just totally wrong.' He invariably has a reason that he considers good enough. In short, *an abuser's core problem is that he has a distorted sense of right and wrong.*"[15]

This is true on the larger social scale. Clearly, a culture killing the planet has a distorted sense of right and wrong. Clearly a police department that arrests treesitters yet neither deforesters nor rapists has a distorted sense of right and wrong.

Bancroft asks his clients whether they ever call their mothers a bitch. When they say they don't, he asks why they feel justified to call their partners that. His answer is that "the abuser's problem lies above all in his belief that controlling or abusing his female partner is justifiable."[16]

Once again, the connections to the larger cultural level should be obvious. In some ways this is a restatement of premise four, but it's different enough and important enough to become the nineteenth premise of this book: *The culture's problem lies above all in the belief that controlling and abusing the natural world is justifiable.*

It all comes down to perceived entitlement. As Bancroft states, "*Entitlement* is the abuser's belief that he has a special status and that it provides him with exclusive rights and privileges that do not apply to his partner. The attitudes that drive abuse can largely be summarized by this one word."[17]

This same attitude applies on the larger social scale. Of course humans are a special species, to whom a wise and omnipotent God has granted the exclusive rights and privileges of dominion over this planet that is here for us to use. And of course even if you subscribe to the religion of Science instead of Christianity, humans' special intelligence and abilities grant us exclusive rights and privileges to work our will on the world that is here for us to use. And of course among humans, the civilized are especially special, because we are such a high stage of social and cultural development, with especially exclusive rights and privileges to use the world as we see fit. And of course among civilized humans, those who run the show are even more special, and so on.

The flattering belief that one is entitled to exploit those around him is a major reason abusers so rarely stop their abuse. Although this is, according to Bancroft, "rarely mentioned in discussions of abuse," it "is actually one of the most important dynamics: the *benefits* that an abuser gets that make his behavior *desirable* to him. In what ways is abusiveness rewarding? How does this destructive pattern get reinforced?"[18]

He also states, "When you are left feeling hurt or confused after a confrontation with your controlling partner, ask yourself: What was he trying to get out of what he just did? What is the ultimate benefit to him? Thinking through these questions can help you clear your head and identify his tactics."[19]

My father tells my sister to do the dishes. She complains that she has never seen him do them. He stares at her. She does them. He points out a place she missed on a plate. He hits her. Never again will she suggest he do dishes, unless she is willing to accept the consequences.

My father wants sex. My mother tells him no. He stares at her. He pouts. Later that day he hits her because of something unrelated. But this happens again later that week, and again the next week, and the week after, until finally she makes the connection. Never again will she tell him no, unless she is willing to accept the consequences.

As Bancroft writes, "Over time, the man grows attached to his ballooning collection of comforts and privileges."[20]

This takes us right back to William Harper's 1837 defense of slavery: "The coercion of Slavery alone is adequate to form man to habits of labour. Without it, there can be no accumulation of property, no providence for the future, no taste for comforts or elegancies, which are the characteristics and essentials of civilization."[21]

On the larger scale, too, each time we are left confused or hurt by the lies or other tactics of those in power—as ExxonMobil changes the climate, as Boise Cascade deforests, as Monsanto poisons the world, as BP lies about its practices, as politicians lie about everything—we need to ask Bancroft's questions: What are those in power trying to get out of what they just did? What is the ultimate benefit to them?

⊕ ⊕ ⊕

One of the bad things about abusers as compared to other sorts of addicts is that at least substance abusers sometimes "hit bottom," where their lives become painful enough to break through their denial. No such luck with those who abuse others.

Bancroft states that partner abuse "is not especially self-destructive, although it is profoundly destructive to *others*. A man can abuse women for twenty or thirty years and still have a stable job or a professional career, keep his finances in good order, and remain popular with his friends and relatives. His self-esteem, his ability to sleep at night, his self-confidence, his physical health, all tend to hold just as steady as they would for a nonabusive man. One of the great sources of pain in the life of an abused woman is her sense of isolation and frustration because no one else seems to notice that anything is awry in her partner. *Her* life and her freedom may slide down the tubes because of what he is doing to her mind, but *his* life usually doesn't."[22]

❁ ❁ ❁

Many Indians have asked these questions about the civilized. I have asked these same questions about CEOs, corporate journalists, politicians. How do these people sleep at night?

Soundly, in comfortable beds, in 5,000 square foot homes, behind gates,with private security systems, thank you very much.

❁ ❁ ❁

It is others who lose sleep over their activities.

❁ ❁ ❁

Within an abusive family dynamic, everything—and I mean everything—is aimed toward protecting the abuser from the physical and emotional consequences of his actions. All members are enculturated to identify more closely with the family structure and its abusive dynamics than with their own wellbeing and the well-being of their loved ones and other victims. Because the dynamic is set up to foster the well-being of the perpetrator, every action, then, by every member of the family—and more to the point every member's every thought and non-thought and feeling and non-feeling and way of being and not-being—has as its goal the protection of the abuser's well-being. This "well-being" is a particular sort, devoid of relationship and accompanying emotions, heavy on the kind of external rewards abusers reap because of their abuse (and of course precisely the kind of external rewards emphasized by a grotesquely materialistic culture), and most especially focused on allowing the perpetrator to avoid confronting his own painful emotions, including the pain he inflicts, the pain he received as a child (and adult) that caused him to separate from his own emotions (to identify not with himself but with an abuser and an abusive dynamic), and the pain of living in an abusive dynamic where rewards gained through abuse never quite compensate for the emptiness of living a "life" devoid of real relationship.

In my book *A Language Older Than Words* I detailed among other things the importance of amnesia or selective memory to the survival of abused children. If you are powerless to prevent being harmed or in any way even to defend yourself, it serves no purpose to consciously remember the atrocities. In fact it can be lifesaving to read and then identify more closely with the perpetrator's emotions and state of being than one's own. After all, the child's emotions don't matter, but the child needs to be capable at all times of reading and if possible placating the powerful adult's emotions. But I did not mention the function this induced amnesia serves for the perpetrator: it allows him to confront neither the emotional consequences nor the emotional motivations for his abusive behavior.

Everyone at every moment acts to protect the abuser. Think about it in your own life. How many times has someone abused you and you did whatever was necessary to make sure the other person did not feel bad? What did you do to take care of the other person? Here is a story a woman just told me. She was sitting in a bar with her sisters, drinking coca cola. A man struck up a superficial conversation with her. Soon she walked into the bathroom. When she emerged from her stall, he was waiting for her. She asked what he was doing. He forced her against the wall, pushed his hips hard into her. She somehow slipped from his grasp, and returned to the main room. He followed. He remained within ten feet of her. She stayed for another hour. Now here's the point: Not only did she not make a scene, but she did not even leave. Even as she was slipping away from his attempted rape and all through the next hour she was thinking, *I don't want to hurt his feelings*.

I cannot tell you how many times I have similarly betrayed myself to protect an abuser.

Years ago, in the midst of one of those abusive relationships I mentioned earlier, a friend was counseling me through the latest incident of abuse. At one point I said, "I don't think she meant to hurt me. Here's what I think she was thinking—"

My friend cut me off: "If I was interested in what she was thinking, I would talk to her. But I'm not, so I won't. I'm interested in what you were thinking, and feeling."

I didn't have an answer. I had no idea. I was too busy taking care of the other person's feelings.

To care about another, to have compassion for another, is beautiful and life-affirming. To care about and have compassion for another who is

abusing you is a toxic mimic of real compassion, and is one of the obsceni-
ties spawned by a culture of abuse.

The same thing happens all the time on the larger scale. I also cannot
tell you how many times I have been told that I must have compassion for
CEOs, who are human too, and who once were children. We must never
hurt their feelings, nor especially their person. We must always be polite
to those who are killing us. If we insist on using any hint of violence, we
are told, if we absolutely must kill them back, we must kill them only with
kindness. This is supposed to somehow be effective at something. But the
only one it helps is the perpetrator.

Bancroft states that one of the most common forms of support for
abusers is the person "who says to the abused woman: 'You should show
him some compassion even if he has done bad things. Don't forget that
he's a human being, too.'" Bancroft continues, "I have almost never
worked with an abused woman who overlooked her partner's humanity.
The problem is the reverse: *He* forgets *her* humanity. Acknowledging his
abusiveness and speaking forcefully and honestly about how he has hurt
her is indispensable to her recovery. It is the *abuser's* perspective that she
is being mean to him by speaking bluntly about the damage he has done.
To suggest to her that his need for compassion should come before her
right to live free from abuse is consistent with the abuser's outlook. I
have repeatedly seen the tendency among friends and acquaintances of
an abused woman to feel that it is their responsibility to make sure that
she realizes *what a good person he really is inside*—in other words, to stay
focused on his needs rather than her own, which is a mistake."[23]

We have all been trained to identify more closely with the abusive personal and social dynamics we call civilization than with our own life and
the lives of those around us, including the landbase. People will do anything—go to any absurd length—to hide the abuse from themselves and
everyone around them. Everything about this culture—and I mean everything—from its absurd "entertainment" to its equally absurd "philosophy"
to its politics to its science to its interspecies relations to its intrahuman
relations is all about protecting the abusive dynamics.

R. D. Laing named three rules that govern abusive family dynamics,
that allow the family to not acknowledge the abuse:

Rule A: Don't.

Rule A.1: Rule A does not exist.

Rule A.2: Never discuss the existence or nonexistence of rules A, A.1, or A.2.[24] These rules hold true for the culture. We see them every day in every way, from the most intimate to the most global. This culture collectively and most of its members individually will give up the world before they'll give up this abusive structure.

⊕ ⊕ ⊕

This culture won't change on its own. The demands it makes on the natural world and on the humans it exploits won't diminish until the culture is destroyed. As Bancroft writes, "An abusive man expects catering, and the more positive attention he receives, the more he demands. He never reaches a point where he is satisfied, where he has been given enough. Rather he gets used to the luxurious treatment he is receiving and soon escalates his demands."[25]

The same is true on the larger scale, as no comforts or elegancies, no feeling of power over another, no accumulation of property can make up for a failure to participate in the great liturgy. It's an attempt to use increasing amounts of emptiness to plug a great void (or, as R.D. Laing wrote, "How do you plug a void plugging a void?"[26]). It's an attempt to cure loneliness through power. But loneliness can only be cured through relationship,[27] and relationship is precisely what exploitation and abuse destroy.

There can be no compromise with the insatiable. They'll ask, then negotiate, then demand, their threat of violence informing all interactions, and in the end they'll take. But that will not be the end, because they'll not be satisfied. They'll begin again, by asking, then negotiating, then demanding, then taking. And then they'll ask, negotiate, demand, take, until there's nothing left. And yet they'll keep on pushing.

Because Bancroft's book is in some ways self-help, he puts all this slightly differently: "*Objectification is a critical reason why an abuser tends to get worse over time.* As his conscience adapts to one level of cruelty—or violence—he builds to the next. By depersonalizing his partner, the abuser protects himself from the natural human emotions of guilt and empathy, so that he can sleep at night with a clear conscience. He distances himself so far from her humanity that her feelings no longer count, or simply cease to exist. These walls tend to grow over time, so that after a few years in a relationship my clients can reach a point where they feel no more guilt over degrading or threatening their partners than you or I would feel after angrily kicking a stone in the driveway."[28]

Or perhaps he means that abusers would feel no more guilt over threatening their partners than civilized humans would feel blasting stones from a quarry, or damming a river, or deforesting a hillside.[29] Stones, rivers, trees, forests, their feelings, far beyond not counting, have within this culture long since ceased to exist.

⊛ ⊛ ⊛

Unfortunately, abusers don't particularly care about what they're losing. Bancroft writes about this, too: "It is true that partner abusers lose intimacy because of their abuse, since true closeness and abuse are mutually exclusive. However, they rarely experience this as much of a loss. Either they find their intimacy through close emotional connections with friends or relatives, as many of my clients do, or they are people for whom intimacy is neither a goal nor a value (as is also true of many nonabusers).You can't miss something that you aren't interested in having."[30] This transposes easily to the larger scale with only a few substitutions. "It is true that the civilized lose intimacy with their landbase because of their exploitation of it, since true closeness and exploitation are mutually exclusive. However, they rarely experience this as much of a loss. Either they find their intimacy through close emotional connections [*sic*] with other humans, or they are people for whom intimacy with the land is neither a goal nor a value (as is true of nearly all of the civilized).You can't miss something that you aren't interested in having."

I've heard many environmentalists state that if only they could get CEOs and politicians out of their boardrooms and legislative halls (or out of their penthouses and vacation homes) long enough to breathe clean forest air and to feel duff beneath their feet, long enough to stop thinking about stock prices and start thinking about spotted owls, that the CEOs would undergo magical transformations and suddenly no longer want to destroy the homes of their newfound forest friends.

It ain't gonna happen. This false hope ignores many things. It ignores the fact that when Europeans first encountered a wildly fecund North America, they were not entranced by it, they did not fall in love with it, they feared and hated it, and they began to dismantle it, a dismantling that continues its acceleration to this day. It ignores the fact that many loggers spend much of their adult lives in forests, claiming to love these forests they're destroying. It ignores the fact that CEOs and politicians, like other abusers, are financially and socially wellrewarded for maintaining their disconnected state. It ignores the fact that if some individual does have an epiphany, he will simply be replaced and the destruction will continue apace. And most of all it ignores the fact that, as mentioned before, the culture's problem lies in the belief that controlling and abusing the natural world is justifiable.

▨   ▨   ▨

Where does this leave us?

Well, if you agree with my thesis—which I think I've more than amply supported—that the motivations, dynamics, and damage of abuse play out not only in the bedrooms of little girls and boys, not only in the black eyes and bruised and torn vaginas of women, not only in the fragmented and fearful psyches of the traumatized; but also in blasted streams and dammed rivers, poisoned oceans and extirpated species; and in enslaved, domesticated, or destroyed humans (and nonhumans, and landscapes), then it means that asking, cajoling, or even sending lovingkindness™ to abusers is at best a waste of time. Bancroft again: "You cannot get an abuser to change by begging or pleading. The only abusers who change are the ones who become willing to accept the consequences of their actions."[31] And yet again: "You cannot, I am sorry to say, get an abuser to work on himself by pleading, soothing, gently leading, getting friends to persuade him, or using any other nonconfrontational method. I have watched hundreds of women attempt such an approach without success. The way you can help him change is to demand that he do so, and settle for nothing less."[32]

Let's apply this on the larger scale: We cannot get large-scale abusers to stop exploiting others by pleading, soothing, gently leading, getting people to persuade them, or using any other nonconfrontational method. It won't work.

But you knew that already.

Bancroft continues, "It is also impossible to persuade an abusive man to change by convincing him that *he* would benefit, because he perceives the benefits of controlling his partner as vastly outweighing the losses. This is part of why so many men initially take steps to change their abusive behavior but then return to their old ways. There is another reason why appealing to his self-interest doesn't work. The abusive man's belief that his own needs should come ahead of his partner's is at the core of the problem. Therefore when anyone, including therapists, tells an abusive man that he should change because that's what's best for *him*, they are inadvertently feeding his selfish focus on himself: *You cannot simultaneously contribute to a problem and solve it.*"[33]

Let's once again explicitly make the connection to the larger scale. It is impossible to persuade the civilized to change by convincing them that they would benefit and simultaneously allowing them to remain within the framework and reward system of civilization, because the civilized perceive the benefits of controlling those around them (including humans and nonhumans; including the land, air, water; including genetic structures; including molecular structures) as vastly outweighing the losses. This is part of why so many of the civilized initially take steps—or at least mouth rhetoric and pretend to take steps—to change their abusive behavior but then return to their exploitative ways. There is another reason why appealing to the self-interest of the civilized doesn't work (apart from the fact that the entire economic system, indeed all of civilization, is based on this limited and unsustainable sense of self which leads people to believe it's in one's self-interest to exploit others, indeed, which causes it to be, within this limited sense of self, *actually in* one's self-interest to exploit others): the belief of the civilized that their own needs should come ahead of the landbase's is at the core of the problem. Therefore when people, including activists, tell a civilized person—for example, a CEO or politician—that he should change because that's what's best for *him*, they are inadvertently feeding his selfish focus on himself: *You cannot simultaneously contribute to a problem and solve it.*

Let's go one more time. Bancroft: "An abuser doesn't change because he feels guilty or gets sober or finds God. He doesn't change after seeing the fear in his children's eyes or feeling them drift away from him. It doesn't suddenly dawn on him that his partner deserves better treatment. Because of his self-focus, combined with the many rewards he gets from control-ling you, an abuser changes only when he has to,[34] so the most important element in creating a context for change in an abuser is placing him in a situation where he has no other choice. Otherwise, it is highly unlikely that he will ever change his behavior."[35] Pay careful attention. No other choice.

No, really. Pay careful attention. No other choice.

No, now *really* pay attention. No other choice.

None.

Let's transpose this to the larger scale. Those who are killing the planet won't change because they feel guilty or drop their addiction to consum-erism or find God, or Nature. They don't change after seeing the fear in factory farmed or vivisected animal's eyes (or in the eyes of the poor) or feeling wild creatures drift away from them. It doesn't suddenly dawn on them that the landbase deserves better treatment. Because of their self-focus, combined with the many rewards they get from controlling those around them, these abusers change only when they have to,[36] so the most important element in creating a context for change in those who are killing the planet is to place them in situations where they have no other choice. Otherwise, it is highly unlikely that they will ever change their behavior.

No other choice.

None.

Let's say you want to take out a big dam. Let's say you have the full power of the state behind you, which means you don't have to worry about pesky cops coming to drag you and your trusty sledgehammer away for knocking down this illegal structure. Of course when the structure was put up, the cops were nowhere in the area. In fact the Law Enforcement Officers were probably off arresting protesters who were trying to block access and stop the dam from being illegally built in the first place. We should change cops' title to Selective Law Enforcement Officers.[37]

How are you going to bring it down?

There are five major ways the state takes out dams, with the most common by far being the last one.

The first consists of digging around the dam to divert the river, then using heavy equipment to dismantle the dam. An example of this would be the twenty-three foot high and nine-hundred foot long Edwards Dam on the Kennebec River in Maine, which was taken down this way in just a few days in 1999.

The second method, usually used on huge earthen dams, is to breach the dam using heavy machinery, and then let the river flow around the rest of the structure, which you allow to remain standing, presumably as a monument to this culture's arrogance and stupidity. I'm still not sure how you breach a dam with water behind it. I'd like to learn, since this is a relatively inexpensive method.

The third method is an easy one. If you've got a barrage-type dam with radial gates, you can just open the gates and pretend the dam isn't there. Two examples of this would be the Nagara Estuary Dam in Japan and the Pak Mun Dam in Thailand.

The fourth method is the one we've all been waiting for: the big blast. Explosives are sometimes used to take out concrete dams. A few examples of this would include dams on the Clearwater (1963), Clyde (1996), Loire (1998), and Kissimmee (2000) rivers. Of course even with explosives it still helps if you've got a friend with some heavy equipment.[38]

The fifth method of dam removal used by those with the full power of the state behind them is the one we've come to expect from those who are able to get the full power of the state behind them, which is to do nothing at all. Although they either don't know or won't admit it, this is a form of dam removal, too, because eventually every dam will fail. The only question will be what's left of the river when that finally happens.

❀ ❀ ❀

If you're like me, you're probably wondering how many explosives it takes to knock out a big dam. The answer may make you as happy as it made me. It doesn't take many at all.

Read that again. It doesn't take many at all.

Imagine the possibilities!

On February 23rd of this year, a hundred foot breach was blasted into the Embry Dam on the Rappahannock River near Fredericksburg, Virginia. The dam was a bit over twenty feet high and nearly eight hundred feet long.

Divers placed explosives, and just a little after noon the signal was given to set them off. Only 10 percent ignited, making a burst of smoke and water but leaving the dam standing. Ninety minutes later they tried again. This time it worked. The river rushed through the broken dam.

All it took was six hundred pounds of explosives.

That's it.

And that was a pretty big dam.

What are you waiting for?

⊛   ⊛   ⊛

Many hundreds of pages ago, and now for me many years ago, I wrote that this book was originally going to be an exploration of when counterviolence is an appropriate response to the violence of the system. In fact what has become this book was supposed to be nothing more than a pamphlet in which I took the main arguments normally presented by pacifists and examined them to see if they make any sense. Here now is that pamphlet.

Here are some standard lines thrown out by pacifists. I'm sure you, too, have heard them enough that if we had a bouncing red ball we could all sing along. Love leads to pacifism, and any use of violence implies a failure to love. You can't use the master's tools to dismantle the master's house. It's far easier to make war than to make peace. We must visualize world peace. To even talk about winning and losing (much less to talk about violence, much, much less to actually do it) perpetuates the destructive dominator mindset that is killing the planet. If we just visualize peace hard enough, we may find it, because, as Johann Christoph Friedrich von Schiller tells us, "Peace is rarely denied to the peaceful." Ends never justify means, which leads to Erasmus saying, and pacifists quoting, "The most disadvantageous peace is better than the most just war." Gandhi gives us some absolutism, as well as absolution for our inability to stop oppressors, when he says, "Mankind has to get out of violence only through non-violence. Hatred can be overcome only by love." Gandhi again, with more magical thinking, "When I despair, I remember that all through history the way of truth and love has always won. There have been tyrants and murderers and for a time they seem invincible but in the end, they always fall—Think of it, ALWAYS."[39] Violence only begets violence. Gandhi again, "We must be the change we wish to see." If you use violence against exploiters, you become like they are. Related to that is the notion that violence destroys your soul. If violence is used, the mass media will distort our message. Every act of violence sets back the movement ten years. If we commit an act of violence, the state will come down hard on us. Because the state has more capacity to inflict violence than we do, we can never win using that tactic, and so must never use it. And finally, violence never accomplishes anything.

Let's take these one by one. Love leads to pacifism, and any use of violence implies a failure to love. If we love we cannot ever consider violence, even to protect those we love. Well, we dealt with this several hundred pages ago, and I'm not sure mother grizzly bears would agree that love implies pacifism, nor mother moose, nor many other mothers I've known.

You can't use the master's tools to dismantle the master's house. I can't tell you how many people have said this to me. I can, however, tell you with reasonable certainty that none of these people have ever read the essay from which the line comes: "The Master's Tools Will Never Dismantle The Master's House," by Audre Lorde (certainly no pacifist herself). The essay has nothing to do with pacifism, but with the exclusion of marginalized voices from discourse ostensibly having to do with social change. If any of these pacifists had read her essay, they would undoubtedly have been horrified, because she is, reasonably enough, suggesting a multivaried approach to the multivarious problems we face. She says, "As women, we have been taught either to ignore our differences, or to view them as causes for separation and suspicion rather than as forces for change. Without community there is no liberation, only the most vulnerable and temporary armistice between an individual and her oppression. But community must not mean a shedding of our differences, nor the pathetic pretense that these differences do not exist."[40] We can say the same for unarmed versus armed resistance, that activists have been taught to view our differences as causes for separation and suspicion, rather than as forces for change. That's a fatal error. She continues, "[Survival] is learning how to take our differences and make them strengths. *For the master's tools will never dismantle the master's house.*"[41]

It has always seemed clear to me that violent and nonviolent approaches to social change are complementary. No one I know who advocates the possibility of armed resistance to the dominant culture's degradation and exploitation rejects nonviolent resistance. Many of us routinely participate in nonviolent resistance and support those for whom this is their only mode of opposition. Just last night I and two other non-pacifists wasted two hours sitting at a county fair tabling for a local environmental organization and watching the—how do I say this politely?—supersized passersby wearing too-small Bush/Cheney 2004 T-shirts and carrying chocolate-covered bananas. We received many scowls. We did this nonviolent work, although we accomplished precisely nothing. But many dogmatic pacifists refuse to grant the same respect the other way. It is not an exaggeration to say that many of the dogmatic pacifists I've encountered have been fundamentalists, perceiving violence as a form of blasphemy (which it is within this culture if it flows up the hierarchy, and these particular fundamentalists have never been too picky about reaping the fiscal fruits of this culture's routine violence down the hierarchy), and refusing to allow any mention of violence in their presence. It's ironic, then, that they end up turning Audre Lorde's comment on its head.

Our survival really does depend on us learning how to "take our differences"—including violent and nonviolent approaches to stopping civilization from killing the planet—"and make them strengths." Yet these fundamentalists attempt to eradicate this difference, to disallow it, to force all discourse and all action into only one path: theirs. That's incredibly harmful, and of course serves those in power. The master's house will never be dismantled using only one tool, whether that tool is discourse, hammers, or high explosives.

I have many other problems with the pacifist use of the idea that force is solely the dominion of those in power. It's certainly true that the master uses the tool of violence, but that doesn't mean he owns it. Those in power have effectively convinced us they own land, which is to say they've convinced us to give up our inalienable right to access our own landbases. They've effectively convinced us they own conflict resolution methods (which they call *laws*), which is to say they've convinced us to give up our inalienable right to resolve our own conflicts (which they call *taking the law into your own hands*). They've convinced us they own water. They've convinced us they own the wild (the government could not offer "timber sales" unless we all agreed it owned the trees in the first place). They're in the process of convincing us they own the air. The state has for millennia been trying to convince us it owns a monopoly on violence, and abusers have been trying to convince us for far longer than that. Pacifists are more than willing to grant them that, and to shout down anyone who disagrees.

Well, I disagree. Violence does not belong exclusively to those at the top of the hierarchy, no matter how much abusers and their allies try to convince us. They have never convinced wild animals, including wild humans, and they will never convince me.

And who is it who says we should not use the master's tools? Often it is Christians, Buddhists, or other adherents of civilized religions. It is routinely people who wish us to vote our way to justice or shop our way to sustainability. But civilized religions are tools used by the master as surely as is violence. So is voting. So is shopping. If we cannot use tools used by the master, what tools, precisely, can we use? How about writing? No, sorry. As I cited Stanley Diamond much earlier, writing has long been a tool used by the master. So I guess we can't use that. Well, how about discourse in general? Yes, those in power own the means of industrial discourse production, and those in power misuse discourse. Does that mean they own all discourse—all discourse is one of the master's tools—and we can never use it? Of course not. They also own the means of industrial religion production, and they misuse religion. Does that mean they own all religion—all religion is one of the master's tools—and we can never use it? Of course not. They own the means of industrial violence production, and they misuse violence. Does that mean they own all violence—all violence is one of the master's tools—and we can never use it? Of course not.

But I have yet another problem with the statement that the master's tools will never dismantle the master's house, which is that it's a terrible metaphor. It just doesn't work. The first and most necessary condition for a metaphor is that it make sense in the real world. This doesn't.

You can use a hammer to build a house, and you can use a hammer to take it down.

It doesn't matter whose hammer it is.

I'm guessing that Audre Lord, for all of her wonderful capabilities as a writer, thinker, activist, and human being never in her entire life dismantled a house. Had she done that, she could never have made up this metaphor, because you sure as hell can use the master's tools to dismantle his house.[42] And you can use the master's high explosives to dismantle the master's dam.

⊛   ⊛   ⊛

I've heard too many pacifists say that violence only begets violence. This is manifestly not true. Violence can beget many things. Violence can beget submission, as when a master beats a slave (some slaves will eventually fight back, in which case this violence will beget more violence; but some slaves will submit for the rest of their lives, as we see; and some will even create a religion or spirituality that attempts to make a virtue of their submission, as we also see; some will write and others repeat that the most disadvantageous peace is better than the most just war; some will speak of the need to love their oppressors; and some will say that the meek shall inherit what's left of the earth). Violence can beget material wealth, as when a robber or a capitalist[43] steals from someone. Violence can beget violence, as when someone attacks someone who fights back. Violence can beget a cessation of violence, as when someone fights off or kills an assailant (it's utterly nonsensical as well as insulting to say that a woman who kills a rapist is begetting more violence).

Back to Gandhi: "We must be the change we wish to see." This ultimately meaningless statement manifests the magical thinking and narcissism we've come to expect from dogmatic pacifists. I can change myself all I want, and if dams still stand, salmon still die. If global warming proceeds apace, birds still starve. If factory trawlers still run, oceans still suffer. If factory farms still pollute, dead zones still grow. If vivisection labs still remain, animals are still tortured.

I have worked very hard to become emotionally healthy, to heal from this culture, my childhood, and my schooling. I'm a genuinely nice guy. But I don't do that emotional work to try to help salmon. I do it to make life better for myself and those around me. My emotional health doesn't help salmon one bit, except insofar as that health leads me to dismantle that which is killing them. This is not cognitively challenging at all.

Next: If you use violence against exploiters, you become like they are. This cliché is, once again, absurd, with no relation to the real world. It is based on the flawed notion that all violence is the same.[44] It is obscene to suggest that a woman who kills a man attempting to rape her becomes like a rapist. It is obscene to suggest that by fighting back Tecumseh became like those who were stealing his people's land. It is obscene to suggest that the Jews at who fought back against their exterminators at Auschwitz/ Birkenau, Treblinka, and Sobibór became like the Nazis. It is obscene to suggest that a tiger who kills a human at a zoo becomes like one of her captors.

Related to that is the notion that committing an act of violence destroys your soul. A couple of years ago I shared a stage with another dogmatic pacifist. He said, "To harm another human being irretrievably damages your very core."

I didn't think Tecumseh would have agreed. I asked, "How do you know?"

He shook his head. "I don't know what you're asking."

"How do you know that violence irretrievably damages your very core?"

He looked at me as though I had just asked him how he knows that gravity exists.

I asked, "Have you ever killed anyone?"

"Of course not."

"So you don't know this by direct experience. Have any of your friends ever killed anyone?"

Disgust crossed his face. "Of course not."

"Have you ever even spoken with anyone who has killed someone?"

"No."

"So your statement is an article of faith, unsupported, based not on direct experience or conversations with anyone who would know."

He said, "It's self-evident."

Nice rhetorical trick, I thought. I said, "I have friends at the prison who've killed people, and I'm acquaintances with many others who've done the same. Because I've heard so many pacifists make this claim before, I asked these men if killing really changed them."

He didn't look at me. He certainly didn't ask about their answers.

I told him anyway." The answers are unpredictable, and as varied as the people themselves. A few were devastated, just as you suggest. Not many, but a few. A bunch said it didn't fundamentally change anything. They were still the exact same people they were before. One said he'd been stunned by how easy it is, physically, to take someone's life, and that made him realize how easily he, too, could be killed. The act of killing made him feel very frightened, he said. Another said it made him feel incredibly powerful, and it felt really, really good. Another said the first time was hard, but after that it quickly became easy."

The pacifist looked like he was going to throw up.

I thought, *This is just reality, man. Reality is a lot more complex than any dogma could ever be. That's one of the problems with abstract principles: they're always smaller and simpler than life, and the only way to make life fit your abstractions is to cut off great parts of it.* I said, "A few told me their answers depended entirely on who they were killing: they regretted some of their murders, but wouldn't take back others even if it meant they could get out of prison. One man, for example, overheard a rapist bragging how he'd made his victim tell him she liked it, and made her beg for more so he wouldn't kill her. The man I spoke with invited the rapist into his cell for a friendly game of chess, and strangled him to death because of what he did to that woman. That murder had felt right at the time, he said, and he knew it would feel right for the next fifteen years till he got out. And one man told me that the thing he was most proud of in his entire life was that he killed three people."

The pacifist shook his head. "That's really sick," he said.

"Let me tell you the story," I responded. "He was a migrant farm worker, from a large Mexican family. He was fifteen. One day he didn't go to the fields but to town. That day three men killed his father. Soon there was a family meeting, and he violated family tradition by interrupting his elders. He insisted that because he was the youngest, the only one without a family relying on him, that he be the one to avenge their father. For the next few years he worked hard to establish a business that would support his mother later on, and when the time came he killed the three men who had killed his father. The next day he went to the police station and turned himself in. He's now serving life."

"He should have let the law handle it."

"I cannot blame him for his actions. They were human." I paused a moment, then said, "And I have known others who killed because they were human. I have known women who killed their abusers. They had no regrets. Not one. Not ever."

"You cannot sway me," he said. "They should let the law handle it."

"The law," I replied. "The law. Let me tell you another story. A woman killed her mother's boyfriend, who had battered her mother for years and finally murdered her mother. And—surprise of all surprises—the district attorney refused to charge him with murder. I suppose this was because women aren't people whose lives actually count. So the woman did a sit-in at the DA's office. For three days, she just kept saying over and over 'You're going to call it murder.' The DA finally had her arrested for trespassing. Having gotten no satisfaction from the system, she bought a gun, tracked the boyfriend down and shot him dead. Because of her sit-in stunt, the lawyers were able to argue temporary insanity. She served two years in prison and didn't regret a single day of it."[45]

The pacifists who say that fighting back against those who are exploiting you or those you love destroys your soul have it all backwards. It is just as wrong and just as harmful to not fight back when one should as it is to fight when one should not. In fact in some cases it may be far more harmful. The Indians who spoke of fighting, killing, and dying—and who fought, killed, and died—to protect not only their land but their dignity from theft by the civilized understood this. So did Zapata. So did the Jews who rose up against the Nazis. Of those who rose up against their exterminators at Auschwitz/Birkenau, and who were able to kill seventy SS, destroy one crematoria, and severely damage another, concentration camp survivor Bruno Bettelheim[46] wrote that "they did only what we would expect all human beings to do: to use their death, if they could not save their lives, to weaken or hinder the enemy as much as possible; to use even their doomed selves for making extermination harder, or maybe impossible, not a smooth running process. . . . If they could do it, so could others. Why didn't they? Why did they throw their lives away instead of making things hard for the enemy? Why did they make a present of their very being to the SS instead of to their families, their friends, even to fellow prisoners; this is the haunting question."[47]

But remember, the Jews who participated in the Warsaw Ghetto uprising, even those who went on what they thought were suicide missions, had a higher rate of survival than those who did not fight back. Never forget that.

Instead of saying, "If we fight back, we run the risk of becoming like they are. If we fight back, we run the risk of destroying our souls," we must say, "If we do not fight, we run the risk of not just acting like but *becoming* slaves. If we do not fight back, we run the risk of destroying our souls and our dignity. If we do not fight back, we run the risk of allowing those who are exterminating the world to move ever faster."

⊕ ⊕ ⊕

To reach the middle of the ocean, those in power must have oil to run their ships and metals to build them. To deforest, they must have gasoline to power their chainsaws and metals to build the saws. No oil, and the ships have no capability to reach the center of the ocean, which means that the oceans can begin to live again. No oil, and chainsaws sputter—actually they don't even do that—and forests must only contend with local use. If those in power have no oil, they cannot rebuild the dams we remove. Part of taking down civilization is the destruction of the oil economy.

Of course in the longer run we must remember how to live in place with what the land willingly gives, but before we can even seriously think about doing that we must remove this threat to the entire planet. To do otherwise is the equivalent of trying to decide how we shall live next summer as we ignore the upraised butcher's knife in front of us.

The first step in taking down civilization is to realize in our own hearts and minds that the dictionaries lied to us, that civilization is not "a high stage of social and cultural development,"[48] or "a developed or advanced state of human society."[49] I am not talking about convincing some hypothetical mass movement of people, which will not happen within this culture. As I said earlier, when fathers are raping daughters, when lovers are beating those they purport to love, there is no hope for the salmon. I am talking about me realizing this in my own heart, and you realizing it in yours.

The next step in taking down civilization is finding a few other people who feel the same. It is hard enough to take on this entire abusive social structure—where everything is set up to protect the abusers—without having to fight our friends as well. It can be lifesaving to have friends who will say, and mean, with courage, love, and determination glistening in their eyes, "Yes, it is unacceptable to me that salmon be exterminated from this river. I will do what it takes to save them." I am talking about small groups of people—small enough to know and trust each other with your very lives—coming to this understanding, and beginning to act upon it.

Next, taking down civilization means understanding that we are in the midst of a war, that war was long ago declared on the natural world, including on humans, and that we must fight back. I am not speaking metaphorically.

Next, taking down civilization means understanding that very few wars are won on actual battlefields. Economic production allows governments to win wars. And the destruction of the means of economic production causes them to lose them. Recall the U.S. military analysis that determined that World War II attacks on German railroads were "the most important single cause of Germany's ultimate economic collapse."[50] This is not to belittle the sacrifices of the soldiers who beat back the Nazis at Stalingrad and elsewhere, but to remind people of the truism that an army fights on its stomach. This includes an army of consumers.

Taking down civilization means acting. It means committing ourselves to defending our landbases, which means committing ourselves to removing the economic and transportation infrastructures, which means committing ourselves to hitting them, and hitting them again, and again, and again. This may be, as we shall see in a few pages, easier than it seems.

Once the economic and transportation infrastructures have been taken down, our fights over how to live sustainably in our own landbases will be local, and face to face, which means they will be human, which means they are eminently winnable, through discourse or violence or some other means.

　　　　　　⊛　⊛　⊛

The final argument I've often heard from pacifists is that violence never accomplishes anything. This argument, even more than any of the others, reveals how completely, desperately, and arrogantly out of touch many dogmatic pacifists are with physical, emotional, and spiritual reality.

If violence accomplishes nothing, how do these people believe the civilized conquered North and South America and Africa, and before these Europe, and before that the Middle East, and since then the rest of the world? The indigenous did not and do not hand over their land because they recognize they're faced with "a high stage of social and cultural development." The land was (and is) seized and the people living there were (and are) slaughtered, terrorized, beaten into submission. The tens of millions of Africans killed in the slave trade would be surprised to learn their slavery was not the result of widespread violence. The same is true for the millions of women burned as witches in Europe. The same is true for the billions of passenger pigeons slaughtered to serve this economic system. The millions of prisoners stuck in gulags here in the US and elsewhere would be astounded to discover that they can walk away anytime they want, that they are not in fact held there by force.

Do the pacifists who say this really believe that people all across the world hand over their resources to the wealthy because they enjoy being impoverished, enjoy seeing their lands and their lives stolen—sorry, I guess under this formulation they're not stolen but received gracefully as gifts—by those they evidently must perceive as more deserving? Do they believe women submit to rape just for the hell of it, and not because of the use or threat of violence?

One reason violence is used so often by those in power is because it works. It works dreadfully well.

And it can work for liberation as well as subjugation. To say that violence never accomplishes anything not only degrades the suffering of those harmed by violence but it also devalues the triumphs of those who have fought their way out of abusive or exploitative situations. Abused women or children have killed their abusers, and become free of his abuse. (Of course, often then the same selective law enforcement agencies and courts that failed to stop the original abuse now step in to imprison those who sent violence the wrong way up the hierarchy.) And there have been many indigenous and other armed struggles for liberation that have succeeded for shorter or longer periods.

In order to maintain their fantasies, dogmatic pacifists must ignore the harmful and helpful efficacy of violence.[51] Years ago I was asked by a publisher to review a book-length manuscript they had just received from a householdname pacifist activist. The document was a mess, and they said they might want me to help edit it. I was younger then, and far less assertive, so my comments were fairly minor throughout, until I came to a statement that made me curse and hurl my pen across the room, then get up and stalk outside for a long walk. The activist claimed that the American movement against the war in Vietnam was a triumph for pacifist resistance, and that it showed that if enough people were just dedicated enough to nonviolence they could bring about liberation in all parts of the globe. He mentioned the four dead at Kent State as martyrs to this nonviolent campaign, and also mentioned "our unfortunate soldiers who lost their lives fighting for this unjust cause," but never once mentioned the millions of Vietnamese who outfought, outdied, and outlasted the invaders. My point is not to disparage or ignore the importance of nonviolent protests in the United States and elsewhere, but rather to point out what the pacifist pointedly ignored: the antiwar movement didn't stop the U.S. invasion—it *helped* stop the invasion. The primary work—and primary suffering—was done by the Vietnamese.

Oddly enough, the publisher didn't hire me to edit it.

I am just being honest when I say that I have talked to hundreds of people who are ready to bring the war home. I've talked to those who went down to assist the Zapatistas but were told, "If you really want to help, go home and start the same thing there." I've talked to family farmers, prisoners, gang members, environmentalists, animal rights activists, hackers, former members of the military who have had their fill of their own enslavement and the destruction of all they love, and who are ready at long last to begin to fight back. I have spoken to Indians who have said their people are ready to bring back out the ceremonial war clubs they have now kept buried or hidden for so long. I have spoken to students and other men and women in their teens, twenties, thirties, forties, fifties, sixties, seventies, eighties who know the world is being killed, and are ready to fight and to kill and if necessary to die to stop this destruction, who, like me, are not willing to stand by while the world is destroyed.

▧   ▧   ▧

How do you win? Someone once asked the Confederate General Nathan Bedford Forrest how he won so many battles. His response summed up the essence of military strategy in six words: "Get there first with the most."

Let's break it down. *Get there.*

You choose where you fight. The person or force who chooses the battlefield has a better chance of winning. Indeed, much military strategy consists of attempting to get your enemy to attack you where you're strong and to not attack you where you're weak, while simultaneously probing for your enemy's weak spots to attack. This is true on battlefields, it is true in antagonistic discourse,[52] it is true in all areas of conflict.

Right now, what are the battlefields on which we are encouraged—allowed—to fight? We are encouraged to vote.[53] But of course we all know the old Wobbly saying: If voting made a difference it would be illegal. And in any case, our choices of whom/what we can vote for are strictly limited. No matter whether a Republican or Democrat wins,we lose.We are encouraged—allowed—to use the courts, and while of course we may get the occasional win there, we must never forget by and for whom the courts are set up. We must never forget that their authority ultimately comes from the ability of the state to inflict violence. We are encouraged—allowed—to write, so long as we never mention social change and violence in the same paragraph. We are encouraged to recycle, to shop green (so long as we shop!), and so on.

Much more interesting are the fields we are not encouraged—allowed—to choose for our battles.Who chooses for us? What fields are off-limits, and why? Who has declared them off-limits? Why have we ceded this territory?

What do we want? How will we accomplish it? I return to the salmon (you can of course return instead to what you love). As I already mentioned, for salmon to survive, dams, industrial logging, industrial fishing, industrial agriculture must go, the oceans must survive, and global warming must cease. Choose one of these, say, dams. What do we need to do to remove dams? (And notice the difference in implication even between using the verb "would" and "do," as in "What would we need to do to remove dams?" *Would* implies theory, which means we're not really going to do it, while *do* implies reality; the choice has been made, and now we're asking *how*.) What battlefields do we choose?

Try for a moment to think for yourself. I'm not being snide, condescending, or sarcastic. Thinking for yourself is one of the most difficult things to do, especially within a culture where we're inculcated into irresponsibility.How do you know if you're thinking your own thoughts, or if you're thinking the thoughts of the people who produce television programs, or thinking the thoughts of your teachers and preachers in junior high, or thinking the thoughts of some guy who writes books about taking down civilization?[54] But try, really try. If you follow your own thoughts, if you follow your own morals, if you choose to protect those you love most, and to protect your landbase (presuming that you love your landbase, but if you do not then you can choose something else[55]), if you choose your own battlefields, what battles do you choose? What do you do? How do you act? Who are you?

It should be clear by now that I do not care what fields you choose for your battles. I do not know you. I do not know your strengths. I do not know your weaknesses. I do not know your loves, and I do not know your hates. I do not know where or how or over what you should fight. And I would neither dare nor even care to make suggestions as to what you should or should not do when I do not know you or your circumstances.[56]

Here's the point: if you allow your enemy to choose the battlefield, you will probably lose. Choosing the field upon which you will fight is the first step to winning. Choose your battlefields wisely.

⊛ ⊛ ⊛

"Get there first with the most." Next: *With the most.*

When it comes to winning battles, local superiority means almost everything. It doesn't matter who has the most troops all over the world: the important thing is who has the most troops right here right now. The United States can have more than 1.4 million soldiers in 135 different countries, and it can have about a million cops just in this country, but if there are four of you and none of them standing next to a cell phone tower, you have achieved local superiority. You got there first with the most, and you will probably win this particular battle. If the four of you show up and find you have not achieved local superiority, don't fight right here right now.

If Nathan Bedford Forrest encapsulated most military strategy into six words, baseball Hall of Famer Wee Willy Keeler accidentally distilled most guerrilla strategy into only five. Someone asked him what was the secret of his batting success, and he responded, "Hit 'em where they ain't."

Until those in power find ways to put surveillance microchips into each and every one of us—something they're feverishly working toward, by the way—they will not be able to be everywhere. This means that so long as we do not identify with them, so long as we have driven them from our hearts and minds, so long as we identify with our own human bodies and the land where we live, we will be able to hit 'em where they ain't.

Their security often stinks. We have been so long and so deeply pacified that for the most part we don't strike back, which means for the most part they do not have to defend the lands they've seized, nor even much that could very easily be attacked.

A report a few years ago revealed that security was so lax at the Grand Coulee Dam that local teens used the dam's interior as a skateboard park. This shouldn't be surprising. With 2 million dams and a more or less fully pacified populace, why bother with security?

Hitting 'em where they ain't is not the only way to win. But I don't believe our movement is large enough yet to allow us the luxury of pitched battles, which generally favor the larger army.[57] To return to the American Civil War, Federal General U.S. Grant had far more soldiers at his disposal[58] than his enemy, and so knew he could afford to hammer away with assault after doomed assault. At Cold Harbor, for example, men pinned their names and addresses to their shirts before charging, so that later their remains could more easily be identified. In that summer's campaign the Army of the Potomac suffered more casualties than there were soldiers in the entire Army of Northern Virginia. But Grant knew that even though he did not get to any of these battlefields first, he sure had the most. And he knew he would continue to have the most. And that was enough.

There is a sense in which for the foreseeable future we will never have the most. This is a problem everyone who has ever tried to stop civilization has faced. It was a constant complaint of the Indians. The Sauk Keokuk, who was highly esteemed by the whites for his conciliatory attitudes, argued that to fight back was tantamount to suicide, saying, "Few, indeed, are our people who do not mourn the death of some near and loved one at the hand of the Long Guns [pioneers],who are becoming very numerous. Their cabins are as plenty as the trees in the forest, and their soldiers are springing up like grass on the prairies. They have the talking thunder [cannon], which carries death a long way off, with long guns and short ones, ammunition and provisions in abundance,with powerful warhorses for their soldiers to ride. In a contest where our numbers are so unequal to theirs we must ultimately fail."[59] Keokuk's warlike rival Maka-taimeshiekiakiak (Black Hawk) spoke after he was defeated of the civilized in similar terms: "Brothers, your houses are as numerous as the leaves on the trees, and your young warriors, like the sands upon the shore of the big lake which rolls before us."[60] Recall the words of the Santee Sioux Taóyatedúta (Little Crow), who also spoke against fighting back: "See!—the white men are like the locusts when they fly so thick that the whole sky is a snow-storm. You may kill one—two—ten; yes, as many as the leaves in the forest yonder, and their brothers will not miss them. Kill one—two—ten, and ten times ten will come to kill you. Count your fingers all day long and white men with guns in their hands will come faster than you can count. . . .Yes, they fight among themselves, but if you strike at them they will all turn on you and devour you and your women and little children just as the locusts in their time fall on the trees and devour all the leaves in one day."[61] The Wyandot Between The Logs, who also was a friend of the whites (specifically the Americans) dropped the metaphorical language and put it bluntly: "I am directed by my American father to inform you that if you reject the advice given you, he will march here with a large army, and if he should find any of the red people opposing him in his passage through this country, he will trample them under his feet. You cannot stand before him. . . . Let me tell you, if you should defeat the American army this time, you have not done. Another will come on, and if you defeat that, still another will appear that you cannot withstand; one that will come like the waves of the great water, and overwhelm you, and sweep you from the face of the earth."[62] It's important to note that the Indians who cautioned against fighting still lost their land.

Each of these declarations by each of these Indians is in some ways a restatement of Thomas Jefferson's line, with subject and object inverted: "In war they will kill some of us; we shall destroy all of them."[63] I do not know any environmentalist or other type of activist who has not experienced the despair that comes from facing civilization's juggernaut of destruction. Let's substitute some words: "Few, indeed, are our people who do not mourn the death [clearcutting, damming, extirpation] of some near and loved one [forest, river, species] at the hand of the Long Guns [timber corporations, energy corporations], who are becoming very numerous. Their cabins [fellerbunchers, caterpillars] are as plenty as the trees in the

forest, and their soldiers [police] are springing up like grass on the prairies. . . . In a contest where our numbers are so unequal to theirs we must
ultimately fail." And, "See!—the white men [CEOs, clearcutters, developers, police] are like the locusts when they fly so thick that the whole sky
is a snow-storm. You may kill one—two—ten; yes, as many as the leaves in
the forest yonder, and their brothers will not miss them. Kill one—two—
ten, and ten times ten will come to kill you. Count your fingers all day
long and white men [CEOs, clearcutters, developers, police] with guns
[palm pilots, chainsaws, maps] in their hands will come faster than you
can count." And one more time: "Let me tell you, if you should defeat the
American army [timber corporation, developer, police unit, or plain old
American army] this time, you have not done. Another will come on, and
if you defeat that, still another will appear that you cannot withstand; one
that will come like the waves of the great water, and overwhelm you, and
sweep you from the face of the earth."

Civilization has from the beginning devoted itself almost completely
to conquest, to war. It's sometimes hard to say—and I'm not sure I care
anyway— whether the civilized hyper-exploit resources to fuel the war
machine, or need a war machine to seize resources (which are then hyper-
exploited to fuel the war machine). It's probably a bit like asking whether
the dominant culture is so destructive because most of its members are
insane, suffering from a form of complex PTSD; or whether the dominant culture is so destructive because its materialistic system of social
rewards—overvaluing the acquisition of wealth and power and underval-
uing relationship—leads inevitably to hatred and atrocity; or whether the
physical resource requirements of cities necessitate widespread violence
and destruction. The answer is yes.

As George Draffan and I asked in *Welcome to the Machine*, "Why does
our machine culture outcompete and overwhelm real live cultures?"
We answered our own question: "Because machines are more efficient
than living beings. Why are machines more efficient than living beings?
Because machines do not give back. All living beings understand that
they must give back to their surroundings as much as they take. If they do
not, they will destroy their surroundings. By definition, machines—and
people and cultures that have turned themselves into machines—do not
give back. They use. And they use up. This gives them short-term advan-

tages in power over the ability to determine outcomes. They outcompete. They overwhelm. They destroy."[64] The point as it relates to the current discussion is that just as there are functional and systematic reasons we will never be able to outspend civilization, there are functional and systematic reasons we'll never be able to outgun them. *In a pitched battle.* But there are other ways to fight. Hit 'em where they ain't.

⊛   ⊛   ⊛

I'm in St. Petersburg, Florida, and it's hot. It's late November, but it's still hotter than hell. I'm talking with another military man, and I'm marveling at all of the useful knowledge taught to GIs at taxpayer expense. I'm also thinking that this knowledge might be a sparkling example of some of the master's tools coming in handy for dismantling the master's house, or rather his economic system.

We go to the beach. The sand is white, almost blinding. There aren't many people here. That's good. We want to talk.

He says, "We were taught in the Army that when we move into a country one of the most important things we want to do is disrupt the delivery of raw materials. If you can disrupt that flow, you disrupt the entire economy. If you disrupt the economy—and keep on disrupting it—you stand a much better chance of winning. Simple as that."

I think of the American and British bombers pounding the Nazi rail lines, and I think of Russian, Belgian, Dutch, French, Czech and many other partisans doing the same. I think of Federal forces in the American Civil War slowly strangling the Confederacy through a blockade and through cutting rail and river lines for transport of materials. I think of German General Erwin Rommel's complaint, looking back on his loss at El Alamein, the turning point of World War II in North Africa, that "the battle is fought and decided by the Quartermasters before the shooting begins. The bravest men can do nothing without guns, the guns nothing without plenty of ammunition, and neither guns nor ammunition are of much use in mobile warfare unless there are vehicles with sufficient petrol to haul them around."[65]

The military man says, "The vast majority of stuff is delivered via three methods: train, truck, and ship. We can ignore air since the amount transported is trivial. Of these, trains and trucks are the easiest to get to. The U.S. rail system, and I would be amazed if Europe's was any different, is wide open. There are tens of thousands of miles of unobserved track that could be taken out with nothing more than a crowbar. When I was a kid we used to pull spikes all the time, just for fun. It takes no time at all. Similarly, millions of miles of roads could be disrupted temporarily by any number of means."

"You learned this in the military?"

"Absolutely. What did you think they taught us in all those classes? What do you think the purpose of the military is?"

He's smiling, so I don't feel chastised for my naïveté.

He continues, "The purpose of the U.S. military is to fuck up the infrastructure of the countries where the United States wants to steal resources or maintain a military presence to use as a staging area to steal somebody *else's* resources. That's what they taught us how to do. Roads (especially junctions), rail lines, ports, and virtually everything else associated with transporting goods has been a military target since day one. So we talked about them quite a lot. But even more than that we were taught to think about the system as a whole and to analyze it looking for the flow of production and goods: both those required locally and those essential elsewhere. We were taught especially to look for choke points, places that some necessary items *must* pass through."

I think about my term for these places: *bottlenecks*.

He continues, "We were also taught to look for transportation segments that are secluded or otherwise isolated, and to look for ways to disrupt the flow of materials even without overt actions."

"What do you mean?"

"If we couldn't blow a bridge, we could still stage a traffic accident. One of those at the right place at the right time could be very useful."

"You were taught that . . ."

"Yes. And you paid for it."

"What else did they teach you?"

"Probably the most important thing is that for this type of activity to be effective it has to be directed and sustained. You must know your area and what resources it requires to function economically, and you must direct your efforts toward interdicting the flow of those resources. You have to know how to get the most bang for your buck."

I smile, thinking I should have known it would be impossible to talk to two military personnel without at least one of them using that phrase, and thinking also that he's talking about what I call *leverage*.

"Of course we were taught about security as well. Never get caught. Never get caught. Never get caught. That was hammered into us."

We sit on the sand in the shade, still hot, and look at the water. It too, is white, and blinding.

He says, "Although they taught us many technical skills, the main thing the classes and the practice did for me was to shift my way of thinking so that now I am constantly evaluating where are the points of greatest stress in any structures and infrastructures that I see. Once you've made that shift in your thinking and perceiving and once you get some experience you begin to understand that all of this work is much easier than it seems."

It's still hot. I don't know how anyone lives here.

He says, "Natural gas."

"What?"

"That's something else they taught us. You know, folks talk all the time about how the military dehumanizes people, destroys their individuality and creativity, and that may be true in some ways, but there are other ways that they taught us to be very creative. We were repeatedly taught to survey our surroundings and find what commonly available resources we could use to achieve our ends. Our exercises always included—no, required—the innovative use of household materials."

"Like what?" Even theoretically, I find all this stuff incredibly fascinating.

"You'd be amazed at what you can do with gasoline and soap flakes. . . ."

"You gonna tell?"

"Napalm. And you can make pipe bombs out of black powder."

"They taught you how to make pipe bombs?"

"It's the military, Derrick. It isn't Boy Scout camp."

"Where does natural gas come in?"

"The delivery of energy is even more fundamental to the system than the transport of raw materials. And not only is the electrical grid wide open but so is the natural gas supply line. Here's an example of just how easy it would be to safely take out a natural gas pipe. Buy a car battery, a piece of glass tubing, and some plastic gloves. First, pour the acid from the battery into a suitable container, then go to one of the millions of gas pipes or relay stations around the world, attach the tube to the pipe with tape molding putty or whatever, pour the acid into the tube, and then walk away and let the acid eat through the pipe. Your onsite time is maybe two minutes."

"They taught you this in the military?"

"Care of Uncle Sam."

Have I mentioned lately that I'm glad I'm a writer?

@ @ @

There was one final series of questions. It was this: "If twelve or fifty or two hundred people really could bring it down, why hasn't it already happened? I mean, it took millions of Vietnamese to fight off the Americans. And certainly around the world more than two hundred people hate civilization. What makes you think it would only take two hundred people, and not two hundred thousand?"

He said, "My estimate of fifty people is definitely at the lower end. Having more certainly wouldn't cause any problems, and would in fact make everything happen more quickly. But I still think fifty people could do it.

"Here's the logic I learned in the military. First, you have to have small groups where everyone knows and trusts each other; everyone is able to set up simple, low-tech communication systems; and each group as a whole is capable of dispersing and reforming easily. So if you had more people I would suggest more and not larger groups. Second, if these groups are dedicated enough—and when I say fifty people can bring it down I'm presuming they're not messing around—they should be able to pull off an action every ten days. That's three a month. That's sixty per month nationwide, with equivalent numbers of actions by members of the resistance worldwide. Thirty pipelines, twenty major rail lines, and ten power lines. Every month. That's a lot. It's what was expected of me in the military. It's what I *did* in the military. Month in and month out. And it works. Trust me on this one. Third, I'm assuming that all targets are preselected and scouted at least a month in advance. Fourth, I'm assuming that the groups stay uninfiltrated and unpredictable. The larger the group the greater chance of infiltration, and also the more organized it must be, and therefore predictable.

"It's always hard to get people to believe that so much is possible with so little. But that's part of the way we've been conditioned to think. We always have to remember: who does it serve for us to think that way? We've been trained to think in terms of pitting ourselves against them in a showdown against their main force. But if you look at some of the more successful guerilla wars, time and again you see smaller lower tech forces prevailing through smarter tactics and leveraged actions.

"The key we must always remember is that it takes large numbers of troops to take and hold ground. If you don't want to take and hold ground you don't need large numbers at all. You just hit and run, and then hit and run again, and you keep doing that. There is no fighting force in the world that can survive the death by a thousand cuts of a dedicated partisan movement."

❧ ❧ ❧

Just today Radio Canada reported on the lax security at the dams on the massive (and genocidal and ecocidal) James Bay project. The journalists were able to walk entirely unquestioned and certainly unstopped into the main control rooms of the dams they visited. The journalists were surprised at how easily someone could have sabotaged the entire enterprise. The response by Hydro-Quebec, the operators of the (genocidal and ecocidal) dams, was to attempt to place a gag order on Radio Canada to stop the story from being reported.

Yes. It's only a matter of time.

❧ ❧ ❧

Unless it is stopped, the dominant culture will kill everything on the planet, or at least everything it can.

Each holocaust is unique. The destruction of the European Jewry did not look like the destruction of the American Indians. It could not, because the technologies involved were not the same, the targets were not the same, and the perpetrators were not the same. They shared motivations and certain aspects of their socialization, to be sure, but they were not the same. Similarly, the slaughter of Armenians (and Kurds) by Turks did not (and does not) look like the slaughter of Vietnamese by Americans.

And just as similarly, the holocausts of the twenty-first century will not and do not already look like the great holocausts of the twentieth. They cannot, because this society has progressed.

And every holocaust looks different depending on the class to which the observer belongs. The Holocaust looked far different to high ranking Nazi officials and to executives of large corporations—both of whose primary social concerns would have been how to maximize production and control, that is, how to most effectively exploit human and nonhuman resources—than it did to good Germans, whose primary concerns were as varied as the people themselves but probably included doing their own jobs—immoral as those jobs may have been from an outside perspective—as well as possible; may have included feelings of relief that those in power were finally doing something about the "Jewish Problem"; and certainly included doing whatever they could to not notice the greasy smoke from the crematoria (constructed with the best materials and faultless workmanship). The Holocaust then also looked different to good Germans than it did to those who resisted, whose main concerns may have been how to bring down the system. And it looked different to those who resisted than it did to those who were considered *untermenschen*, whose main concerns may have been staying alive, or failing that, dying with humanity.

What will the great holocausts of the twenty-first century look like? It depends on where you stand. Look around.

If you're one of those in power, your post-modern holocausts will be at most barely visible, and at least a price you're willing to pay, as Madame Albright said about killing Iraqi children. The holocausts will probably share similarities with other holocausts, as you attempt to maximize production—to "grow the economy," as you might say—and as when necessary you attempt to eradicate dissent. This means the holocaust will look like a booming economy beset by shifting problems that somehow always keep you from ever reaching the Promised Land, whatever that might be. The holocaust will look like numbers on ledgers. It will look like technical problems to be solved, whether those problems are increasing your access to necessary resources, dealing with global warming, calming unrest on the streets, or figuring out what to do about too many unproductive people on land you know you could put to better use. The holocaust will look like houses with gates, limousines with bullet-proof glass, and a military budget that can never stop increasing.

The holocaust will feel like economics. It will feel like progress. It will feel like technological innovation. It will feel like civilization. It will feel like the way things are.

If you're in the second group, the good Germans, you will continue to be coopted into supporting the system that does not serve you well. Perhaps the holocaust will look like a new car. Perhaps it will look like lending your talents to a major corporation—or more broadly toward economic production—so you can make a better life for your children. Perhaps it will look like working as an engineer for Shell or on an assembly line for General Motors. Maybe it will look like basing a person's value on her or his employability or productivity. Perhaps it will look like anger at Mexicans or Pakistanis or Algerians or Hmong who compete with you for jobs. Perhaps it will look like outrage at environmentalists who want to save some damn suckerfish, even (or especially) if it impinges on your property rights, or if it takes water you need to irrigate, to make the desert bloom, to make the desert productive. Maybe it will feel like continuing to do a job that you hate—and that requires so little of your humanity—because no matter how you try, you never can seem to catch up. Maybe it will feel like being tired at the end of the day, and just wanting to sit and watch some television.

If you're in the subsection of the third group who might some day resist but don't know where to put your rage, the holocaust might look like armed robbery, auto theft, assault. It might look like joining a gang. It might look like needle tracks down the insides of your arms, and might smell like the bitter, vinegary stench of tar heroin. Or maybe it smells minty strong, like menthol, like the sweet smell of crack brought into your neighborhood at the behest of the CIA. Or maybe not. Maybe it's the unmistakable smell of the inside of a cop car, and a vision through that backseat window of a little girl eating an ice cream cone, with the knowledge that never in your life will you see this sight again. Maybe it looks like Pelican Bay, or Marion, or San Quentin, or Leavenworth. Or maybe it feels like a bullet in the back of the head, and leaves you lying on the streets of New York City, Cincinnati, Seattle, Oakland, Los Angeles, Atlanta, Baltimore, Washington.

If you're a member of the subsection of group three already working against the centralization of power, against the system, then maybe from your perspective the holocaust looks like rows of black-clad armored policemen, and it smells like tear gas. Maybe it looks like lobbying a congress you know has never served you. Maybe it looks like the destruction of place after wild place, and feels like an impotence sharp as a broken leg. Maybe it looks like staring down the barrel of an American-made gun in the hands of a Colombian man wearing American-made camo fatigues, and knowing that your life is over.

For those of the fourth class, the simply extra, maybe it looks like the view from just outside the chainlink fence surrounding a chemical refinery, and maybe it smells like Cancer Alley. Maybe it looks like children with leukemia, children with cancer of the spine, children with birth defects. Maybe it feels like the grinding ache of hunger that has been your closest companion since you were born. Maybe it looks like the death of your daughter from starvation, and the death of your son from diphtheria, measles, or chicken pox. Maybe it feels like death from dehydration, when a tablet costing less than a penny could have saved your life. Or maybe it feels like nothing. Maybe it sounds like nothing, looks like nothing: what does it feel like to be struck by a missile in the middle of the night, a missile traveling faster than the speed of sound, a missile launched a thousand miles away?

Maybe it feels like salmon battering themselves against dams, monkeys locked in steel cages, polar bears starving on a dwindling ice cap, hogs confined in crates so small they cannot stand, trees falling to the chainsaw, rivers poisoned, whales deafened by sonic blasts from Navy experiments. Maybe it feels like the crack of tibia under the unforgiving jaws of a leghold trap.

Maybe it looks like the destruction of the planet's life support systems. Maybe it looks like the final conversion of the living to the dead.

As much as I cannot help but see the similarities between prisons and concentration camps, it seems to me a grave error to count on Zyklon-B-dispensing showers to mark the new holocaust. Perhaps the new holocaust is dioxin in polar bear fat, metam sodium in the Smith River. Perhaps it comes in the form of decreasing numbers of corporations controlling increasing portions of our food supply, until, as now, three huge corporations control more than 80 percent of the beef market, and seven corporations control more than 90 percent of the grain market. Perhaps it comes in the form of these corporations, and the governments which provide the muscle for them, deciding who eats and who does not. Perhaps it comes in the form of so much starvation that we cannot count the dead. Perhaps it comes in the form of all of these, and in many others I could not name even if I were able to predict.

But this I know. The pattern has been of increasing efficiency in the destruction, and increasing abstraction. Andrew Jackson himself took the "sculps" of the Indians he murdered. Heinrich Himmler nearly fainted when a hundred Jews were shot in front of him, which was surely one reason for the increased use of gas. Now, of course, it can all be done by economics.

And this I know, too. No matter what form it takes, most of us will not notice it. Those who notice will pay too little attention. It does not matter how great the cost to others nor even to ourselves, we will soldier on. We will, ourselves, walk quietly, meekly, into whatever form the gas chambers take, if only we are allowed to believe they are bathrooms.

❁ ❁ ❁

Over the years I have been criticized because I do not suggest models by which people should live. "You're only interested in tearing things down," some people say, "not in providing alternatives."

I do not provide alternatives because there is no need. The alternatives already exist, and they have existed—and worked—for thousands and tens of thousands of years.

Over the years I have heard many of the civilized ask how we could possibly live without civilization. It is a question I have never heard any Indians answer publicly. It is a question I have never asked, because I already know the answer. In private many Indians have answered this question I have never asked. They have said, "After civilization is gone from the earth and from your hearts, we will teach you how to live. We will not do it before then because your culture has been trying to kill us, and also because you would try to make money from what we say, or you would try to paste what we tell you onto your unworkable system. So until civilization is gone we will just hold on to our traditions and hold on to our existence. Later, if you come to us, we will help."

What they say is true. And it is true also of the land. Once civilization is gone, once it is only a terrible, terrible memory, the land, too, will teach us how to live.

⊛   ⊛   ⊛

The Christian Cheyenne Chief Lawrence Hart described one tradition of his people, which he called the Cheyenne Peace Tradition. I want to describe another, called the picket pin and stake. Before a battle, a few of the bravest Cheyenne Dog Soldiers would be chosen to wear sashes of tanned skins called "dog ropes." Attached to each dog rope was a picket-pin, normally used to tether horses. During battle, the pin would be driven into the ground as a mark of resolve. Once the pin was driven, the Dog Soldier would remain staked to that piece of ground, even to his death. Retreat was no longer an option. The pin could only be removed when everyone was again safe, or when another Dog Soldier relieved him of his duty.

It is time. I have driven my picket-pin. I am staked out, and willing to give in no more.

Where will you drive your own picket stake? Where will you choose to make your stand? Give me a threshold, a specific point at which you will finally stop running, at which you will finally fight back.

⊛   ⊛   ⊛

Stand with me. Stand and fight. I am one. We would be two. Two more might join and we would be four. When four more join we will be eight. And we will be eight people fighting whom others will join. And then more people, and more. Stand and fight.

⊛   ⊛   ⊛

The questions before each of us now are: What are your gifts and how can you use them in the service of your landbase? What can you do? What does your landbase most need from you? How can you achieve it? What do you want to do?

And right now, perhaps the most important of all: What are you willing to do?

Thought to Exist in the Wild
Awakening from the Nightmare of Zoos

with Karen Tweedy-Holmes

No Voice Unheard, 2007

Karen Tweedy-Holmes approached me with her devastatingly sorrowful photographs of animals as prisoners in zoos. I wanted to write something that would honor her work, honor the suffering of these animals, and attempt to stop the suffering by helping to debunk the notion that zoos help animals. I wanted to help end zoos.

One of the things I'm most proud of in that book is that I was able to show the relationship between zoos and pornography. In both cases, they require a subject who is in control of an object who is the recipient of the subject's gaze, and who is essentially held captive and exploited for the nominally educational or entertainment purposes of the subject. The most important lesson taught by zoos and by pornography is that I, the viewer, have power over you, the one in the cage.

The bear takes seven steps, her claws clicking on concrete. She dips her head, turns, and takes three steps toward the front of the cage. Again she dips her head, again she turns, again she takes seven steps. Another dip, another turn, another three steps. When she gets to where she started she begins all over. Then she does it again. And again. And again.

This is what's left of her life.

Outside the cage, people pass by on a sidewalk. Strollers barely come to a stop before their drivers realize there's nothing here to see. They move on. Still the bear paces. Seven steps, head dip, turn. A pair of teenagers approach, wearing walkmans and holding hands. One glance inside is enough, and they're off to the next cage. Three steps, head dip, turn.

My fingers have wrapped themselves tight around the metal railing out-

265

side the enclosure. I notice they're sore. My breath catches in my throat. Still the bear paces. I look at the silver on her back, the concave bridge of her nose. Seven steps, head dip, turn. I wonder how long she's been here. A father and son approach, do not stop to stand next to me. Three steps, head dip, turn.

I release the rail, turn, and as I walk away I hear, slowly fading, the rhythmic clicking of claws on concrete.

⊛ ⊛ ⊛

A zoo is a nightmare taking shape in concrete and steel, iron and glass, moats and electrified fences. It is a nightmare that, for its victims, has no end save death.

⊛ ⊛ ⊛

Zoo director David Hancocks writes something echoed by many others: "Zoos have evolved independently in all cultures around the globe."[1] Many echo this statement, but it isn't quite true. It is the equivalent of saying that the divine right of kings, Cartesian science, pornography, writing, gunpowder, chainsaws, backhoes, pavement, and nuclear bombs have evolved independently in all cultures around the globe.[2] Some cultures have developed some of these, and some have not. Some cultures have developed zoos, and some have not. Human cultures existed for scores of thousands of years prior to the first zoo's appearance about 4,300 years ago in the Sumerian city of Ur,[3] meaning zoos did not evolve in these cultures. And in the time since the first zoo thousands of cultures have existed—some to this day (until the dominant culture finishes eradicating them)—with no zoos or their equivalents to be found.

Zoos have, however, evolved in many cultures, from ancient Sumer to Egypt to China to the Mogul Empire to Greece and Rome, on up the lineage of western civilization to the present. But these cultures share something not shared by indigenous cultures such as the San, Tolowa, Shawnee, Aborigine, Karen, and others who did not or do not maintain zoos: they're all civilized.

The change of just one word makes Hancocks' sentence true: "Zoos have evolved independently in all *civilizations* around the globe." As Michael H. Robinson, director of the National Zoo, wrote, "The period of civilization accounts for perhaps 1 percent of our history as hominids. With civilization came urbanization. Shortly after we had developed cities on a grand scale, zoos and botanical gardens sprang up in countries as far apart as Egypt and China."[4]

Civilizations are ways of life characterized by the growth of cities.[5] Cities destroy natural habitat and create environments inimical to the survival of many wild creatures. By definition cities separate their human inhabitants from nonhumans, depriving them of the routine, daily, neighborly contact with wild creatures, which until the onset of civilizations—for 99 percent of our existence—was central to the lives of all humans, and to this day remains central to the lives of the non-civilized.

If it can be said that we are the relationships we share, or at least that relationships form us, or at the very least that they influence who we are, how we act, and how we perceive, then the absence of this fundamental daily bond with wild nonhuman others will change who we are, how we perceive wild creatures, how we perceive our role in the world around us, and how we treat ourselves, other humans, and those who are still wild.

🐝 🐝 🐝

Many ancient zoos contained tremendous numbers of animals. Egyptian zoos held thousands of monkeys, wild cats, antelopes, hyenas, gazelles, ibex, and oryx.[6] Some zoo historians suggest that because the creatures in these zoos were considered sacred they were treated well, but as Hancocks points out, "Deification of a species, however, brought dubious honor. Used in ritualistic sacrifices, sacred ibis, falcons, and crocodiles were mummified by the hundreds of thousands in sanctified ceremonies. The temple slaughters were so great they led to extermination of these species in many parts of Egypt."[7] The Chinese, too, built large zoos, as did princes in India: the mogul Akbar had five thousand elephants, one thousand camels, and one thousand cheetahs in his collection.[8] The Aztecs' aviary and zoo in Tenochtitlán was large enough to require almost three hundred keepers just for the ducks, fish, and snakes, and three hundred for the rest of the animals. Five hundred turkeys per day were fed to captive eagles and hawks.[9]

Zoo animals have been kept as pets, as oddities, as objects of study, as entertainment, but mainly—and this is as true today as it was in ancient times—as symbols of prestige and power.

⊗ ⊗ ⊗

One of the great delights of living on this land is getting to know my neighbors—the plants, animals, and others who live here—as they introduce themselves to me in their own time, on their own terms. The bears, for example, weren't shy, showing me their scat immediately and their bodies soon after, standing on hind legs to put muddy paws on windows and look inside, or showing glimpses of furry rumps that disappeared quickly whenever I approached on a path through the forest, or walking slowly like black ghosts in the deep gray of pre-dawn. Though I am used to them being so forward, it is always a gift when they reveal themselves even more, as one did recently when he took a swim in the pond in front of me.

Robins, flickers, hummingbirds, and phoebes all present themselves, too. Or rather, like the bear, they present the parts of themselves they want seen. I see robins often, and a couple of times I've seen fragments of blue eggshells long after the babies have left, but I've never seen their nests. It's the same with the others.

These encounters—these introductions—and so many more are always on terms chosen by those who were on this land long before I was: they choose the time, place, and duration of our meetings. Like my human neighbors, and like my human friends, they show me what they want of themselves, when they want to show it, how they want to show it, and for that I am glad. To demand they show me more—and this is as true for nonhumans as it is for humans—would be unconscionably rude. It would be arrogant. It would be abusive. It would destroy the others' trust. It would destroy any potential our relationship may once have had. It would be downright unneighborly.

⊗ ⊗ ⊗

I'm at a zoo. I'm horrified. All across the zoo I see consoles atop small stands. The consoles, with cartoony designs clearly aimed at children, each have a speaker with a button. When I push the button I hear a voice begin the singsong, *All the animals in the zoo are eagerly awaiting you.* The song ends by reminding the children to be sure and "get in on the fun."

⊛ ⊛ ⊛

I push the button. I hear the song. I look at the concrete walls, the glassed-in spaces, the moats, the electrified fences. I see the expressions on the animals' faces, so different from the expressions of the many wild animals I've seen. And I have seen the similarities between the eyes of imprisoned humans and the eyes of those imprisoned in zoos. If you will only care to look, you will see the differences, and you will see the similarities.

Again I push the button: *All the animals in the zoo are eagerly awaiting you.* The central conceit of the zoo, and in fact the central conceit of this whole culture, is that all these others have been placed here for us, that they do not have any existence independent of us, that the fish in the oceans are waiting there for us to catch them, that the trees in the forests are waiting there for us to cut them down, that the animals in the zoo are waiting there for us to be entertained by them.

It may be flattering in an infantile sort of way to believe that everything is here to serve you, but in the real world where real creatures exist and real creatures suffer, it's pretty pathetic to pretend nobody matters but you.

⊛ ⊛ ⊛

Unfortunately we live in an entire culture suffering from narcissism, or to be more precise, we live in an entire world suffering from this culture's narcissism. Zoo proponents are especially prone to narcissism; they have to be or they couldn't rationalize zoos. In the book *Zoo Culture: The Book About Watching People Watch Animals*, Bob Mullan and Garry Marvin ask, "Why preserve wildlife at all? One might well respond that the world would be impoverished if the animals under threat of extinction were allowed [sic] to die out. But who precisely would be impoverished?" They then answer their own question in a way that makes this narcissism especially clear: "Our answer is that the human world would be impoverished,[10] for animals are preserved solely for human benefit, because human beings have decided they want them to exist for human pleasure. The notion that they are preserved for *their* sakes is a peculiar one, for it implies that animals might wish a certain condition to endure. It is, however, nonsensical for humans to imagine that animals might want to continue the existence of their species." It is obvious that neither of these authors has ever known any real wild animals, and certainly has never bothered to ask these animals—either literally or metaphorically—whether they want to survive. Of course an utter disinterest in the perspective of the other is one of the defining characteristics of narcissism.[11] Far worse than a disinterest, however, is this denial that the other's perspective even exists.

Contrast their words with those of Bill Frank Jr., Chair of the Northwest Indian Fisheries Commission, who stated, "If the salmon could speak, he would ask us to help him survive. This is something we must tackle together."[12] And I would say that the salmon are already speaking, if only we would listen. Mullan and Marvin continue, "Animals other than man [sic] cannot have a sense of species identity; they cannot reflect on the nature of their collective identity; nor can they have a sense that it would be a good thing for them to continue in existence." The authors' assertions are unsupportable, arrogant, and absolutely necessary to justify the continuation of the extermination of nonhumans. Again they continue, "The *desire* for a species to continue is merely a projection on the part of human beings." Once again unsupported, unsupportable, and necessary. Again: "The preservation of the natural world is only a preservation for our benefit."[13]

The authors also argue against giving zoo animals larger cages, saying that because animals generally stay in one part of the cage, they don't need a larger territory. They cite another zoo proponent's quip that cheetahs stay in only one part of their cage because "unlike Jogging Man, they saw no point in needlessly expending all that energy." Mullan and Marvin add, "The desire for space, in other words, is the public's desire, not the animals'. According to Dick van Dam, of Blijdorp Zoo, Rotterdam, 'The animals don't need the space but the public of course wants to see them roam on the big plain.'"[14] Professor H. Hediger of the Zurich Zoo expands on these same ideas: "The cage used to be something in which a wild animal was incarcerated against its will, chiefly to prevent its escape. Wild animals lived in cages like convicts in prison. This led to the idea, largely extinct today but still smouldering among some people with little knowledge of animals, that animals in the zoo were indeed convicts and innocent convicts at that, pining away in grief, sorrow, and resentment at the loss of their 'golden freedom' and frequently dying of homesickness." Heddiger is saying that if we believe that animals feel—and remember, humans are animals, too—then we must have "little knowledge of animals." He continues, "Today the idea that zoo animals are in any way like innocent convicts is just as fanciful as the belief that the voices in the radio emanate from little men imprisoned in the box." Now if we believe animals feel—and remember, humans are animals, too—then according to Dr. Hedigger, we must be crazy. He keeps going: "Wild animals in the zoo rather resemble estate owners. Far from desiring to escape and regain their freedom, they are only bent on defending the space they inhabit and on keeping it safe from intrusion."[15] Need I comment on this, or is it as obvious to you as it is to me that this way of thinking is insane? Time and again we see the same rationale with slightly different words. Here are the words of yet another zookeeper: "If you had to spend a weekend in a superdome without contact with other people, you would be going up the wall with boredom by Monday morning. But if I locked you in this office (a small one) for the weekend, and gave you a radio, books, pencils and so forth, you would keep yourself occupied."[16]

I'm sure you can see the fallacies. First, these animals are not locked in these cages only over a weekend, but for their lives. Second, the options are not solely whether the animals should be locked in a small cage or a large one—an office or a superdome. The zookeeper ignores the third option: to blow up both the office and the superdome, the small cage and the large one, and let the animals go. Or even better: to not capture the animals in the first place. Next, if the animals need only a small space and do not wish to roam—or, as Hedigger put it, the animals have ceased "desiring to escape and regain their freedom"—then surely there would be no need for bars, moats, or electrified fences. Again, I can tell again that none of these zookeepers has ever had a meaningful relationship with— or, for that matter, even truly seen—any wild animal. Have they never seen sea lions surfing, or seagulls playing in the wind? Haven't they seen wolfpacks playing together, and deer prancing and playing from joy? Have they never seen squirrels racing up and down trees, teasing each other and teasing dogs and others who cannot climb to catch them? I used to raise chickens, and on cold nights I would bring the motherless chicks inside. Each morning when I'd take them back into the sun they would leap and dance and turn pirouettes. They would play. Both wild and domestic animals—and this is the birthright of all of us, including humans, though civilized humans have been forced to forget this—spend a tremendous amount of time playing. It's a lot of what we *do*. It's a lot of why we're here.

⊕ ⊕ ⊕

I go to a zoo. I see animals on display. I push a button and hear: *All the animals in the zoo are eagerly awaiting you.*

I come home. I open a newspaper and see an article titled, "Animal Planet: From India's famed camel fair to Indonesia's fierce Komodo dragons—all the world's a zoo." And the subhead, large font, boldface: "**Animal World Awaits**."[17] For whom? For you, of course.

I put down the newspaper and turn on the computer. I go to a porn site. I see women on display. I click a mouse and read, "All my ladies love to undress in front of the camera and have a great time doing all the photo sessions that you get to view totally uncensored." I click again and read, "These girls are *sex crazy, they can't get enough!*" Click again, "Hi guys, I'm Pamela! I just started my freshman year at college. Recently I shot my first uncensored hardcore fuck video and want YOU to see me fucked in all holes. I hope that you come inside to see me soon :)" And again: "Cock it, load it, shoot it inside her. She's got the proper lubrication and she wants to be your target. Unload your big guns all over Christy. There's a battle brewing between her thighs, and she's waiting for you to come extinguish the flames. In the heat of the moment, when her body is raging with desire, she waits for the cavalry. Her body is a battle field. You are her last hope. Unleash your big guns deep inside. Fill her up with all you got."

It's not enough to put these others on display. We must convince ourselves that they are desperately willing participants in their own degradation, that we are not exploiting them but doing them a favor. We are rescuing bears from the wild, saving orphans from sentences of death. The animals in zoos are so happy that we need cages to keep the others out. The animals are rich, even, estate owners leading lives of idle luxury.

I click the computer mouse and read, "There are now reported to be one dozen Gorillas and one dozen Chimpanzees living in this new slice of Ape heaven. *They all want to meet you.*"

I push a button. I hear, *All the animals in the zoo are eagerly awaiting you.*

☺ ☺ ☺

It is often said that one of the primary positive functions of zoos is education. The standard ending to the standard zoo book is a plea in high blown language that because the earth has become a battlefield, with the animals losing the battle and the war, that zoos really are the last hope for beleaguered wildlife. Only through unleashing the full potential of zoos for education will the mass of people ever grow to care enough about wildlife to not destroy the planet. The challenge of zoos, according to one not atypical passage, is, "To allow living, breathing animals to inspire wonder and awe of the natural world; to teach us that animal's place in the cosmos and to illuminate the tangled and fragile web of life that sustains it; to open the door to conservation for the millions of people who want to help save this planet and the incredible creatures it contains. To enrich, enlighten and empower the people who care, so that through huge numbers and sheer willpower we save the beetle and the snail and the alligator along with the panda and the rhino and the condor."[18]

Let's parse this out. The author Vicki Croke's use of the word "allow" carries with it the same old implication of willingness on the part of the encaged, ignoring that they are forcibly imprisoned: we must capture and imprison these others so we can then allow them to teach us. Her word *allow* would be entirely appropriate if we were talking about wild animals in wild circumstances coming to us as teachers. Within many indigenous cosmologies wild creatures are our primary teachers. I think often of the words of Brave Buffalo, "I have noticed in my life that all men have a liking for some special animal, tree, plant, or spot of earth. If men would pay more attention to these preferences and seek what is best to do in order to make themselves worthy of that toward which they are attracted, they might have dreams which would purify their lives. Let a man decide upon his favorite animal and make a study of it, learning its innocent ways. Let him learn to understand its sounds and motions. The animals want to communicate with man, but Wakan´tanka [the Great Spirit] does not intend they shall do so directly—man must do the greater part in securing an understanding."[19]

And what, according to Vicki Croke, will these incarcerated animals—oh, sorry, these estate owners—teach us? They will "inspire wonder and awe of the natural world."

Have you ever been to a zoo? Zoos consist of row after row, promenade after promenade, of animals in cages—oh, sorry, habitats. Zoos are at their very best weak simulations of the natural world. So what can at best be conveyed is a sense of appreciation for the cleverness of those who attempt these simulations as well as befuddlement that anyone would even try (why try—and fail miserably—to replicate nature when nature does it for free?).

And have you seen the people at zoos? The pacing grizzly bear did not elicit any response at all from those who passed by, much less wonder and awe. And what feelings are inspired by the hippopotamus drifting in a concrete tank of turds and water; the chained elephant; the lonely giraffe? Awe and wonder would be entirely inappropriate, unless they are at the resilience of these creatures in the face of these horrors. Zoos do not inspire in me a sense of awe and wonder in me. They inspire a sense of loneliness and deep sorrow.

I see no awe and wonder on the faces of other zoo patrons. I hear children laughing at the animals. Not the sweet sound of children's laughter that we so often read about in bad poems, but the derisive laughter of the schoolyard, the laughter at someone else's misfortunes, the laughter that gives voice to the same contempt manifested in flippant newspaper headlines and in "Jogging Man" jokes. I see mothers with their young children, laughing with the children, pointing at the silly animals, laughing at the fat orangutan, laughing at the pacing wolf, making scary faces at the snake, ignoring the pacing bear, laughing at the anteater walking back and forth, back and forth, endlessly back and forth. And the women with their strollers, with their young children crying for cotton candy, crying for plush bears, never stop walking, never stop talking, never stop pointing and laughing. They enter the monkey house. They shriek at the silly monkeys, the silly chimpanzees who pick their noses and who stare straight through the glass at the women, at the children. The children laugh and pound on the glass. They put their faces close to it, stare back at the animal on the other side. Make faces. I turn away. I hear the mothers shriek again, say, "Oh, look, the monkey is doing a poopie." I close my eyes, find myself again gripping a rail. The children laugh and scream. The mothers shriek yet one more time, and say, "Oh, look, the monkey is smearing his poopie on the glass." The women and children laugh and laugh.

I think, "Don't you know what that chimpanzee just said to you? Are you so insensate you do not even know when you've been insulted?"

What are these women and children learning? What "awe and wonder" are they "allowing" the animals to inspire?

Croke continues that a purpose of zoos is "to teach us that animal's place in the cosmos and to illuminate the tangled and fragile web of life that sustains it." This makes no sense. Zoos teach us that a hippo's place is in a turd-filled concrete pool, a monkey's place is behind glass so he can't smear shit on our faces—which at this point he would surely love to do—and a grizzly's place is in a 10,000 square foot "habitat." How could a zoo teach us an animal's place in the cosmos when the creature's presence in the zoo requires it or its ancestors to have been forcibly removed from that rightful place? And how could a zoo illuminate a tangled and fragile web when all of the component parts are split apart and caged? The web consists of the relationships between the different animals and plants and soils and weather and cannot be simulated in a concrete box, no matter how much "enrichment" is added.

Vicki Croke has a lot of company. David Hancocks uses similarly messianic language to once again convey the notion that zoos are nature's last hope: "Zoos have the marvelous potential to develop a concerned, aware, energized, enthusiastic, caring, and sympathetic citizenry. Zoos can encourage gentleness toward all other animals and compassion for the well-being of wild places. Zoos can cultivate environmental sensitivity among their hundreds of millions of patrons. Such a populace might then want to live more lightly on the land, be more careful about using the world's natural resources, and actually choose to vote for politicians who care about the wild inhabitants of the Earth and the health of the wild places that remain. To help save all wildlife, to work toward a healthier planet, to encourage a more sensitive populace: these are the goals for the new zoos."[20]

We could parse this out the same as we did for Vicki Croke to find the same unfounded assumptions and magical thinking, but it might be better if you just go to a zoo for yourself and watch the patrons.

Even if we accept their claims for the educational potential of zoos at face value, study after study has shown that zoos fail miserably at this task. As one author notes, "A study of observation periods at Regent's Park in 1985 revealed that spectators stood in front of the monkey enclosure for an average of 46 seconds, and spent 32 minutes in a pavilion containing a hundred cages. Rather than indicating thorough examination, this is reminiscent of the speed at which television programmes, and even works of art in museums, are 'consumed.'"[21] This 46 seconds includes time spent reading—or rather skimming—the information posted about the animals. Further, while 80 percent of zoo visitors claim to have learned something at the zoo, studies have shown that even after their visit patrons remain "less sensitive to the need to respect nature" than hikers.[22] Further still, inquiry after inquiry has revealed that even while patrons are in the zoo, standing directly in front of the animals in question, they consistently fail even rudimentary nomenclature questions: they still call gibbons and orangutans *monkeys*; vultures *buzzards*; cassowaries *peacocks*; toucans *fruitloop birds*; tigers *lions*; otters *beavers*; and so on.

Peter Batten comments on the educational value of zoos: "Whether anyone derives lasting benefit by seeing wild animals from other countries in enclosures which inhibit their natural behavior must be evaluated without bias. Should one learn that the chimpanzee, for example, is a neurotic humanoid that cadges food from humans, and throws tantrums and excreta should this not materialize? Or that the orangutan, which [who] by nature seldom descends to the soft forest floor, is a pathetic bundle of matted red fur in the corner of a tiled cell? Must the alert, gregarious California sea lion be represented by an animal, half blinded from filthy unsalted water, that [who] spends its life begging for rotten fish?"

All this said, I actually think zoos overwhelmingly succeed at teaching visitors about nonhumans. But the question remains: what are they teaching?

☙ ☙ ☙

While it's true that, as Berger wrote, "The capturing of animals was a symbolic representation of the conquest of all distant and exotic lands," and it is true that zoos are symbols of wealth and power, we must never forget that there is much more at stake here than mere symbols, especially to those most intimately involved.

In the age of the Roman Empire, pits and traps were traditionally favored for capturing most animals. Injuries were common, often fatal. Even animals not physically injured did not emerge unscathed. In addition to forever losing their freedom, obviously, Baratay and Hardouin-Fugier report, "The shock of being captured was such that, according to tamers, 'a big cat [would] be almost mad upon arrival.'"[23] Historically, about fifty percent of animals died on ships bound for Europe or America. Baratay and Hardouin-Fugier writes that "Deaths before embarkation cannot even be guessed at. For most monkeys and for some other animals, the destruction of mothers and, effectively, of their descendants must also be counted. James Fisher, an assistant manager of London Zoo, estimated that one captured orang-utan eliminates four in the wild, of whom three would be potential mothers. Domalain reckoned the number of animals killed for every one visible at a zoo to be ten. Even in the twentieth century, mortality rates for legitimate air transport continued to be high: between 1988 and 1991, they were between 10 and 37 percent for baboons and long-tailed monkeys from Africa, around 10 percent from the Philippines, and 18-54 percent from Indonesia."[24]

The traditional method for capturing many social creatures, including elephants, gorillas, chimpanzees, and many others, was—and remains—to kill the mothers. About elephants it was said, "The only way to capture a living animal was to kill the suckling females or the herd's leaders. The account of the Tornblad expedition to Kenya tells of the slaughter of adult giraffes that enabled the capture of a calf, who was immediately welcomed into the group, cared for and given a name, Rosalie. Hagenbeck found himself 'too often obliged to kill' elephants who were protecting their young by using their own bodies as shields."[25]

Just keep telling yourself: they're only animals. They don't feel. They don't care. They don't grieve. The mothers and fathers do not love their children. The children do not love their parents. Keep repeating that it is a peculiar notion to believe that animals might wish a certain condition to endure.

✥ ✥ ✥

I think it best to describe the capture of zoo animals in the words of those humans most directly involved. I cannot improve on their language.

Hans Dominik was a German living in Africa around the start of the twentieth century. He extensively captured and traded many types of animals, including human animals. Here he describes capturing elephants for transport to zoos. "There was little activity among the animals. The calls of the working humans which carried clearly through the quiet forest hardly appeared to bother them. One bull stood apart, preoccupied with tearing twigs off branches with his trunk and consuming the leaves. Closest to us stood a cow using her trunk to lovingly caress her baby, which was barely larger than a pig and stood between her legs. A few animals ate—sweeping together and ripping up low-growing grasses and using their trunks like sickles—most of them appeared to be sleeping. . . . We seemed so small, so insignificant when compared to the mighty animals in the mighty wild."[26]

How does it go? "You show power by keeping an animal captive; how much more powerful are you if you kill it?"[27]

That night Dominik and his servants built a fence to prevent the animals' escape. The "hunt" began the next morning. "One after the other, a head turning to the left to pull up something green from here and there, the elephants came slowly toward us. The safeties were released. 'You, the second,' I whispered to Zampa. Now we had the animals ready. I fired at the right ear of the foremost animal. At the sharp crack the elephant threw its trunk into the air and trumpeted loudly. The short tail stretched out far, he turned upon himself like a top. In this moment Zampa also fired. Close before me the second animal buckled at his knees, but quickly stood up again and followed the incessantly bellowing and bleeding lead bull which pushed up the hill."[28]

Dominick followed the wounded animals, continuing to fire as he went. He found them. "There lay one of the animals; apparently the spine had been hit because the elephant had only collapsed in the rear and was in a sitting position. Like columns, the forelegs projected from out of the ground, the head and trunk swung left and right: a muffled moan sounded, thick clumps of blood flowed at the side, a sign that the lungs were also wounded. The other stood next to him, motionless except for his trunk. He blew frequently, and with his trunk threw soil on himself. Our approach didn't seem to bother the animals. We crept around them. I had the eye of the sitting giant exactly in the rifle sight, when beside me Zampa fired. The standing elephant trumpeted loudly. Now I squeezed the trigger and the animal collapsed onto its side. The other elephant was still standing; finally with the first shot from my second chamber he collapsed. Close beside one another lay the two giants in a massive pool of blood. Amba and Balla were already there; with their sharp machetes they cut through the trunks, which were half the thickness of a man. The animals were still breathing. As if from a fountain, the red blood sprayed up from the thick arteries onto our clothes as we stood beside the animals examining our guns and discussed how we should proceed with the hunt."[29]

In his indispensable book *Savages and Beasts*, Nigel Rothfels details the rest of Dominik's story: "The fascination with grisly detail which permeates this story continues as the hunt progresses. Soon Dominik encountered a female with a young calf; after several shots, also graphically described, the female was dispatched with a shot in the left eye. The calf was roped to a tree, where it 'churned up the soil with its small tusks, bellowed and moaned, charged backwards, stood on its head, and foamed at the mouth in rage as bloodshot eyes protruded from its head.' Three remaining calves were soon captured as well, one dying of suffocation after having its trunk pulled between its forelegs and tied to its rear legs so that it 'breathed with difficulty and lay on the ground like a large gray sack.' Another calf died during the night of wounds sustained in the capture, but Dominik had still managed to secure two calves from the herd and soon added three more to his collection. Two died a month later, but the remaining three apparently thrived [sic] in their new environment,[30] and one found its way through Hagenbeck to the Berlin Zoo, where it was seen by literally thousands of Berliners who lined up to view the newest acquisition from the colonies."[31]

And how does it go? *All the animals in the zoo are eagerly awaiting you.*

Heinrich Leutemann clarified the priorities of those who capture animals for zoos: "For the animal trader, the method of capture is, from a business point of view, a trivial issue."[32] He gives examples: "Without exceptions lions are captured as cubs after the mother has been killed, the same happens with tigers, because those animals, when caught as adults in such things as traps and pits, are too powerful and untamable, and usually die while resisting. . . . The larger anthropoid apes can, in addition, only be captured—taking into account occasional exceptions—quite young beside the killed mother. The same is the case with almost all animals; in the processes, for example, giraffes and antelopes, when hunted, simply abandon their young which have fallen behind, while in contrast the mother elephant more often defends her calf and therefore must [sic] be killed, as is the case with hippopotamuses. . . . Also in the case of the rhinoceros, the young are captured from the adults, which [sic] are usually killed as a result."[33]

Perhaps the most famous elephant of the nineteenth century was Jumbo. He was captured in a similar fashion. A hunter, Hermann Schomburgk, shot his mother. He describes it himself: "She collapsed in the rear and gave me the opportunity to jump quickly sideways and bring to bear a deadly shot, after which she immediately died. Obeying the laws of nature, the young animal remained standing beside its [sic] mother. . . . Until my men arrived, I observed how the pitiful little baby continuously ran about its mother while hitting her with his trunk as if he wanted to wake her and make their escape."[34]

🜨 🜨 🜨

What do we really learn from zoos? What do we learn looking at the pathetic, dejected, angry, or insane animals? What do we learn beyond the platitudes on the plaques in front of the bars, moats, or electrified fences?

We learn that humans are not animals. We learn that we are here and they are there. We learn that they are there for us, for our pleasure, our entertainment, our education: us. We learn that they have no existence independent of us. We learn that our world is limitless and their worlds are limited, constrained, constricted. We learn that we are more clever than they, or they would outwit us and escape. Or maybe that they do not want to escape, that the provision of bad food—the grizzlies in the San Francisco Zoo are now being fed commercial dog food—and concrete shelter within a cage is more important than freedom (the importance of having humans internalize this lesson for their own lives cannot be overstated). We learn that we are more powerful than they, or we could not confine them. We learn that it is acceptable for the technologically powerful to confine the less technologically powerful (once again, the importance of having especially less technologically powerful humans internalize this message cannot be overstated). We learn that each and every one of us, no matter how powerless we may feel in our own lives, is more powerful than the most mighty elephant or polar bear. Why? Because we can come, and we can go.

We learn that "habitat" is not unspoiled forests and plains and deserts and rivers and mountains and seas, but concrete cages with concrete rocks and the trunks of dead trees. We learn that "habitat" has sharp, immutable edges: everything inside the electrified fence is "bear habitat" and everything outside the fence is not. We learn that habitats do not meld and mix and flow back and forth over time. We learn that humans can make "habitat," and from that can come to believe that humans can make habitat.

We learn that you can remove a creature from her habitat and still have a creature. We see a sea lion in a concrete pool and believe that we're still seeing a sea lion. But we are not. That is all wrong. We should never let zoologists define for us what or who an animal is. A sea lion *is* her habitat. She is the school of fish she chases. She is the water. She is the cold wind blowing over the ocean. She is the waves that strike the rocks on which she sleeps, and she is the rocks. She is the constant calling back and forth between members of her family, this talking to each other that never seems to stop. She is the shark who eventually ends her life. She is all of these things. She is that web. She is the process of being a sea lion, in place. She is her desires, which we can only learn by letting her show us, if she wants; not by encaging her.

We could and should say the same for every other creature, whether wolverine, gibbon, macaw, or elephant. I have a friend who has spent his life in the wild, and ecstatically reported to me one time that he saw a wolverine. I could have responded, "Big deal. I've seen plenty in zoos. They look like a big weasel." But I have never seen a wolverine in the wild, which means I have never seen a wolverine.

Zoos teach us that animals are meat and bones in sacks of skin. You could put a wolverine into tinier and tinier cages, until you had a cage precisely the size of the wolverine, and you would still, according to what zoos implicitly teach, have a wolverine.

Zoos teach us that animals are like machine parts: separable, replaceable, interchangeable. They teach us that there is no web of life, that you can remove one part and put it into a box and still have that part. But that is all wrong. What is this wolverine? Who is this wolverine? What is her life really like? Not her life constrained by moats and walls, but her life in the forest, surrounded by that life, doing what wolverines *do*.

Zoos cannot teach us anything true about the lives of animals—not even human animals. They teach us that a wolverine/elephant/giraffe/anteater/grizzly bear/lion is a ridiculous animal pacing past its own shit in a cement cage. Zoos teach us implicitly that animals need to be managed, that they can't survive without us. They are our dependents, not our teachers, our neighbors, our betters, our equals, our friends, our gods. They are ours. We must assume the interspecies version of the white man's burden, and out of the goodness of our hearts we must benevolently control their lives. We must "rescue them from the wild."

Here is the real lesson taught by zoos, the ubiquitous lesson, the inescapable lesson, the overarching lesson, and really the only lesson that matters: a vast gulf separates humans and all other animals. It is wider than the widest moat, stronger than the strongest bars, more certain than the most lethal electric fence. We are here. They are there. We are special. We are separate.

⊕ ⊕ ⊕

Zoos commit at least four unforgivable sins. First, they destroy the lives of those they cage. Second, they destroy our understanding of who and what animals and habitats really are. Third, they destroy our understanding of who and what *we* really are. And fourth, they destroy the potential for mutual relationships, not only with those particular encaged animals but also with those still wild.

Zoos—like pornography, like science—substitute superficial relationships based on hierarchy, based on dominance and submission, based on a detached consumer manipulating and observing another who may or may not have given permission to be the object of this gaze, for deep relationships based on mutual respect and the giving of gifts.

Think of a pornographic picture. Even in cases where women are paid and willingly pose for pornography, they have not given me permission to see their bodies—or rather images of their bodies—right here right now. If I have a photograph, I have it forever, even if subsequently the woman withdraws her permission. This is the opposite of relationship, where the woman can present herself to me now, and now, and now, always at both her and my and our discretion (and of course I can present myself to her now, and now, and now, also always at her and my and our discretion). What in the latter case is a moment by moment gift becomes in the former case my property, to do with as I choose. This is, of course, true of all photographs.

And it is true of zoos. I do not and cannot command the bear whose home I share to appear before me. Nor the gray jays, the slender salamanders, the slugs. They are willful and independent.

Everything is far worse than I am making it seem. Zoos—like pornography, like science, like other toxic mimics—take a very real, necessary, creative, life-affirming, and most of all relational urge and turn it—pervert it—until it furthers not fully mutual relationships at all but instead superficial relationships based on domination and control. Indeed, zoos—like pornography, like science, like other toxic mimics—can cause people to forget those original relational urges, to forget mutuality is possible, to forget depth is possible, to believe control is natural and desirable. Pornography takes the creative relational need for sexuality with willing partners—and the intimacy this can imply—and simplifies it to the relationship of watcher and watched. Science takes the creative relational need for understanding and the gaining of wisdom and simplifies it to that same dynamic; watcher and watched; dominator and dominated; subject and object.[35] Zoos take the creative need for participating in relationships with wild nonhuman others and simplify it until our "nature experience" consists of spending a few moments looking at—or simply walking by—insane bears and angry chimpanzees in concrete cages.

It's actually worse than this. Incarcerating animals in zoos is to entering into relationships with them in the wild as rape is to making love. The former in each case requires coercion, limits the freedom of the victim, springs from, manifests, and reinforces self-perceived entitlement to full access to the victim on the part of the perpetrator. The former in each case damages the ability of both victim and perpetrator to enter into future intimate relationships. It distorts the notion of what *constitutes* a relationship. It is based on the dyad of dominance and submission. It closes off any possibility for real and willing understanding of the other.

The latter in each case is a dance among willing participants who give what they wish as they wish when they wish. It inspires present and future intimacy, present and future understanding of the other and the self. It nourishes those involved. It makes us more of who we are.

⊡ ⊡ ⊡

Early on in this book zookeepers asked the question, "What do grizzly bears want? Their question, however, was a toxic mimic of a real question. It was a rhetorical device to lead themselves toward a predetermined answer. It was a lie, substituting for their real question of, "What do grizzly bears want, given that we zookeepers will forever control their lives, and will keep them forever inside small cages we will call habitat?"

Let's ask that question again, only this time sincerely: What do grizzly bears want? And then let's ask, What do salmon want? What do spotted owls want? What do hamadryas baboons want? What do redwoods want? What do American chestnuts want?

All these lead to the next question: How do we know what they want? And when we ask those questions—What do they want, and how do we know what they want?—honestly ask those questions, ask those questions without preconceptions, ask those questions not as an excuse to incarcerate and exploit them, ask those questions not as "lords of the earth" but as friends and neighbors and loved ones, ask those questions respectfully, ask those questions of our elders, ask those questions of those who have lived on the landbases we share far longer than we have, ask those questions not just of individuals but of families, clans, communities,[36] and landbases, ask those questions as though their and our lives depend on it (because they do), we will find in time—soon, soon—that everything we've ever known will change. The din of the echo chamber will lessen, the isolation-induced hallucinations and delusions of grandeur will begin to fade. The loneliness—the devastating, soul-breaking, heart-numbing loneliness—will crack, rip, crumble, and get rushed away in a flood of newfound neighbors who have been here from the beginning, until the loneliness no longer pains, until the loneliness moves from all-consuming present to traumatic memory to the realm of cautionary tale told to future generations who literally cannot imagine how anyone could have been so absurd as to fail to listen.

As the World Burns
50 Simple Things You Can Do to Stay in Denial

with Stephanie McMillan

Seven Stories Press, 2007

When I encountered Stephanie McMillan's work, I knew that I wanted to work with her because she is such a great cartoonist, she is so funny, and her politics are so good. We both absolutely despise books like 100 Simple Things You Can Do to Save the Planet *because they're wrong and they trivialize tremendous suffering. They trivialize the murder of the planet. They point away from the real problems, which are capitalism, civilization, and this entire exploitative way of life. It's absurd to think that you can inflate your tires and that's going to stop global warming. Someone actually did that math on all the suggestions that Al Gore makes in* An Inconvenient Truth, *and even if every person in the United States did everything he suggests, it would only reduce carbon emissions by about 20 percent. Since Gore doesn't speak against a growth economy, all of that would disappear in a few years anyway. So we wanted to write a book about resistance, about what it would really take to stop this culture from killing the planet.*

A piece of trivia: for both of us, our favorite character to write was the president. That kept us from working together again for a few years. We were afraid that if we worked together immediately, we would just make The Adventures of the Stupid President, Volume Two.

Another piece of trivia: we thought that the main character was going to be Kranti, and we were both entirely surprised through the writing of the book to learn that in the sense of a standard novel, where the protagonist has to undergo a dramatic shift, the real protagonist and the real hero of the book is Bananabelle.

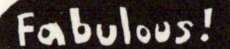

In response to the recent terrorist act of blowing up a dam, today the FBI and local police around the country arrested 700 rabbits and rabbit supporters as suspected terrorists.

An anonymous eyewitness provided this description of the terrorist mastermind: "I saw a bunny slinking around, not hopping like most of them do. So I looked real close and saw he had one eye gone. Then I saw him plant the charges and tamp them down with his back legs."

When asked why he did not stop this terrorist rabbit from carrying out this terrorist act, the anonymous witness replied: "I did what anybody would do if they saw a one-eyed rabbit tamping down explosives next to a big dam. I ran like hell."

What We Leave Behind

with Aric McBay

Seven Stories Press, 2009

Ever since I was a child, I've been fascinated by decay. In eighth grade, I was supposed to do a science experiment. I bought a fish at a fish market and cut it in half and put half of it on the ground and half of it in water. Several times every day I would go out and observe the differences in how the decay happened. The dry fish became a crawling, pulsating mound of maggots. The one in the water essentially looked like chocolate milk. I was in eighth grade, which means by definition I had no common sense, which means I didn't do this experiment away from the house. Nor did I tell my mother. I put the fish in the garage and let it rot. After a couple of days my mom asked, "What is that smell?" That put an end to the experiment.

I've always wanted to know what happens to our waste products. End-game describes how you cannot take more than the land gives willingly, and What We Leave Behind describes how all of our waste products have to be food for someone else. For example, plastic is a really, really bad idea because no one in the history of the planet has ever eaten plastic. Plastic as a waste product is not food for someone, which means it does not decompose. By using plastic you are not participating in the cycle of life, but instead circumventing it and destroying it.

The health of the land begins with shit, with dead bodies, with body parts that fall to the ground. It begins with death, decomposition, decay. It begins with eating, metabolizing, excreting. That's how it has always been, since the beginning of life. You feed me, I feed the soil, the soil feeds everyone, the soil feeds me, I feed you, you feed the soil, and so on.

❁ · ❁ · ❁

Here's another way to look at it: I eat you, the soil eats me, everyone eats the soil, I eat the soil, you eat me, the soil eats you.

It's all the same.

❁ · ❁ · ❁

Our relationship—both personal and collective—with shit, and more broadly with our waste products, reveals much about our relationship with the land—with our habitat—and much about why and how this culture is killing the planet. In the case of shit, this culture has turned what was a gift from us to our habitat—a gift of fertile soil, given in response to the nourishment our habitat gives us—into something toxic, something harmful. Something shameful. And that is a terrible shame.

❁ · ❁ · ❁

We're called to this book, and to this discussion, by many driving questions. What is waste, really? Is there good waste, or ways to waste well? When and how did shit become waste? More to the point, when and how did waste become waste? How is "waste" dealt with in the natural world? How are cultural relationships with shit and waste mirrored in attitudes toward wasted people, castes, and classes? How did the dominant culture come to cause such massive wastage of materials, people, and lands, and what will happen to that waste when this culture is gone?

Answering these questions is not a mere matter of academic or intellectual curiosity. These questions go to the root of our relationships with our biological selves, with the future, with the land, with sex, with death, with the sacred, and with all other living creatures. If we want to live in a world that is not being laid to waste—where living creatures are not viewed as garbage or lives callously wasted—we have to find good answers to these questions, and soon.

❁ · ❁ · ❁

Here's something about toilet paper. The toilet paper I leave in a clearing lasts relatively intact through the summer, not breaking down until the rains come. If, however, I leave toilet paper in the forest and return even one or two days later, I see that the stained part of the paper is gone, eaten by slugs, who clearly prefer to live in moist forests over open sunlight where trees have been removed.

I mention this for a few reasons. The first is that it means if I got desperate I suppose I could use slugs as toilet paper cleaners and reuse the toilet paper (er, maybe not). The second and far more important is that it returns us to the question of how long it takes for something to break down. Now, it may not matter much whether my toilet paper breaks down in six days or six months, but when you have a society producing toxic wastes lasting thousands, tens of thousands, or in some cases hundreds of thousands of years, the question of how long it takes for something to break down becomes far more important. But there's an even more important reason, which is that it leads to the question of whether and how our "garbage," our "refuse," our "waste," is useful or harmful to others in our landbase, and to our landbase in general. And this, of course, leads to the most important question of all: Does our presence and do our actions help or harm the land who supports us, the land who is our home, the land on whom our own survival ultimately depends? It's very simple: any way of living that doesn't help the landbase—any way of living that doesn't feed the soil what it needs to survive and thrive—will not last. This means that any way of living that produces waste that doesn't help—or worse, harms—the plants and animals and fungi and bacteria and land and water and air upon which (or rather, whom) that way of living—in fact the species' survival—depends, will not last.

⊕ ⊕ ⊕

The questions we need to ask ourselves about every action—as we live in the midst of a culture killing the planet—are these: Is this action sustainable? Why or why not? How would this action need to be different for it to be sustainable?

Before we can answer these questions, however, we have to define *sustainable*. Many politicians, business people, "green" architects, land managers, foresters and other resource specialists, and so on throw that word around a lot in meaningless or deceptive ways, labeling as "sustainable" many manifestly unsustainable actions that most often make a lot of money for them or for the corporations to whom they are beholden. We hear about sustainable buildings, sustainable agriculture, sustainable forestry, sustainable this and sustainable that, and of course, within this culture, little or none of it is even remotely sustainable.

For an action to be sustainable you must be able to perform it indefinitely. This means that the action must either help or at the very least not materially harm the landbase. If an action materially harms the landbase, it cannot be performed indefinitely: any line sloping downward eventually reaches zero.

Central to sustainability is the landbase itself. What may be beneficial to one landbase may be harmful or lethal to another. I feel good shitting outside and dropping pieces of toilet paper willy-nilly across the rainforest floor, comfortable in the knowledge that it will all break down at most within a year. Would this be appropriate behavior in a desert? Certainly beings in deserts still have to defecate, and certainly deserts have developed ways to turn shit into something they can use. But in a desert, I might have to spread my shit and paper over a larger area, and maybe not use paper at all, to ensure I won't negatively impact the land. A nonhuman example may help make this a little more clear. Cows did not evolve in a desert. Their poop makes big patties, which in moist climates break down into potent packages of food for scavengers and soil alike. In dry climates, however, these patties can ossify, turning into a sort of fecal asphalt that smothers and harms the soil. Antelope, bighorn sheep, and others who evolved in deserts do not poop in big mounds, but rather tiny pellets that are more easily convertible to food the desert can use.

Here's another way to look at sustainability's dependence on context. I live on Tolowa land. Prior to conquest, the Tolowa lived here without materially harming the place for at least 12,500 years. By any reasonable definition they lived here sustainably. Their homes were made of wood. This means that here in their rainforest this particular use of wood was sustainable. But people who live where trees are sparse may not be able to sustainably use wood as a building material.

Any working definition of sustainability must emerge from and conform to a particular landbase—to what that landbase can freely give forever—and not be an abstract set of principles, or rationalizations, imposed upon the landbase. The landbase is primary, and what we do to it (or far more appropriately, with and for it) must always follow the landbase's lead.

What actions are sustainable is determined not only by context, but also by scale. One human shitting in the forest near my home is probably a good thing. Two, three, or four might make an even bigger bonanza for slugs. But let a thousand people shit right here and the slugs will quickly say *no mas*. Likewise, the fact that the Tolowa took out a few trees does not mean this forest (or any forest) can survive industrial (or even extractive) forestry (which should more accurately be called deforestation). And the fact that the Tolowa took salmon to eat doesn't mean salmon can survive industrial (or even extractive) fishing (which once again should really be called de-fishing).

Similarly, I don't shit in the pond by my home. I don't believe the pond is big enough to take in my shit without materially harming it. If I lived near the Amazon River, on the other hand, and were the Amazon not horribly stressed by the various activities and wastes of this culture, I'd gladly defecate into it, knowing that I was helping the river. I remember years ago reading a story by Herbert R. Axelrod, an expert on tropical fish, in which he was wading waist-deep in a sluggish backwater of the Amazon when a big fish started nudging him in the butt. His native guides told him the fish was begging for food, hoping it could induce him to drop his pants and defecate.

At least my dogs have the patience to not begin their begging till I've removed my pants.

The land is a living entity who like any other living entity requires certain foods to survive. Certain other foods can be toxic. But even nourishing foods can be toxic out of scale.

My own shit is just one small bite for this land. But just as I might find one apple delectable and conducive to good health, and just as ten apples at once might make me sick to my stomach and give me the runs, and just as dropping ten tons of apples on my head will kill me, so too the land could find one person's shit delicious and beneficial, more shit harmful, and more shit than that all at once lethal.

Time is as important as scale to an action's sustainability. Indeed they are related. The Tolowa, Yurok, Karuk, and other tribes probably killed more salmon in this region over the last 12,500 years than this culture has in the past 180. But they did it over 12,500 years (they also didn't dam or poison rivers, deforest hillsides, murder oceans, change the climate, and so on). Killing that many salmon to eat (and dying yourself on this same land, feeding the land and thus eventually the salmon in the same manner as they are feeding you) could have continued in human terms forever. But killing as many salmon as the dominant culture has in such a short time has not been a mutual feeding, but instead a slaughter, and has decimated—and in many cases extirpated—salmon.

🐟 🐟 🐟

The real world does not resemble our compartmentalized version of it. Where do I end and you begin? Are you still in the air you just exhaled, the air I now take in, the air that carries with it the sweetness of your breath? Do you end where your fingers touch my skin, or do you follow these sensations into my body? Where do I end when I move inside of you, and where do you end when you move inside of me? And when you leave are you still inside of me? Am I still inside of you? Where do you end and where do I begin?

🐟 🐟 🐟

A week ago my dog Amaru died. He was fourteen. He had been sick for more than a year, from Cushing's, arthritis, and most of all age. He had lost about 30 percent of his weight, and even more of his strength. When he was younger he gladly chased bears, but those days were gone. He was losing his sight and hearing. And then, the night before he died, he began to throw up. I stroked him, spoke with him, tried to reassure him, but I did not know how to take away his pain and nausea.

He finally fell asleep around midnight, and I fell asleep sometime later. I awoke near dawn knowing something was wrong. I had to find him. He wasn't near my bed, wasn't in the house. He was outside collapsed in shallow water at the edge of a pond. He had only been there a short while (I could tell because the water had not wicked up his body), but still he was very cold. I brought him in, warmed him up. He got up once, hours later, to stagger outside to relieve himself, taking care of me by not soiling the house even as he was dying. He couldn't make it back inside. I carried him in, laid back down. He never got up again.

I held him as he died, as I'd held him all through the day, talking to him, rubbing him, not quite believing he was dying. But he did. His breathing grew forced, and then stopped altogether, although his heart still beat. I pushed on his chest, and pushed again, and again, but he was dead.

I did not bury him that night. He wasn't ready to go, and I wasn't ready to let him go. I let him sleep or maybe he let me let him sleep next to my bed one last time, near the head, where for so long I had reached down to pet him in the middle of the night.

The next day I buried him, where he will over time become more and more a part of the forest.

I miss him terribly, though he is even now only thirty feet from me.

I have always had difficulty falling asleep, often lying awake for hours.

I've written elsewhere that it sometimes feels as though I've forgotten how to sleep, or like I never learned. In my late thirties I finally experimented with various herbs—valerian, chamomile, and so on—but none seemed to help, until finally I came upon a combination that put me to sleep: kava kava, 5-htp, and melatonin.

For years then I marveled at the deeply sensuous pleasure of slipping so seamlessly between waking and dreaming worlds. But a few months ago the herbs stopped working, and once again I would lie awake each night till near dawn, thinking and not thinking, writing and not writing, meandering sometimes into shallow dreams but rarely going deeper, instead nearly always bubbling back a few moments later to and through that permeable surface, back to this side.

I tried different herbs, and I tried changing other variables in my life.

Nothing worked.

Nothing worked, that is, until the night after I buried Amaru. Ever since then I've slept soundly, and deep.

It took me only a couple of days to understand that Amaru had been keeping me awake. Not by snoring (my other dog Narcissus does that, sometimes driving me to wear earplugs or to carry him gently into another room) but because he was in so much pain and carrying so much psychic distress that it hung in the air, seeped through my skin.

Where does Amaru end, and where do I begin?

I often have dreams that are not my own, but rather they come from one friend or another. I may, for example, understand everything in a dream except a strange blue car that drives by too fast and scares me. I try to make some interpretation fit. I think about every blue car I can recall, think about their owners, try to assign some meaning. Nothing works. Then a friend calls, tells me she was nearly run over the day before by someone driving a blue car. This happens often enough that my friends and I have a name for it: *leakage*, because their realities leak into my dreams.

Where do my friends end, and where do I begin?

You and I touch. We make love. I touch the middle of your chest with two fingers. You kiss my forehead, my lips, my shoulder.

One of us gets up, leaves. What is left behind in this room? Where do you end, and where do I begin?

I live in a forest. What is this forest? Is it redwood trees? Or is it soil? Or is it frogs? Slender salamanders? Huckleberries? Shrews? Steller's jays? Or is this forest all of these and so much more? Where do frogs end and where does water begin? And when the small streams taste like tea from flowing past and through and with all this vegetation—living, dead, whatever—where does the forest end, and where does the water begin?

Here is the bottom line: the world is being killed. It is being murdered. And one of the ways it is being murdered is that it is being poisoned: the waste products of this culture do not help landbases—as waste products are *supposed to do*, and *have always done*—but instead they harm and kill them. In nature there is no waste. Waste in this sense is a modern invention. And it's a rotten one.[1]

As epidemiologist Rosalie Bertel has pointed out, the probable fate of our species is extermination by poisoning. We could at this point add that this is the probable—in fact looming—fate for the oceans, the air, the soil, the bodies of most every living being.

All of the fancy talk of sustainability—by us and others—is just dancing around the central issue: this culture is killing the planet.

This culture is killing the planet.

This culture is killing the planet.

This culture is killing the planet.

If we repeat this enough times, perhaps we will start to comprehend even the tiniest terrifying bit of *what this means*, and we will begin to act as if any of this matters to us.

I have in front of me a photograph. It is a photograph of a turtle. Or what would be a turtle. Or what could or should be a turtle, and of course still is a turtle, but is a turtle who got a plastic ring caught around the shell's middle.[2]

The turtle grew. The shell surrounding the ring did not. The turtle—and I wish I were making this up—looks like an hourglass. I first saw this photograph a few weeks ago, and have not been able to get it out of my head (nor my heart). But of course there is a difference between feeling empathy for another and actually having to live the life of that other. I can walk away. I can live my life pretending nothing is wrong. The turtle cannot do that. The plastic ring deformed the turtle, changed the turtle's life for much the worse.

Now I have another picture before me. It is of a river, or so I am told. I cannot tell, because there's too much trash. As the accompanying article states: "It was once a gently flowing river, where fishermen cast their nets, sea birds came to feed and natural beauty left visitors spellbound. Villagers collected water for their simple homes and rice paddies thrived on its irrigation channels. Today, the Citarum is a river in crisis, choked by the domestic waste of nine million people and thick with the cast-off from hundreds of factories. So dense is the carpet of refuse that the tiny wooden fishing craft which float through it are the only clue to the presence of water. Their occupants no longer try to fish. It is more profitable to forage for rubbish they can salvage and trade—plastic bottles, broken chair legs, rubber gloves—risking disease for one or two pounds a week if they are lucky."[3]

And now another picture. It is of the skeleton of a sea gull. Inside the rib cage is a mound of plastic. There's a sense in which this picture is less horrible than the previous ones, since it could have been staged: someone could have placed the plastic inside the skeleton.[4] But I know that a study of fulmars—a type of seagull—in the North Sea revealed that the gulls had an average of forty-four pieces of plastic in their stomachs (weighing what in a human would be the equivalent of five pounds).[5] One animal had consumed and retained more than 1,000 pieces of plastic.

Now I see a picture of a sea turtle with a plastic bag hanging out of its mouth: creatures in the ocean often mistake plastic bags for jellyfish. Sometimes they eat them. Sometimes they die.

Now I see a picture of a "ghost net," a plastic fishing net that was cut loose from a commercial fishing vessel. The net hangs in the water, fills with fish, turtles, sea mammals, sea birds—anyone captured by it—and eventually sinks to the bottom. When the bodies decompose sufficiently to fall apart, the net floats again toward the surface, where it begins this process anew.

I'm sure by now you know the numbers. Marine trash kills more than a million seabirds and 100,000 mammals and turtles each year, as well as unimaginable numbers of fish[6]—each and every one of these an individual worthy of consideration. There is at least six times more plastic in the middle of the Pacific Ocean than phytoplankton. (Imagine trying to eat and six out of every seven swallows bring only plastic; it is no wonder that so many sea creatures are starving to death, bellies full of plastic. Others die of constipation brought on by plastic blocking their intestines; having suffered a blocked intestine, with its pain at least an order of magnitude worse than a broken bone, I can tell you that I cannot imagine many more excruciating ways to die). This plastic is not degrading, but merely breaking into smaller and smaller pieces, until by now it is routinely found inside the cells of phytoplankton.

Plastic is everywhere in the oceans—and I mean everywhere—but it also accumulates where currents carry it. In these places the essence and endpoint of this culture could not be more clear. As one author wrote, "It began with a line of plastic bags ghosting the surface, followed by an ugly tangle of junk: nets and ropes and bottles, motor-oil jugs and cracked bath toys, a mangled tarp. Tires. A traffic cone. Moore could not believe his eyes. Out here in this desolate place, the water was a stew of plastic crap. It was as though someone had taken the pristine seascape of his youth and swapped it for a landfill."

The article continues, "How did all the plastic end up here? How did this trash tsunami begin? What did it mean? If the questions seemed overwhelming, Moore would soon learn that the answers were even more so, and that his discovery had dire implications for human—and planetary—health. As Alguita glided through the area that scientists now refer to as the 'Eastern Garbage Patch,' Moore realized that the trail of plastic went on for hundreds of miles. Depressed and stunned, he sailed for a week through bobbing, toxic debris trapped in a purgatory of circling currents. To his horror, he had stumbled across the 21st-century Leviathan. It had no head, no tail. Just an endless body."[7]

That particular "Garbage Patch" is nearly the size of Africa. And there are six others. Combined, they cover 40 percent of all of the oceans, or 25 percent of the entire planet.[8]

It's not merely river and ocean creatures—and rivers and oceans themselves who are being murdered by plastic. So are land dwellers, including us. And frankly, although I love humans, at this point I feel even worse for those like the turtle whose species have done nothing to deserve this than I do for most especially the rich humans whose lifestyles are causing these murders. At least rich humans get to drink from plastic cups and play with plastic Barbies and watch televisions housed in plastic (using electricity flowing through wires insulated with plastic) before these plastics poison and suffocate us all.

And plastics *are* poisoning and suffocating us all.

⊗   ⊗   ⊗

Plastics are of course not the only persistent pollutants made by this culture. The list is long. There are plutonium-239 and other radioactive wastes, many of which will persist essentially for the life of the planet (of course given the velocity and acceleration of this culture's destructiveness, the death of the planet will probably come sooner than we think, insofar as most of us think about it at all). There are heavy metals removed from the earth and put into living bodies. There is carbon dioxide, which used to simply be part of the air but is now a pollutant, significant quantities of which will not be metabolized by plants or otherwise removed from the biosphere for an extremely long time. Look around your own life—and get out into the land where you live—and find the long-term pollutants produced or released by this culture.

I grew up in Colorado, and spent a lot of time in the mountains, exploring ghost towns and the fringes of old mines. I say fringes because I was forbidden by my mom from entering mines for fear they would collapse or I would fall down a shaft. So I spent those days wandering around the tailings: great mounds of crushed rock that a hundred years later still supported no life. In retrospect, the mine shafts may have been the least of my safety worries: the tailings are probably more dangerous than the mines themselves.

Mine tailings are what's left over after miners dig up the ground and extract whatever they're going to sell. These previously-buried, now-exposed rocks have normally been broken, and range from much larger than a softball down to coarse sand or even fine powder. Some minerals commonly found in tailings include (but are certainly not restricted to) arsenic (especially in gold mine wastes), barite, calcite, fluorite, many radioactive materials (that the earth had previously stored where the earth wanted them: underground), sulfur (and many sulfide compounds), cadmium, zinc, lead, manganese, and so on.[9]

Many of these minerals are toxic. That explains the dead zones so familiar to anyone who has had the misfortune to go near a mine.

Many of these minerals don't stay in place, but weather away—their rapid weathering facilitated by their relatively small particle size, which results in greater surface area available for wind, air, and water to affect—into surrounding soil, and more ominously, water. This happens most frequently with the sulfide minerals—especially pyrite (iron disulfide: one molecule of iron and two of sulfur), the most common mineral in the majority of mine tailings[10]—which are oxidated by chemical processes and by being metabolized by certain bacteria.[11] These processes produce sulfuric acid, which often reduces the pH of afflicted streams and groundwater to less than 3.0. Of course acidifying streams and groundwater kills the streams, as well as the plants and animals who live in them. As even one pro-mining website puts it: "The presence of these toxic heavy metal ions and acidic pHs has an adverse affect on every aquatic species found in the stream. In many instances, streams are almost completely void of life for many miles downstream of a mine drainage source."[12]

As the acidified water slowly gets diluted, the pH rises back above 3.0. You'd think this would be a good thing, but making the stream less acidic causes iron ions (from that original iron disulfide) to, as that same website says, "precipitate out of the water and coat the stream bottom with a slimy orange sludge (iron III hydroxide, $Fe(OH)_3$, and related compounds). This unsightly sludge, called 'Yellow Boy,' tints the stream water an unnatural reddish-orange color and smothers the organisms that thrive on the stream bottom."[13]

Great.

And it gets worse yet. Once miners have dug ore from the ground they extract from it the gold, platinum, silver, lead, coal, or whatever else they want to sell. Frequently, and it gets ever more frequent with each passing year, this extraction (or "milling") uses toxic or otherwise harmful chemicals: in other words, the valued minerals are no longer simply separated mechanically, but are separated through processes often involving chemicals. For example, because these days gold remaining in its native state typically occurs at concentrations of less than a third of an ounce per ton—meaning mining corporations dig up and later dump three tons of rock and soil for every ounce of gold—it's not economically feasible to mechanically separate gold from ore. Instead, the gold is dissolved into a liquid, adsorbed from this liquid onto activated carbon, and then washed away from the carbon using solvents. Because gold isn't soluble in water, it requires both a complexant and an oxidant in order to dissolve. Many of the chemicals used in these extraction processes are toxic or otherwise harm the land and water, and of course also harm those who require land or water in order to live. In other words, they harm all of us.

Even though the mining industry claims these chemicals are recycled—and they often are recycled to the best of their technological capability, for what that's worth—the chemicals still often end up in mine tailings or tailing ponds, and often from there the chemicals move into streams or groundwater. From there they end up in living beings.

You may recognize some of these chemicals. Let's start with cyanide. Yes, that cyanide. The one from those World War II movies where a character bites down on an ampoule of hydrogen cyanide and immediately goes into convulsions and dies. The one that, so we hear in all these movies, smells like bitter almonds. The one that prevents cellular respiration, and kills at concentrations of 100 to 300 parts per million (and kills fish and other aquatic life at concentrations in the range of parts per billion). The one that went by the name Zyklon B and was used by the Nazis as an efficient form of mass murder. The one used in gas chambers. The one responsible for many of the deaths at Bhopal.[14] The one commonly listed as a chemical warfare agent. The one allegedly used by Hussein's Iraq regime against both Iran and the Kurds (with ingredients provided with the help of the United States). The one the Aum Shinrikyo cult attempted to use to commit mass murder in a Tokyo subway in May 1995. The one Al Qaeda allegedly planned to use to commit mass murder in a New York subway in 2003. Yes, that cyanide. The same cyanide produced routinely—1.4 million tons per year—for use in the production of plastics, adhesives, cosmetics, pharmaceuticals, and so on. The same cyanide used in mass quantities—182,000 tons per year—to facilitate the extraction of gold from surrounding ores.[15] The same cyanide used to extract about 90 percent of the gold mined each year.

It seems that those who put small amounts of cyanide in subways are terrorists. Those, however, who produce it in mass quantities and contaminate broad reaches of soil, water, and air, killing countless living beings, are not terrorists, but rather capitalists, and are counted among the finest and most powerful people on the planet.

Mining industry representatives—whose relationship to truth is as strained as that of representatives of other industries—like to tell us that using cyanide to extract gold from ore is reasonably safe. But at this stage in the corporate destruction of both the planet and honest discourse, it should not surprise us that they are lying.

Above, when I wrote that "gold is dissolved into a liquid, adsorbed from this liquid onto activated carbon, and then washed away from the carbon using solvents," and also wrote that since "gold isn't soluble in water, it requires both a complexant and an oxidant in order to dissolve" my description failed to get to the essence of the process: this articulation makes the process seem clean, even if that "complexant" is something as toxic as cyanide. If the miners are careful enough—and we know they will be—then this could and should be a safe process. Right?

Well, let's try this description. First, when we're talking about miners, we're not talking about Humphrey Bogart, Tim Holt, and Walter Huston protecting their hard-earned treasure by fighting off "cops" who have no stinking badges (nor, and this is more to the point, are we talking about miners who say, and more importantly actualize, the understanding that they've "wounded this mountain. It's our duty to close her wounds. It's the least we can do to show our gratitude for all the wealth she's given us").[16]

Instead, we're talking about huge transnational mining corporations that often bribe local, regional, or national officials, politicians, dictators—whoever has the power to sign pieces of paper that legalize their activities—to give them permits to dig up entire mountains. They often buy off, bully, beat, capture, or kill those who oppose them; or they get their money's worth from the local, regional, national officials, politicians, or dictators who send in the military or police (with stinking badges) to bully, beat, capture, or kill those who oppose their mines. In the United States and Canada the preferred tactic is to buy off the opposition (except especially in the case of those of the indigenous who cannot be bought off), whereas in the colonies it's often more cost-effective for these transnational corporations—and the military and police who serve them—to skip directly to the latter three options: in both the short and long-run, bullets are cheaper for them than bread, and bullets are certainly (fiscally) cheaper for them than not digging up the earth.

Next, when these huge corporations mine and mill ore, these processes are not quite so spick and span as the technical descriptions make them seem. First, picture a living mountain, its base covered in trees, its top above the tree line. Now, picture this mountain flattened. Picture its guts removed. Picture pits so large that, as one pamphlet puts it, "they could swallow cities."[17] Picture heaps of extracted ore several hundred feet high and several times larger than a football field. Now, picture spraying a solution containing cyanide over those heaps. Picture the cyanide trickling down, chemically bonding with microscopic bits of gold. Picture this solution draining to a huge rubber blanket beneath this heap. Picture this blanket channeling the cyanide solution toward a large holding pond, where the gold is stripped away, and as much of the cyanide as possible is recovered to be reused.[18]

Picture birds landing in this pond. Picture birds dying. Picture every living being who comes in contact with this pond dying.

Picture these heaps being dumped on plains. Picture them being dumped in valleys. Picture them filling these valleys, until the valleys no longer exist. Picture these heaps being dumped anywhere these transnational mining corporations have permits to dump them. Picture governments handing out these permits to transnational corporations. Picture transnational corporations handing money to local, regional, or national officials, politicians, dictators—whoever has the power to sign pieces of paper that legalize these activities. Picture streams below turning to sulfuric acid. Picture all creatures dying. Picture the bottoms of these streams coated with Yellow Boy, and picture the water itself a sickly orange. Picture those humans who live by these streams dying. Picture paid representatives of these mines telling us that these processes are safe. Picture newspapers repeating these claims. Picture paid local, regional, or national officials, politicians, dictators repeating these claims as well. Picture paid local, regional, or national officials, politicians, dictators passing laws making it illegal or impossible (at least through "proper channels") to effectively stop the gutting of these mountains, the poisoning of these streams, the poisoning of this water, this land, this air, these nonhumans, these humans. Picture paid local, regional, or national officials, politicians, dictators buying off some of those who resist, and bullying, beating, capturing, or killing the rest. Picture the owners and CEOs of these corporations living nowhere near the tailings piles or the acidic orange streams.

Picture this cycle being repeated.

Picture the planet being killed.

Now picture a catastrophe. Picture the Baia Mare gold mine in northwest Romania. Picture a dam holding back a tailings pond. A large tailings pond. Full of water polluted with cyanide (120 tons just of cyanide) and other toxins. Now, picture this dam breaking. Picture a twenty-five-foot tall wave of toxic mud and water rushing into the Lapus River, and from there into the Somes and the Tisza Rivers. Picture this wave crossing the border into Hungary, hitting the Danube, and crossing another border, this time into Yugoslavia. Picture this wave killing nearly everything in its path. Picture the Tisza River being habitat and home to nineteen of Romania's twenty-nine protected fish species. Picture one hundred tons of dead fish being collected from the surface of this river. Picture otters on the Tisza and Somes Rivers eating cyanide-laced fish. Picture these otters disappearing, presumed dead: casualties gone MIA in this culture's war on the world. Picture the Tisza and Somes rivers without otters. Now picture another spill higher on the Tisza, and this time picture 20,000 tons of sediment toxified by lead, zinc, copper, aluminumand cyanide. Picture a river being systematically murdered.[19]

Picture the owners living elsewhere, and not being poisoned by the effects of their decisions.

Now, picture the Zortman-Landusky mine in Zortman, Montana. Picture spill after spill of cyanide-laced water. Picture once again a tailings pond. Picture a heavy rain. Picture a tailings pond in danger of overflowing. Picture mine operators deciding to pump this cyanide solution out of the tailings pond and onto another plot of land. Picture them not telling anyone about this. Now, picture living in Zortman. Pretend you are thirsty. You walk to your tap. You turn it on. You smell bitter almonds. You've seen enough World War II movies to know what this means. You turn off the tap. Your water has been poisoned with cyanide,[20] and not by Aum Shinrikyo or al-Qaeda. Instead by capitalists (who, by the way, kill a hell of a lot more people than terrorists). By a transnational mining corporation. Of course.

Now, picture these mine owners. Pretend that one of them is thirsty. He gets a drink of water. Do you think it smells like bitter almonds?

Of course not.

Now, pretend you are a rancher in southern Colorado. Pretend your cattle are on the range. You check on them. You notice they will not drink. You look more closely at the Alamosa River, and more closely still. You notice something different. You can't quite puzzle it out. And then you understand. You see far less aquatic life than normal.

Life goes on. You keep checking the river. And over time a pattern becomes clear to you. You realize you are seeing a river being murdered. You see no aquatic life: no fish, no insects, nothing.

Somehow this does not surprise you, because you know about the Summitville Mine, a nearly-two-square-mile open pit mine: nearly two square miles of toxified landscape high in the mountains. You know that this mine tore open the mountainside, and you know that the absentee mineowners—mainly European investors, although Bank of America invested $20 million when it learned that the huge corporation Bechtel would be involved—did not follow the ethic laid out in *The Treasure of the Sierra Madre*, but rather walked away after wounding the mountain. You know that this landscape will remain toxified long after you are dead, long after your own bones have become no longer yours but have rejoined the earth, have become soil, become trees, become those others who live on this land (presuming they still live).

You want to know how severely these absentee mineowners have wounded not only the mountain—which you know they have wounded gravely—but the river. Studies are done. The government—always a better friend to the absentee mine owners than to you (after all, who pays the policymakers to make their policies?)—declares that all aquatic life was killed in a seventeen-mile stretch of the Alamosa River. You know that is not correct. You believe the local alfalfa farmer—who is in no way beholden to the mining corporation—who says, "There were 55 miles killed. Usually, the papers just mention the top of the watershed and not the residential areas that were contaminated. It affected our entire watershed including the river, its laterals and the stock ponds where all life was killed."

You know, don't you, that none of the absentee mineowners was poisoned. None of them were killed.

On the contrary, they, as always seems to happen, increased their wealth, at the expense of the mountain, the river, you, and your community.

Oh, and by the way, the total value of gold and silver taken from this mine is less than half of what the cleanup has cost *so far*.

Somehow this doesn't surprise you.[21]

🐝 🐝 🐝

The most intimate, fundamental gift we can and must and do give each other is our bodies. Our bodies are the most ancient, most vital of all gifts.

This gift is older by far than language, far older and deeper than words. It is as old as life itself, if not older. It is older than birth, perhaps older than death, as old as metaphor, possibly even as old as dreams.

This giving, and taking, of bodies, of flesh on flesh, flesh on stone, stone on root, root in stone, stomach on stone, skin, soil, is within and before all life, if there is a before all life. It is older than sex, older than that joining and unjoining then joining to become a third: different, new. It is older than this creativity. It is older than the bee, the pollen, the nectar, the pistil. It is older in this way than the wind. It is older than trees, older than ferns, older than algae. It is older than animals, older than mushrooms, older even than mycelia.

This gift is old. Old, like waves falling onto rocks, and rocks falling into waves. It is as old as mud, old as clouds, or even older. It is old, like water, like air, like everything.

Bodies shared, through sex, through touching, through breathing, through absorbing, eating, being eaten, becoming one or becoming another: all this is the gift of life, in all physical truth.

Bodies sustain us. They are us. They support us. We consume them. They become us, are us, as we become them, now, and later.

This gift, this support, this becoming, is where we come from, and where, whether we wish or not, we go.

▨ ▨ ▨

Think about the prominent "solutions" suggested to help curb the worst of global warming. What do they have in common? I'm talking about *every* major "solution," from those proposed by Al Gore (compact fluorescents, inflating tires, reducing packaging, and so on); to James Lovelock (nuclear energy); to Newt Gingrich (giving polluters tax credits to lean them toward voluntarily reducing their carbon emissions); to the various ideas proposed and promoted by scientists, such as the idea of dumping tons of iron, or alternatively, tons of agricultural waste—how conveeeenient!—into the ocean in the hope that this will cause algae to flourish, absorbing CO_2 into the algae's bodies and, by the way, doing god knows how much damage to the already-being-murdered oceans; or that of injecting sulfur particles high into the atmosphere to reflect sunlight back into space; or a further refinement of this idea, put forward officially by the United States government, to put giant mirrors in outer space to reduce the sunlight that arrives here; or (and I can hardly believe I'm not making up these obscenely and insanely stupid ideas, but each one has come from a "respected" source and received a lot of mainstream media attention) an idea pushed by NASA scientists to move the Earth farther from the Sun.[22] I never thought I would see solutions presented that would make me pine for the relative sanity of plants on Ford truck factories.

What all of these "solutions"—and of course the same is true for the "solutions" presented by people like William McDonough, Paul Hawken (who wrote *Natural* [sic] *Capitalism* and *The Ecology* [sic] *of Commerce*[23]), Al Gore, and nearly all of the so-called environmental intelligentsia (or "Bioneers" as some call them)—is that they all suffer a stupid and insane reversal of what is real. They all take industrial capitalism as a given, as that which *must* be saved, as that which must be maintained at all costs (including the murder of the planet, the murder of all that is real), as the independent variable, as primary; and they take the real, physical world—filled with real physical beings who live, die, make the world more diverse—as secondary, as a dependent variable, as something (never someone, of course) which (never who) must conform to industrial capitalismor die. Even someone as smart and dedicated as Peter Montague, who runs the indispensable *Rachel's Newsletter,* can say, about an insane plan to "solve" global warming by burying carbon underground (which of course is where it was before some genius pumped it up and burned it), "What's at stake: After trillions of tons of carbon dioxide have been buried in the deep earth, if even a tiny proportion of it leaks back out into the atmosphere, the planet could heat rapidly and civilization as we know it could be disrupted."[24] No, Peter, it's not civilization we should worry about. Disrupting civilization is a good thing for the planet, which means it's a good thing. Far more problematic than the possibility that "civilization as we know it could be disrupted" is the very real possibility that the planet (both as we know it and as we have never bothered to learn about it) could die. Another example: in a speech in which he called for "urgent action to fight global warming," and in which he called global warming "an emergency," UN Secretary-General Ban Ki-moon gave the reason he wants urgent action to combat this emergency: "We must be actively engaged in confronting the global challenge of climate change, which is a serious threat to development everywhere."[25] Nevermind it being a serious threat to the planet. He's worried about "development," which is in this case code language for industrialization.

This is the same perspective of those who do not hide the fact that they are grotesquely anti-environmental. Just recently, Bjorn Lomborg, the latest in a long line of writers who are paid well to deny or understate the damage this culture causes to the natural world[26] finally acknowledged that global warming is happening, and that it is caused by industrial civilization. But his next move was mind-numbingly predictable: he immediately shifted to the fallback position of saying that nothing can (or should) be done about it, stating that it is "somewhat silly" to think that this culture can change.

And it's not silly to harm or destroy the planet you live on?

As always, it is this culture which is primary, permanent, immutable; and the real world which is secondary, and which (rather than who) must bend to this culture's will.

Trying to force sustainability onto a functionally unsustainable culture causes severe cognitive dissonance, and makes people suggest absurd solutions. No solution can be too absurd so long as it fulfills its primary purpose of keeping us from seeing that the culture can never be sustainable, and that to attempt to sustain this culture is to harm the world. The real, physical world.

Any solution that springs from the (most often entirely unconscious) belief that the culture is more important than the world (or that the culture is real and the real world exists only as a backdrop and a source of raw materials) will not solve the problem.

⊕ ⊕ ⊕

Unless you're ideologically blinkered, irredeemably selfish, or just plain stupid, it's pretty easy to recognize that every action involving the industrial economy is destructive. And because the continued existence of the industrial economy cannot be questioned, much less threatened, and because we must *always* be disallowed from realizing that the problem is the culture, not us (just as in any abusive situation all people must always be disallowed from realizing that the problems are caused by the abuser, not the victims), many of us choose to "fight back" by decreasing our involvement in the industrial economy, by "living simply so that others may simply live."

So we eat less. We drive less. We do not own a car. We take shorter showers. We live more and more simply. We feel more and more pure.

We're doing what we know we can control.

Living simply is a good thing to do. Sadly, it in no way stops this culture from killing the planet. In no way is it a sufficient response to this culture's destructiveness. In no way is it a substitute for actively and effectively resisting actions and policies that harm our (and others') habitat.

I want to be clear. I'm not saying we shouldn't live simply. I live reasonably simply myself, but that's primarily because I only buy stuff I want, and I don't really want a lot (except I'd love to buy a lot of land to protect it, which would of course be analogous to buying individual slaves to free them, which doesn't alter the fact that I want to do it). But I don't pretend that me not buying much (or me not driving much, or me not having kids) is a powerful political act, or that it's deeply revolutionary. It's not. Personal change doesn't equal social change. It's not a significant threat to those in power, nor to the system itself.

Besides being ineffective at causing the sorts of changes necessary to stop this culture from killing the planet, there are at least five other problems with perceiving simple living as a political act (as opposed to living simply because that's what you want). The first is that it's fundamentally as narcissistic and as much a product of magical thinking as BaringWitness or orgasms for peace in that it substitutes private personal actions that accomplish very little in the real world, and a whole lot of wishing ("But if everybody lived simply . . ." they say, to which we can respond, "If we're going to fantasize about everybody doing something, let's fantasize about them demolishing the oil infrastructure to slow carbon emissions") for organized (or solo) resistance. Once again, I'm not dissing simple living. This book started with me shitting in the forest because it makes food for slugs. But I'm not going to trumpet that act as particularly political. Although it does help those particular slugs and the frogs who eat them, it's not going to slow global warming or stop plastics from being dumped in the ocean. Ultimately it won't even help these slug and frog communities, because unless the industrial economy is stopped, global warming and global poisoning will kill them.

The second is that it's predicated on the flawed notion that humans inevitably harm their landbase, in that it consists solely of harm reduction. The world is still a worse place than had you never been born, only this time it's not quite as bad as it would have been had you not been so pure. But humans can help the earth as well as harm it, and simple living as a political act ignores this. There are other things we can do as well. We can rehabilitate streams, we can get rid of noxious invasives, we can remove dams, we can shut down gold mines that are poisoning water sources, we can destroy the industrial economy that is destroying the real, physical world.

The third problem is that it incorrectly assigns blame to the individual (and most especially to individuals who have no particular power in this system except their ability to consume) instead of to those who actually wield power in this system and to the system itself.

The fourth problem is that it fundamentally accepts capitalism's redefinition of us from citizens to consumers, such that the "political acts" of the simple living "activists" are not the acts of citizens, with all the responsibilities citizenship implies, but are explicitly the acts of consumers. This redefinition is as wrenching, alienating, demeaning, disempowering, and wrong as this culture's previous redefinition of us from human animals in functioning communities to citizens of nation-states. Each of these redefinitions gravely reduces our range of possible forms of resistance. Human animals in functioning communities perceive themselves as having a wider range of forms of resistance to threats (both internal and in the case of functioning communities primarily external) available to them than citizens of nation-states, who perceive themselves as having a wider range available to them than consumers.

The fifth problem is that the endpoint of the logic behind simple living as a political act is suicide. If every act within an industrial economy is destructive; and if we want to stop this destruction; and if we are unwilling (or unable) to question (much less destroy) the intellectual, moral, economic, and at least as importantly physical infrastructures that *cause* every act within an industrial economy to be destructive, then we can easily come to believe that we will cause the least destruction possible if we are dead. Partly because it's true. The world would be better off without humans who do not actively attempt to stop industrial civilization from killing the planet.

No, that's not true. Whether or not we "attempt" to stop this culture is irrelevant. Results matter, in this case. The world would be better off without humans who do not actively and *successfully* stop industrial civilization from killing the planet.

Because the industrial economy is based on omnicide (and you thought it would never get around to consuming you?), to participate in this economy without proactively shutting it down is to be thrust into a double-bind, in fact into the double-bind to end all double-binds (in fact the double-bind to end all life). A double-bind is a situation where you are presented with two (or more) options, and no matter which option you choose, you lose, with the additional constraint that you cannot leave. If we avidly participate in the industrial economy, we may in the short term think we win because we may accumulate wealth, the marker of "success" in this culture. But we lose, because in doing so we give up our empathy, our animal humanity. And we really lose because industrial civilization is killing the planet, which means everyone loses. If we choose the "alternate" option of living more simply, thus causing less harm, but still not stopping the industrial economy from killing the planet, we may in the short term think we win because we get to feel pure and self-righteous, and we haven't even had to give up *all* of our empathy (only enough of it to not stop the horrors), but once again we really lose because industrial civilization is still killing the planet, which means everyone still loses. And unless you've found a way to leave the planet—which would be an odious abrogation of responsibility anyway—you can't leave. Except by dying.

The good news is that there are other options. The option I perceive as the most real, fundamental, necessary, and most importantly *life-affirming*, is not to die, but to get rid of the industrial economy.

⊛   ⊛   ⊛

Think about it. Your bathtub is overflowing. The water is running full-blast. What's the first thing you do? Do you begin to carry water outside in a teaspoon? Do you use energy-intensive machines invented at great taxpayer expense by your buddies at huge high-tech corporations and fueled by oil refined at great taxpayer expense by your buddies at huge oil conglomerates to suck up the water, transport it elsewhere, and dump the now-polluted water into what were once salmon-bearing streams? Do you plant native grasses on the bathroom floor, and call it a "new model for sustaining industry"?

Or do you turn off the tap?

If you're not insane, you turn off the tap.

But the problem is that the greedy abusive assholes—okay, so they're sociopaths, too, and immature, and so on—at the top do their damnedest to make sure the bathtub always overflows on someone *else's* floor. And because they don't care about others, they have no real incentive to turn off the tap.

This means someone else has to turn off the tap.

This analogy of an overflowing bathtub is true for carbon dioxide, plastics, and other waste products of this culture.

We need to turn off the fucking tap.

And while we're at it, we should pull the fucking plug.

❧ ❧ ❧

All this brings to mind Thoreau's observation that "There are a thousand hacking at the branches of evil to one who is striking at the root." Most incentives that motivate people aren't from programs, they're systemic. If you want to change a system that's profoundly broken, you don't tinker with programs, you get a different system.

If you really want to remove the incentives for being wasteful, you have to change this culture at the deepest level you can find, and then deeper than that, and deeper still. How deep can we go? Yes, individual people are wasteful because it's cheap, convenient, and they don't know any better. Corporations make a profit because you make more money by selling things that shortly thereafter become garbage. Governments benefit when corporations do this because it keeps the economy going and offers taxes.

These things only persist because the consequences of wastefulness are exported, like so many trash-laden barges, to seemingly distant places and times. If you want to stop the incentives for wastefulness, you have to stop the systems that export and postpone the disincentives. You have to bring the consequences home. And most of all, you have to stop the industries that manufacture waste in the first place.

❧ ❧ ❧

Industrial society has grown so much because people are systematically rewarded within it by increasing production. Within society that means more money and power. Between societies, this means that a society that produces more can outcompete or even conquer another. Labor efficiency is central to maximizing production, but land and energy are only important to the extent that they boost or interfere with production in the short term. All decisions are made in the short term, because in such a highly competitive society, any group that can't compete in the short term will not exist to make decisions later on. This is strangely appropriate, seeing as a society that makes all its decisions in the short term will not exist in the long term.

◉ ◉ ◉

There are also specific questions we can ask if we want to evaluate any particular technology for its neutrality.

First, there are questions about the prerequisites for the technology. How many people are required to make and run the infrastructure? Does the use of the technology presuppose the existence of other infrastructure like roads or energy distribution grids? Can the materials be obtained sustainably? Can human-scale communities with minimal hierarchy implement the technology? Or is a large, hierarchal society required to utilize it?

And second, there are questions about who can direct and make decisions about the technology. Is the infrastructure distributed so that it is under the control of many people, or is it centralized under the control of an elite? Does the technology allow people to become more autonomous and fulfilled, or do people surrender their autonomy to meet the needs of the technology? And who gets veto power? Can the technology be repealed or stopped?

And third, there are questions about the inherent effects of the technology. Are any benefits widely available? Are there barriers to accessing them? Is the technology useful to all, or only to an elite? And are the costs paid by the same people who get the benefits, or are the costs exported? Does the technology affect the degree of stratification and hierarchy in society?

Many of the technologies that we take for granted clearly fail the neutrality test and fall into the camp of authoritarian technics. Nuclear power and nuclear weapons are especially strong examples of this.

For nuclear technologies even to be developed requires the existence of a megamachine. Nothing less could marshal together the large numbers of people and raw materials required for an endeavor like the Manhattan Project. For example, as one of several massive complexes created for the Manhattan Project, the US government built an entire secret city called Oak Ridge to manufacture enriched uranium. Oak Ridge was home to the largest building in the world (which housed the first nuclear reactor), and by itself used *one sixth* of all electricity generated in the US.[27] Constructing the infrastructure for nuclear power is simply beyond the capability (and likely even further beyond the desires) of decentralized democratic communities.

Clearly all decision-making about the technology is highly centralized. That certainly goes for today, when much of the nuclear energy in the world is under military control. And it definitely went for the Manhattan Project—anything that is a complete state secret can hardly be called democratic. Furthermore, nuclear infrastructure requires a regimented hierarchy to function, and requires the people involved to sacrifice their autonomy to meet the needs of the technology and the social structure implementing it. In one illuminating case, a World War II–era Oak Ridge employee wondered, "Why on earth did they have all these high-school girls running this machinery? We could have blown up the whole of Tennessee!" Visiting the still active facility six decades after the bombs were dropped on Hiroshima and Nagasaki, she got her answer. "I was told that they wanted young women who would do what they were told and not ask questions. Really, we were just robots."[28]

What about the effects of the technology, the distribution of benefits and costs? It's clear that the purpose of nuclear weaponry, from the begin- ning, has been to cement and increase the power of the countries that created them—more accurately, to cement and increase the power of megamachines, a primary goal of all megamachines being to increase their own power and extent. In that regard, nuclear weaponry has so far been a success. So it certainly has benefits for those in power, though I think few people would say that the world is better off with nuclear weapons than it would be without them. And the costs, as usual, have been imposed on those with relatively little power. The Japanese civilians who were killed or injured at Hiroshima and Nagasaki. The Micronesian islanders (and other indigenous peoples) who were displaced or poisoned by nuclear testing on their territory. The many people in Iraq and other countries who have been poisoned by the use of non-fissile nuclear weapons like depleted uranium munitions. And, of course, all of the (human and non-human) people displaced, oppressed, exploited and killed by the megamachine so that it could create such weapons in the first place.

⊛ ⊛ ⊛

Many people would probably admit that nuclear weapons don't fall into the category of neutral technology, but might insist that it's an exception to the general rule. So let's choose another, less obvious example. In the interests of fairness, I'll simply go to the Technology portal on Wikipedia and choose a random page. (Oddly enough, the current featured article is "Nuclear Weapon.") And our next example? The steam locomotive.

All right, prerequisites. Metal is the single greatest infrastructural requirement for the steam locomotive. And metal means mining, both for ore and for fuels like coal to operate smelters and forges. In a number of very real ways, the steam locomotive—and all rail transit—is a product of mines. The metal that forms a locomotive, the steam engine that drives it, and the rails it travels on all originate from mines.

The first practical steam engines were invented only about three hundred years ago. Because of their large size and weight, they were stationary. The first and most important application of the steam engine was in mines. Since deep mine shafts are essentially gigantic wells, large machines were needed to continuously pump out water to allow miners to work. It was in this application that early steam engines were refined and developed. Steam engines weren't used in self-propelled vehicles until some seventy years later.

Rails themselves, originally made of wood, also originate in mining. Because metals are usually extracted from much larger amounts of ore, miners must find a way to move the ore to a place where it can be melted and refined. Almost five hundred years ago, miners in Europe started to use rails for this, though carts were hauled by hand or draft animals rather than by machines. Again, it was for mining that the railway was gradually developed—to carry ever-heavier loads of ore—until it reached the solid metal tracks we see today.

These different mining technologies converged to create the steam locomotive only two hundred years ago.

The steam locomotive also has social prerequisites, especially because of the extensive systems of railroads put in place for locomotives. Railroad construction in the North American West in the late nineteenth century involved tens of thousands of laborers at a time, most of them poorly paid and poorly treated Chinese immigrants, who had horrendous injury and mortality rates because of the dangerous work. Those railroads also presupposed the dispossession, displacement, and genocide of the many indigenous peoples whose territories they were built on. It seems fairly clear that building railway networks is the job of a large, hierarchical, and expansionistic society.

Question two, decision-making. Who controls steam locomotives? As a massive machine requiring extensive mining (smelting, etc.) and transportation infrastructures to build and maintain equipment, the steam locomotive is not a community-scale device. In North America a century ago, steam locomotive manufacturing was essentially limited to three specialized companies.[29] At the same time, it's not as centralized as our previous nuclear example. Even though locomotives and trains are themselves under the control of an elite, the infrastructure is very spread out. Individual communities cannot use that infrastructure for themselves, except by buying passage on a train. However, they can use—and have used—the distributed nature of railroads as a means of political leverage in disputes with empire. It's pretty easy to block the tracks. And when the going gets rough, railways—especially bridges—have been a favorite target of resistance and guerilla groups, since they are often unprotected and fairly easy to disable.

Category three, effects. Steam locomotives—and railroads—are a basis of an industrial empire. In a sense, they shrink the world, and allow governments and corporations to extend their influence, especially in military and industrial terms. They make possible and speed up the transport of large volumes of raw materials and other goods across long distances over land, for which railroads are still used today. And they allow the movement of heavy equipment and machinery over great distances, and to places that previously could only be reached by hauling over dirt roads or wagon trails.

Railways allow those in power to increase and concentrate their power ever more, both materially and socially. A railway brings distant resources within reach of the engines of industry, and can ship the engines of industry to those resources. By dramatically increasing the geographic scale and convenience of resource extraction, railways make possible the creation of much larger and more powerful megamachines.

Let's talk briefly about two other consequences of steam locomotives, to show how deep and wide can—and will, and often must—be the consequences of a piece of technology. The invention and use of any piece of technology will of necessity affect many other parts of any culture, because for better or worse, technology and culture are intertwined: certain attitudes and social mores and reward systems will lead to certain technologies; and certain technologies will lead ineluctably to certain changes in perception, experience, behavior, and personal and social (cultural) mores and systems of rewards. The first of the two social consequences I want to mention here is that railroads led to the standardization of time zones. Prior to railroads, different communities kept their own time. Standardization wasn't necessary because by the time someone walked three days from one town to another, whether the time was now 8:15 or 8:34 in the evening rarely was of consequence. But with railroads came railroad timetables, and with timetables came the necessity for one community's 8:15 to be another community's 8:15. Now, superficially, this may not seem important to you, but the point is that the existence of railroad locomotives requires—as do so many technologies—standardization, even in parts of our lives where we might not expect.

The second consequence to mention here is that railroads really were the prototypical modern corporation. The invention of steam locomotives was tightly tied to the rise of corporations. Does anyone still want to argue that technology is inherently neutral?

A common thread in both the nuclear and locomotive examples is that both require a large, hierarchical society in order to develop. And in both cases, the effects of the technology are to benefit those at the top in their efforts to make megamachines even larger, to gain more power. Now, we can get into an argument about whether making bigger megamachines is good or bad (though if you're still reading, that's probably not going to be an issue), but it's clear that such an act is far from amoral.

The issue is complicated slightly by the fact that the development of technology is an iterative process: new technologies are built upon old ones, often stacked so deep that the foundational technologies become invisible and unquestioned. When people talk about technology they almost invariably mean something from the past few decades, not the technological bases of industry or agriculture.

Because of this stacking, we're sometimes put in confusing and morally complex situations—technics that could be somewhat democratic may be based on deeper authoritarian technics. While I like the idea that the internet can be used for rapid, long-distance communication between equals, the internet and all electronic communications are based on

deeper, authoritarian technics like mining and industrial manufacturing. And despite the tendency of some academics to refer to this age as "post-industrial," the character of modern civilization is determined much more by its clanking industrial foundation than its digital veneer. Not only did the manufacture of my computer require the ecological and social damage caused by mining and manufacturing, the social conditions required for its manufacture emerge from and lead to the dominance of authoritarian technics. Because of this, even the most theoretically liberatory technologies, like the internet, end up being dominated by corporations and governments that use them to advertise to, propagandize, spy on, and track people. We should also never forget that although we may be able to use computers to quickly communicate with others of those who oppose the destruction of life on Earth, computers are used more efficiently by those who are killing the planet. Indeed, much modern commerce and war (and surveillance) would be impossible without computers.

None of this is to say we should shun potentially democratic—or even potentially useful—technics that are based on authoritarian ones. We can use them judiciously while keeping our end goals in mind. We can disillusion ourselves of myths of a magical and painless conversion to a non-authoritarian system using identical technologies. We can rid ourselves of the narcissistic notion that our personal purity (spiritual or otherwise) is more important than conditions in the real, physical world, including life on earth. And we can use the most effective tools and techniques to which we have access in order to systemically dismantle the underlying authoritarian system.

⊕   ⊕   ⊕

A great many books on environmentalism written in the last few decades conclude with a listing of things that you, as a citizen, or more likely as a consumer, can do to address the problems. In your capacity as a consumer, you can reduce your consumption or buy products that are allegedly more eco-friendly in order to convince corporations to enact change. As a citizen, you can write to "your" congresspeople or other governmental "representatives," and ask *them* to enact change. If you want to do one better, you can donate to a non-profit organization that will lobby governments and corporations on your behalf. Authors may offer a plethora of different vicarious solutions involving various ways to try to persuade large, entrenched institutions to act against their underlying drives.

We're not going to do that.

If you've gotten to this point and you are yearning for a way to reduce your personal use of disposable packaging or compost more of your household waste, there are already hundreds of books with tips on exactly those subjects. It's been done. And we're not saying you shouldn't try to reduce your production of household waste. Minimizing waste is certainly a good thing to do, and we don't want to insult that group of people (which includes ourselves) who have taken steps to reduce their waste. And obviously we aren't saying you shouldn't donate to non-profit organizations, especially local ones. We *will* say that in general you shouldn't give a dime to big corporate "green" organizations. One example why: Jay Hair, former head of the National Wildlife Federation, immediately went from there to becoming a spokesperson for Plum Creek Timber Company, a timber company so nasty even a Republican called it the Darth Vader of the timber industry. Such is business as usual among the big corporate "green" organizations.

What we're are saying is this: we aren't going to insult your intelligence by asserting that such solutions are even remotely sufficient to address the problem. They are a drop in the plastic-suffocated ocean in terms of real change. We don't have the time or patience to immerse ourselves in a fantasy world where corporations and governments act in ways that contradict their own fundamental imperatives and immediate self-interest because we send them politely-worded and well-researched letters. And we aren't going to blindly swallow the premise that you, the reader, are a mere consumer, taxpayer, or even citizen. Your identity, your being, is not limited to your economic function in relation to some vast bureaucracy. You are a human being, an animal: whether you recognize it or not, you are a living creature embedded in a network of trillions and trillions of other living creatures, all interdependent.

In a way, that's really what this entire book is about. Is your identity that of a consumer, or a person? When you say "the real world" are you talking about wage slave capitalism, or are you talking about a living breathing world of trees and rivers and lakes and deserts and forests and mountains and seas? And if you identify as a living member of that much larger community, a community that is being systematically destroyed by a toxic mimic of the real world, what are you going to do about it? How are you going to defend your community?

⊛ ⊛ ⊛

There are those who disagree that the world needs defending. I'm not talking about the people who think everything is fine, or those who wouldn't understand or care if every fish in the ocean were killed. I'm talking about people who recognize the scope and seriousness of our predicament, who recognize that the problems we face are deeply rooted and systemic, but who don't think that they in particular need to do anything other than "walk away." Of course, as was already discussed, this is not nearly enough, and we should not pretend it is. Those who would truly walk away, who aim to abandon the dominant culture completely, clearly recognize that this culture is fundamentally irredeemable. And they presumably recognize that civilization's voracious industrial appetite is eating up the planet at an ever increasing rate. So why do they view walking away as an adequate strategy? If this culture is not going to change, where do they expect to be safe? What do they expect will happen as industrial society exhausts its last remaining resources? If this monstrosity is not stopped, the carefully tended permaculture gardens and groves of lifeboat ecovillages will be nothing more than after-dinner snacks for civilization.

I think the problem is partly a lack of historical perspective. It's not as though living outside of civilization is new, after all. A little more than five centuries ago, North and South America (and Africa, Oceania, and Asia; if we go back 2,000 years we can add Europe, and if we go back 6,000 years we can add the Near and Middle East) were filled with tens of thousands of uncivilized communities. Any five-year-old child among them would have been better at finding wild edible foods than I will ever be. The Confederacy of the Haudenosaunee was a proven and effective participatory democracy that I can only dream of emulating in my community. These continents abounded with warriors who were skilled and courageous beyond my conception. And yet they were all but wiped out by the insane civilized (whom they significantly outnumbered, at least at the beginning) using technologies that are hopelessly crude by modern standards of conquest and genocide. Do civilized people who walk away, many of whom are essentially novices both to living in a healthy community and living with the land, believe they can survive so much better than entire indigenous nations with countless millennia of uninterrupted experience?

I don't mean to sound overly pessimistic. I don't think we're completely doomed. We do have some advantages that those living centuries ago did not. We recognize how deeply pathological this culture is, we recognize the need for it to be dismantled entirely, and we can understand that in a way for which indigenous Americans of the fifteenth through nineteenth centuries (and others of the indigenous) simply didn't have the context.

They were taken off-guard by an enemy more hideous, insatiable, and cruel than they had ever encountered, could reasonably anticipate, or even imagine. We no longer have to imagine. We need merely pay attention. We also have the benefit that modern society is more monolithic, more dependent on a small number of centralized industrial and economic systems, more brittle, and more vulnerable to collapse. And we can use those systems to our own (and the world's) advantage, both by employing and disrupting them.

But if we choose solely to walk away, we give up these and other advantages. If we choose not to fight back, we concede most of the few slim possibilities we have for success, let alone survival.

Like most decisions people make, especially life decisions based on complex or unpredictable situations, the decision to just walk away is not one based on reasoned analysis. And because of this, I think the motivation has much to do not just with identity, but with what Dietrich Bonhoeffer wrote: that action comes from a readiness for responsibility. To succeed in stopping the destruction of the planet, you have to be ready to take responsibility. Not to belittle responsibility by pretending that solely personal actions aimed primarily at protecting you and yours can solve vast problems. Not to surrender responsibility to governments and businesses which claim to act on your behalf or in your best interest. Not to renounce responsibility by pretending that walking away from the destruction will somehow cause it to stop.

You have to be ready to take responsibility to defend your community. And when your community (by which we mean your landbase, the living

Earth, your human community, and your own body) is in danger—no, not just in danger but actively under attack—that means fighting back.

⊛ ⊛ ⊛

There's no doubt, we're in a serious situation that requires a serious response. And though we shouldn't unnecessarily provoke those in power, we must recognize that effective political strategy will meet with reprisals from those in power regardless of the specific tactics used. It can be frightening to think of those reprisals targeting you—that's the whole point after all: it's a form of government-sponsored terrorism (and one could easily argue that civilized governments *are* a form of terrorism)—but in the long term (and by now the short-term), those in power are destroying the world. What do we have to lose? If we make them really mad, what are they going to do, destroy the earth twice?

Some people will not resist. Some will actively collaborate. Perhaps they benefit, at least temporarily, from civilization's hierarchy. Or perhaps those in power have determined the precise measure of empty promises that most people will tolerate, calculated the "exact measure of injustice and wrong" that they can get away with. But for the rest of·us, our job is to devise and enact a plan—many plans, actually—that will make those in power afraid, not just for themselves, but for the entire wretched system that keeps them in power.

So what do we need in order to do that? First, to quote Lierre Keith: "What any movement needs is an effective strategy. That means identifying two things: where is power weak, and where are you strong? The overlap is where you strike. One problem with nonviolence is that it depends on huge numbers of people to be effective. Rosa Parks on her own ended up in jail. Rosa Parks plus the whole Black community of Montgomery ended segregation on the public transportation system. Without a mass movement, the technique doesn't work."

So that's actually two things we need so far: an effective strategy, and an effective strategy that's actually congruent with the numbers of people we have, the resources we have, and the time we have. That helps to narrow things down. Though a wholly nonviolent mass movement bent on systematically uprooting the fundamental causes of human exploitation and ecocide would be wonderful, it falls short in our case as a valid strategy. Currently, there simply aren't enough people willing to address the issue. And worse, there isn't enough time to build that movement—each day that passes means hundreds of species wiped out forever, means more landbased cultures destroyed by the industrial onslaught, means global fossil energy consumption brings us closer and closer to a runaway greenhouse effect, means more plastic and fewer fish in the oceans, means fewer amphibians, and so on, ad omnicidium.

A third asset we need is a collective recognition of the real systemic roots of the problem. As discussed earlier, the garbage problem will not go away because you or I stop producing garbage. And global warming will not go away because you or I stop using gasoline. Those problems will not go away as long as there is a global industrial system that produces waste and burns gasoline. In fact, it's conceivable that in the coming decades a focus on reducing personal consumption could even make things worse. I'm not talking about worsening caused by a focus on more symbolic action at the expense of more effective action, although that's certainly a valid argument. Rather, I'm talking about the fact that we're entering a post-peak period for oil, and for many other commodities. If demand for oil far outstrips supply, and we all decide to get together to reduce our consumption of gas for altruistic reasons, we will reduce demand for the finite supply of oil. The net effect of this "green" action will simply be to make the remaining oil cheaper and more readily available for militaries, corporations, and other institutions which lack our scruples. Which, again, isn't to say we shouldn't reduce our consumption, but we should do it because it's the right thing and not because we expect it to topple those in power.

A fourth prerequisite for effectiveness is a culture of resistance. This should not be confused with an "alternative culture." Instead, a culture of resistance is an explicitly oppositional culture. An effective culture of resistance does not seek a "cultural revolution": cultural change is not the objective, but material change, accomplished though the organized work of a large and diverse group. A culture of resistance is, collectively, that group of people with an understanding of the root causes of their predicament, and a willingness to work together in opposition to authority to address those causes.

A culture of resistance is not the same as an organized resistance movement, but is necessary for the success and growth of such a movement. In every country where a successful revolution has taken place, there has been a culture of resistance. In every occupied nation with an ongoing resistance movement, there is a culture of resistance. If a resistance movement is a sturdy tree, a culture of resistance is the soil from which it grows, a soil itself enriched by the growth of the tree.

There is a story about a member of an Irish resistance group in the early twentieth century. One night, while carrying out resistance activities, he was discovered by the British and shot as he escaped. The man was wounded, but managed to hide in an alley and avoid discovery by the British. Later that night, a group of men on their way home passed through the alley and found the injured man. Though not active members of the resistance, they immediately recognized what had happened and brought the man to a doctor and safehouse. They didn't need to be told to do this—they knew, because their culture was a culture of resistance.

Such a culture benefits from shared goals and group norms that allow the culture to propagate and persist, and gives rise to effective tactics and strategies. Solidarity and mutual aid, such as in the above example, are one important characteristic. Further, those in the culture acknowledge and support a broad diversity of tactics and involvement, with the understanding that they're all working toward the same goals with the same general strategy. This permits individuals and small groups to focus on the projects and tasks they're best suited for, as well as limiting risk to the entire group while supporting those in the most high risk positions.

Here's what I mean by that. In any given army, only a tiny percentage of the army is actually involved in fighting. (For example, in 1918, just before the Irish War of Independence, the IRA had about 100,000 enlisted members, but only about 3,000 of them were actually fighters at any given time.) The rest of those involved participate in logistical and support roles, doing recruitment and training; communication; logistics like obtaining, manufacturing, and moving materiel; medical support; and even things as basic as feeding the troops and maintaining equipment. And many of those who fight aren't professional soldiers. An army commonly consists of a core or skeleton of professional officers and non-commissioned officers. When war is declared, that skeleton is fleshed out by conscripts, reservists, or civilian militia.

Guerilla or resistance movements are likely to have a similar dispro-portion between the number of people who actively carry out operations, and the people who support them. Relatively few members of an armed resistance movement actually take up arms as active guerillas. But in order for them to succeed they require a much larger support network of sym-pathizers, fund raisers, above-ground political agitators, reconnaissance workers, and those who offer direct material support such as food and shelter. Even nonviolent movements are likely to have a parallel struc-ture. Those people who put themselves in harm's way through civil dis-obedience or direct action—be they forest defenders in tree sits, Project Ploughshares activists smashing military hardware, or indigenous people blockading loggers or miners in their homeland—ultimately rely on those who can offer support for prisoners and their families, medical aid, aware-ness raising, and material support. (For example, when people think of treesitters, how many of them think about the people who bring them food and water, and carry away their shit buckets, people without whom the treesitters would not be able to last more than a few days.)

When the issue of fighting back comes up, I sometimes hear people argue that we mustn't *fight* back, because those in power will only rebuild, or because they'll only increase their repression and violence. That this concern is considered by some to be a valid reason for inaction tells me many things. One of the most important things it tells me is that the people asking that question do not live in a culture of resistance. The question of how those in power will respond to different actions is certainly strategi-cally valid. But in a culture of resistance, it's not a reason to not resist by whatever means are most appropriate and effective. Of course those in power will inflict reprisals on those who resist them. Of course they will try to frighten and terrorize dissidents into accepting their authority. Of course they will try to harm even those who do not directly participate in actions against power. *This is not a reason to hold back—this is why we fight them.* In a culture of resistance, reprisals and state terrorism are certainly not trivial, and they are not ignored. Instead, they underscore the impor-tance of resistance, and strengthen the resolve of those who fight back.

☙ ☙ ☙

Here are some questions that anyone contemplating serious action should ask themselves.[30] What are the risks if you take action? (Loss of status? Status? State reprisals? Prison? Torture? Murder by the state?) What are the risks if you don't? (A freefall slide into fascist dystopia? Runaway global warming? The collapse of the biosphere? Loss of self-respect?) What would you need from yourself, from your friends, your family, your community, your institutions to make action more possible? (Moral support? Material support? Familial support? Collaboration?)

Where do your loyalties lie? Where do you end, and other creatures begin? What will be your legacy? What do you want to leave behind?

What do you need, and what do you have to give up, to make that happen? And if you don't do it, who will?

Knowing the answers to those questions, having discarded the paralyzing mythologies of those in power, choose your future, and fight for it.

Songs of the Dead
A Novel

PM Press, 2009

Ten years before I wrote Songs of the Dead, I woke up one morning to see three paragraphs on a piece of paper. Evidently, I'd woken up in the middle of the night and written them. I had no recollection of doing so. I didn't know where they came from, or what they meant, but I knew I had to structure a book around them. They became the first three paragraphs of Songs of the Dead. I didn't even edit them.

It took ten years for the plot to develop. The line that really called to me in those opening paragraphs was about falling through time. I lived right near a place in Spokane, Washington, called Hangman Valley, called that because during the white theft of the area from the Indians, the Indians were fighting back and a white colonel—Colonel Wright, who became a hero—called Indians in to parlay under a flag of truce and then hanged them. There was a golf course named after one of the hanged Indians. This same valley also had streams that used to be salmon bearing and no longer were. When I lived there, a serial killer dumped the bodies of women onto this golf course. So, in this same place, I saw everything that is wrong with this culture: genocide, ecocide, and gynocide . . . and golf. I had to write a book that took on all of that.

Each night, I walk the line that wends between unconsciousness and terror, between forgetting and remembering, between present and past. Each night I do not fall asleep but instead stumble through time, falling into deep impressions—like five-pointed handprints on soft clay—of past on present, living in house after house after house of imagination, each one an edifice of events uncompleted. Does the land dream so, too, carrying with it the weight of thousands of years of nights on nights, remembering salmon that were and are not, caressing them in the infancy of their evolution and caring for them in their absence? Does the land mourn these losses as I mourn my own, and does she—it, he, pieces of moist soil between my fingertips, the orange bellies of ponderosa pine four arm lengths around—dream as well of times unwounded, and of woundings? Does time wind and unwind for her—for I know now it is her—each night as she sleeps beneath snow, stars, cold wind, trees sighing sadly or giving up their own ghosts before meeting what we have become, beneath a moon that night after night sees all, yet keeps remembering?

I know now that there is and always has been a heart that beats beyond the grasping of our mechanical fingers, unfound in the claws of our braced backhoes, slipping away in the face of our too-coarse bulldozers. The past resides in the soil, and though we believe it blows away and is lost, that is not true. It is there all the time, though we do not see it.

Our dreams carry with them the perfume of this soil, and will not without a fight let go of that which beneath it all makes each of us who we are. So each night I walk that fine line, and sometimes awaken to freeze before all that has happened to me, to her, to each of us, and to wish that things could be different than they are.

There's a fire somewhere. I can smell more than see it, but my eyes trick me, with a slight sting, into pretending that I see the smoke. I don't, of course, except when I do, and even then, like all of us, I'm never sure if what I see is what I see.

There *is* a haze in the distance, but it's just the sky settling back to earth at the end of the day. It's July, and it's hot. I'm sweaty, wet beneath my arms, on my lower back where my shirt touches my skin, and under the elastic band of my underwear.

I'm in Hangman Valley, in the western part of Spokane. I'm walking, as I often do, near Hangman Creek, which used to be Latah Creek before any of this began, and certainly long before any of it began with me.

Or maybe not. That's one of the things I often have difficulty with. Before. During. After. Sometimes I don't understand what any of it means.

But it's hot. I understand that. It's hot enough that the leaves on the trees hang limp, except when a hot breeze makes the air quake with their paper rattling. Edges of these leaves are turning brown, and the grasses beneath have long since died or gone dormant, used up for and by the summer, and dry as tinder. Even the needles of the pine trees seem to have lost their strength and their shine.

It's cooler by the creek, though not as much cooler as I'm sure it once was, back when the creek was a creek, deeper, wider, stronger. I go there often. It's a reasonably long walk from my home—probably a couple of hours—and a longer walk back since I have to go so much uphill.

I sit by the creek, take off my shoes and socks, roll up my pants, and put my feet in. I lean forward to search for tiny fish. None. I close my eyes, then open them again quickly, just to see if this will make the salmon appear. I know that's not how it works, but it's never stopped me from hoping. And sometimes I do see them. They haven't been here since the Grand Coulee Dam was built back in the thirties, but sometimes I still do see them.

      🐾 🐾 🐾

There is a woman. She takes a shortcut through an alley. She is thinking, or not thinking, but seeing inside of her what she saw that morning, which was a puppy she gave her son for his birthday four days before. When the puppy wagged his tail he did not so much wag his tail as wag his whole body when he squirmed toward her son, who in turn did not smile so much with his lips and teeth as he, too, smiled with his whole body. This is what she is seeing when she hears the sound that is not a sound but the movement of a sound throughout her whole body, the sharp cracking of lightning as it strikes inside her brain, but does not stop after the bolt has gone; it keeps expanding outward until there is nothing left of her skull and of what was inside her skull, and she is flying, having been struck, and there is nothing but the sound that keeps expanding, and no longer can she see the puppy or her son or anything but the sound that is no longer a sound, but everything she knows.

That is what I hear. When I walk where the car struck her, that is what I hear.

Not every time. But often. And if the truth is that while I see salmon not nearly often enough, this I see far too often.

⊛ ⊛ ⊛

I haven't always seen like this, and even now I often do not. I used to not see anything more than anyone else, or maybe I should say not more than any of my neighbors, or maybe I should be even more precise and say not more than any of my human neighbors. I think nonhumans—and some humans—see this all the time.

For example, just a few days ago a huge submarine earthquake caused a tsunami that rocked parts of Indonesia, Sri Lanka, Thailand, Malaysia, and India, killing more than a hundred thousand humans. Just today I read a news report saying, "Wildlife officials in Sri Lanka expressed surprise Wednesday that they found no evidence of large-scale animal deaths from the weekend's massive tsunami—indicating that animals may have sensed the wave coming and fled to higher ground. An Associated Press photographer who flew over Sri Lanka's Yala National Park in an air force helicopter saw abundant wildlife, including elephants, buffalo, deer, and not a single animal corpse." The response by one person was, "Maybe what we think is true, that animals have a sixth sense."

I'm not saying I have a sixth sense. Sometimes I'm not even sure about the other five, and my girlfriend Allison will tell you I sure don't have much of the common one. But I see things, and hear things. No, I see places, and I hear places. Places where I'm standing. Places where I'm sitting. Places where I'm sleeping. Sometimes I hear what the place says to me.

It's not something I can force, by any means. It just happens. It used to scare me more than it does now, but even now I do not understand it, and even now sometimes it terrifies me.

⊛ ⊛ ⊛

Kristine looks at her watch. Time to go to work. She opens her wallet to look at the mirror inside. Not great, she thinks, but good enough. She runs her hand through her hair, feels the slight stickiness of her scalp and the texture of her hair made thick and brittle, like straw, by dirt, sweat, and hairspray. She looks again at her watch. Yeah, there's time, she thinks, there has to be time. Otherwise she's never going to make it. She rummages through her canvas bag of clothes, but can't find what she's looking for.

"Fuck."

Kristine keeps digging. She sees a black tube top and realizes she hasn't worn it for a few days. She remembers the tip she got the last time she did. She could use the money. Maybe it's a lucky shirt. She puts down the bag, unbuttons and pulls off her fuchsia blouse, stuffs it into the bag, and shimmies into the tube top. She looks again in the mirror, and again she runs her hand through her hair.

Back to the bag. She finds a small black chunk of heroin wrapped in plastic, along with a pocketknife, syringe, bent spoon, and a lighter. She unwraps the heroin, and the stench makes her salivate. She uses the pocketknife to scrape a little into the spoon. Not much, just enough to remove the edge. Then she pours in a little water and stirs the mix with the tip of her needle. She flicks the lighter, holds it under the spoon. The tar dissolves. Using the cotton ball as a filter, she fills the syringe. She sits cross-legged on the ground, then extends her right leg while keeping her hips open so she can see the back of her knee. The needle finds its own way into her vein, and the plunger finds its own way down.

She feels good. Not so good she can't move or do anything but stay here under the bridge—just good enough that now she can go to work.

◈ ◈ ◈

Nika is awake, but the apartment is silent, so she lies in bed with her opened box of memories. So long as she keeps her eyes closed and doesn't move, doesn't hear anything, she can pretend she's in bed at home, that she is somewhere and someone else, a world away from where and who she is now. This is how she gets through each day. She takes each memory out of the box, holds it, turns it around and around in her mind, tries to re-create its feeling in her body. There's her little brother Petya playing with his dog in the field behind their home, and there are the flowers in the field. There is the sun on her shoulders as she watches. Even the sun somehow felt different then: it's hard to believe it's the same sun shining now. There is her mother giving her the pendant cross given to her by her mother, whose mother gave it to her. There is the feeling of her mother's fingers on Nika's neck as she attaches it, the smell of her mother, the smell of the kitchen. There is her father's smile as Nika tells him her marks at Lyceum.

She lies there comfortably, almost drifting, almost smiling, as image after image bubbles up. Blood sausages with her grandmother. Bathing her great-grandmother, cutting her hair, clipping her toenails, listening to her stories of the German occupation and holding her when she got confused over what year it was and thought the Nazis were coming to the door. Nika remembers her first kiss with her boyfriend Osip, how neither had known what to do but had learned so quickly and easily. She remembers watching Petya practice ballet.

She hears steps on the stairs outside the apartment, and starts to put the memories back into their box, starts to shut it up tight and lock it. But then, as so often happens, another memory forces its way before her. An advertisement in a newspaper. She sees the newspaper as though it's before her now, sees the ad circled in red pen. A secretarial job in Vilnius. She remembers begging her mother to let her have this adventure the summer before she begins her college, and her mother and father finally approving. She wishes she would have not chosen that day to look in the newspaper, wishes she would have listened to the dreams that told her not to go.

The steps are closer on the stairs, and she needs to conjure another memory before she can shut the box. She cannot bear to end on this one. She searches, her eyes moving below her eyelids. Seven years old, she thinks. Eight. She needs to find something good. And then she remembers. Six years old. Christmas. A gift from her parents. Normally the gifts were simple and necessary, like pencils or notebooks. But this time she rips apart the paper to find a toy drum. Her parents smile as she bangs on it. She almost laughs now as she wonders whether a few days later they were still smiling, or whether they regretted bringing all that noise into their home.

The footfalls cross the hallway outside the apartment. The front door opens. She locks her box of memories, closes her eyes tight, then opens them wide. She's not in Russia. She's in the United States. Spokane, Washington, in a shitty apartment just off East Sprague. The door slams shut. She hears Viktor's voice, in Russian, as he shouts, "Nika, you lazy slut. Get up. It's time for work."

Kristine gets out of the car, done with her first john of the day, a regular who likes things, as he says, vigorous. She walks to the corner, sees Nika putting her pendant around her neck. That means she must have just finished a job, too: she never wears it around men.

"Hello, Kristine."

Kristine nods, smiles. "How are you doing?"

"I'm making some money."

"Viktor letting you keep any?"

"He says I'm not making enough to keep him happy."

"Fucker." Nika makes more than anyone else Kristine knows. She's what men want. She's young, blond, pretty, slender but not crack-thin. She doesn't use. She's quiet—you have to strain to hear her speak.

Kristine doesn't know much of Nika's history—the woman doesn't open up to anyone, at least to Kristine's knowledge—but she presumes from the accent and the shared apartment that Nika was part of a big shipment of women from Russia by way of Lithuania and then Amsterdam.

Kristine envies Nika. Certainly not her being so far from home, but half a world or half a continent, does it really matter? Besides, "home" was the last place Kristine would ever go again. At least here she gets money for her services.

Nor does she envy Nika's looks or figure. She knows how long they'll last. No, she envies Nika's ready access to a shower. In order for Kristine to bathe, she has to convince a john to rent a room, then afterwards take a quick shower and put on makeup before heading back out to the street. She sometimes fantasizes about a long hot bath, with soap and bubbles and bath oil in those squishy, slippery marbles that slowly dissolve. She could live in a house or apartment—and she has spent a fair amount of time in squats, though of course that doesn't solve the shower problem—but then she'd have to put up with the other women, and especially with the pimp.

Kristine asks, "How much does he say you owe by now?"

"I'll never see the end of this. The more I make, the more I owe."

"You and me both, sister."

"Who do you owe?"

"My dealer. You've got Viktor, I've got heroin."

Nika looks at her for a long moment, then to the ground.

Kristine continues, "At least the heroin makes me feel good."

"And it won't kill you if you run away."

Kristine laughs. "Oh, it will kill me all right if I try to leave. I've done that a couple of times, and it came right after me to bring me back."

Nika is silent.

Kristine says, "I don't know how you do all of this sober."

"The tricks?"

"All of it. Look around. Do you ever actually look at the people? Not just the johns. All of them. They're as dead as we are. Only we've got the sense to know it. And the cars. Do you ever notice the air? It tastes like shit. No, it doesn't. I grew up on a farm, and this smells far worse than shit."

"I grew up . . ." Nika trails off.

Kristine doesn't look at her directly. She wants to know more about her friend, but knows if she says the wrong thing she'll scare her away. The silence stretches longer.

Finally Nika says, "In the country."

More silence. Kristine wants to ask where, what it was like, who was her family, but doesn't know where to start. So she does what she knows is best. She lets the other be.

Nika says simply, "I'm never going home."

Kristine knows better than to disagree directly. She says, "It has happened before. Some women have made it." A pause before she continues, "Do you want to go home?"

"More than . . ."

A car slows, pulls up to the curb. It's one of Kristine's semi-regulars. Kristine says to Nika, "Fuck. I'm sorry. Maybe later?"

The man opens the passenger window, leans across, says, "Hey, Kristine, who's your pretty friend?"

Kristine senses money slipping away, and wouldn't mind if it were slipping to Nika. She would mind it going to Viktor.

Nika comes over to the car. The man looks from her face to her breasts and back to her face. He does the same to Kristine, then says, "I'd forgotten how much your shoulders turn me on. Same price? Get in."

The street is hot, and empty. No people, no cars. Nika paces back and forth, facing then going with the nonexistent traffic. She doesn't see the truck pull up next to her, and jumps a little when she hears it close by. She turns, looks at the man inside. His passenger window is already down.

She walks to his vehicle.

He says, "Would you like to party?"

"What do you have in mind?"

"Depends on the price."

"First," she says, "you've got to show me something I don't have."

The man has done this before, knows the game. Cops can't expose themselves. He unzips his pants, pulls out a nondescript penis.

She licks her lips. "Very nice," she says. Make the sale, she thinks. "Well?"

"Makes me want to drop my price. For you I'll do a blow for twenty-five, a lay for fifty, half and half for sixty, and for a hundred you get me for an hour."

"That's a discount?"

"That's my discount." She pushes back from the truck.

"No, wait, here." He pulls a couple of fifty dollar bills from his shirt pocket.

She puts the money in the front left pocket of her tight shorts, pulls the pendant from her neck, puts that in the other pocket, and gets in.

The man says, "Buckle up. I don't want to get a ticket."

She does. He begins to drive. They make small talk. He asks her name. She tells him. She asks his name, and he gives her one she knows is false. He asks her other questions and she lies, too. He doesn't pull into an alley like she was expecting, but drives around, as though uncertain what he wants to do next.

Finally she says, "You'll need to pull over if you want me to do you."

The man just says, "We've got time."

She thinks, It's your money.

Then the man says, "Do you believe in God?"

She doesn't say anything. She tries to read what he wants, give it to him. It is safest—and makes the most money—if you give the man what he wants before he asks. But he already asked, and she doesn't know how to answer.

"Do you," he repeats, "believe in God?"

She frowns, then says, "Do you want to fuck?"

"Look," he says, "I bought you for an hour. If I want you to answer my question, you'll do it. Do you get it?"

"Yes."

"So . . ."

She remembers that once, long ago, she did believe, and still does enough to wear her mother's cross. But that's in memory of her mother, not Jesus. And it was two years and fifty lifetimes ago that her mother gave it to her, and now both Jesus and her mother are too far away to help. She says, "Yes, I believe—"

He cuts her off. "Oh, I get it. You're afraid I'm some sort of fundy and if you say you don't believe that the Lord Jesus Christ died for you I'll spend the next hour trying to save your soul. Well, I don't believe in souls. I'm a scientist, and so it's against my religion to believe in superstitions." He laughs at his own joke, then says, "It's your body I want."

She doesn't understand his joke. She says, "Should we stop here? We can go down this alley."

He reaches with his right hand into his shirt pocket, pulls out two more bills, and says, "Instead of buying an hour I'm buying two. Let's go somewhere private."

She takes the money. "There's a hotel on North Division, just a few blocks. We can get a room."

"A room? Where other men have fucked you? And even if they didn't fuck you they fucked someone else and left their sperm on the sheets. It doesn't wash out. It leaves traces even after cleaning. Do you think I want some man's DNA all over me?"

She doesn't say anything.

"Do you?"

"Where, then?"

"South of town, a nice little park where we can get out of the car."

Again, she doesn't say anything.

"So like I was saying, I'm a scientist. I look at things from a scientific perspective. That doesn't mean I'm anti-Christian, though. That's a mistake a lot of scientists make. The truth is that science and Christianity are two sides of the same coin."

She tries to look interested. If he wants to spend his money lecturing her, she'll take the money. Maybe he'll buy her something to eat.

He continues, "Both of them are attempts to explain the universe, attempts to explain what is. They're both articulations of systems of power. They both tell us how to live, how to experience the world, how to be in the world. They tell us how to relate to each other. Do you see?"

"Yes," she says, wondering what science and what religion would cause a man to pay to fuck a woman, what science and what religion would cause another man to force a woman to have sex for money and to give that money to him. What sort of science and what sort of religion would cause people to value money over another's freedom or happiness? What sort of science and what sort of religion would cause someone to want to wield such power over another? She says none of this, shows none of this on her face. There are very few men she does not hate.

He says, "There's one line from the Bible I've always especially liked, a line that says everything we need to know about the relationship between men and women. Do you know the Bible?"

"I—" Her great-grandmother used to read the Bible to her.

She no longer remembers much of it.

"I read a lot of books. I want to know everything I can. Because knowledge is power. It really is. The more knowledge you have, the more power you have. Do you see that, too?"

"I understand," she says, but she thinks: no, power is a fist in my face, a knife at my throat, rape after rape after rape until I don't care anymore. You and your books and your science and your religion don't know anything.

He says, "There's a line from the *Malleus Maleficarum* that has always spoken to me. I don't suppose you've ever heard of that? No? Not many people have. It's the Christian response to witchcraft. You could say that's one superstition taking out another, but once again I think that's a mistake. There's a reason they burned those witches . . ."

She has no idea what he's talking about. She hates him. She hates these pompous theories that she knows will somehow—surprise—pretend to prove that men are superior to women and to everything, and that this superiority grants them the right to the lives and bodies of women. She hates all men, except her father and Petya and Osip. Listening to him drone on—no, pretending to listen to him drone—is worse than giving him a blowjob. She wishes he would shut up. It's bad enough that he wants to fuck her for money—to buy her, as he accurately put it—but she wishes he wouldn't try so hard to rationalize it. It is what it is, and he should just be honest about that.

But then he tells her the line from the *Malleus Maleficarum*, "A woman is beautiful to look upon, contaminating to the touch, and deadly to keep."

Nika doesn't understand. The man is starting to scare her. She wishes the car would slow so she could jump out. But he turns onto the on-ramp of the interstate.

He asks, "Are you happy?"

"What?"

"Are you happy?"

"I don't . . . That's not what most men ask."

He looks her straight in the eye: "I am not most men." He looks back to the highway. Then he says, light, casual, "Or maybe I am." A pause, then, "You're one beautiful woman, and I'm sure every man wants to have you." Another pause, then, "You have an accent, where are you from?"

She hesitates, then says, "Russia."

"What makes you happy?"

"It's my job to make you happy."

"It's your job to do what I say."

She doesn't want to think about happiness. That's back in the box. No one knows about the box. No one gets into the box. She asks, "Do you want to take me in the ass? I won't make you pay extra. I like it. Just . . ."

"I want to know where you go when a man takes you. Where do you go when you go away?"

She closes her eyes and then opens them. She thinks, *He will not get inside.* She takes a deep breath, but quietly so he can't hear, and tries to force away the answers to his questions.

They turn south off the interstate onto the Pullman Highway.

He says, "I just want to know."

But she knows that's not true. She knows what he wants. He's a liar and a thief. He doesn't want only her body. Him and his words and all his belief that knowledge is power. He wants those deep places inside no one ever touches, not that Lithuanian man Linas who broke her with his lies, beatings, rapes, not Viktor the pimp who now continues where Linas left off, not even the other girls. No one.

. He doesn't say anything, and she knows why. He knows that she knows, and she can tell he likes it.

They drive. She tries not to think about the box, tries not to think about anyone back at home, all those who surely by now think she is dead.

"We're here," the man says. He turns right onto a two-lane road, then soon left onto a dirt trail that heads sharply down. He stops next to a small creek, turns off the truck. "Should we do it?"

"Where?"

He points to an opening on her right. She nods, unbuckles her seat belt, and gets out.

He reaches behind the seat and says, "I brought something for us to put down on the grass." It's a towel, folded tightly and sealed with duct tape. He opens his door and gets out, walks to her side of the truck. He motions for her to walk ahead, then gestures before them, "Beautiful, isn't it?"

"It's nice," she says. She is concerned about being where no one can hear, but the forest just right here, the sound of the stream, reminds her of home. In the opening she sees three young apple trees. She knows apple trees from home. These trees should begin to bear good fruit this year. The trees make her smile. And the smells. They aren't like the city. Kristine was right: How do we all survive this?

She begins to walk down the path.

She hears him walking behind her. He says, "Did you know that the word vagina is Latin for sheath?"

She doesn't know the English word sheath. She keeps walking.

He says, "I never did tell you my favorite line from the Bible. It is from the thirty-first chapter of Numbers, where God instructed his chosen people to kill every woman who has had intercourse with a man, but spare for themselves every woman among them who has not had intercourse."

For just a moment too long she puzzles over the meaning of what he has said, and when she finally begins to understand, the last voice she hears is his, asking, "Nika, have you had intercourse with a man?"

The ground is tilting and she is trying to run but the ground is moving far too quickly. She doesn't know why the ground is tilting but the sound she heard must have been an earthquake that brings the ground up to meet her face. She sees the tan soil, the small stones, the yellow blades of dried grass and the green that lies beneath, and then she falls through all of these and into the dark inside the earth, and she sees her mother and her father and she reaches out to them as she hears her voice say inside her head, "Oh, mother, mother."

⧆ ⧆ ⧆

I don't know about you, but when I catch a cold, I get psychologically down. It sinks into my experience. I get a little bit crabby. I don't deal with that stress very well. The virus infects my spirit as well as my body. I guess what I'm saying is that if my body is sick, my brain changes. So it would make sense that if I have a spiritual sickness, my brain and my body are apt to change as well.

Why are the only epidemics that we recognize physical? I think it's because we take such great pains to keep our physical and spiritual selves apart. It's crazy that people devastated by physical illness receive all kinds of support—or at least some of them do—while those who become desperately sick mentally or emotionally most often do not. In regular hospitals, patients get flowers and people come to visit. Mental hospital inmates are shamed.

Physical illness I can see and measure and diagnose. So because I feel I can understand it, I can respect it. But because we don't know how to understand mental illness we pretend it doesn't exist, and we shut the ill into mental hospitals far away. Even if they're not physically far away, they are far away from our hearts and our minds.

How much moreso, then, do we fail to acknowledge any disease of the soul? There is a cannibal sickness, which is a sickness just like any other plague or epidemic, highly contagious, with physical vectors, spread by contact, by air, by water, by touch, even also by spoken or written word—spread till it now covers the earth and to a greater or lesser degree infects us all. There are no hospitals for this sickness. If we cannot acknowledge it, how can we attempt to cure it?

Of course I'd heard about rabies from when I was a small child: everyone who has ever bawled through the end of Old Yeller knows that any beloved pet who contracts the disease turns into a vicious monster frothing at the mouth and lunging at anyone who comes too near, and everyone who lives in the country knows that the fear of rabies is why you never pick up injured rodents.

But the implications of rabies didn't hit me until my twelfth year, and to this day I remember where I was and what I was doing when the central question of rabies struck me. I was sitting on a wood bench on our deck on a hot summer day, holding an encyclopedia and thinking about the ground squirrel who had gotten stuck in our garage the day before. I'd caught her and put her in a cage, because that's what I'd been taught you do with wild animals unfortunate enough to come in contact with you: you turn them into "pets," whether they want that or not. Fortunately the cage was rickety, and overnight the ground squirrel escaped.

I wasn't thinking about the ground squirrel's bad fortune of encountering me or her good fortune of the cage being old. I wasn't even feeling guilty or bad for caging her in the first place: the understanding that an other has a life of her own, and is not here solely for my use, didn't come to me until a bit later: I'm grateful it's come at all, since the same cannot be said for most people in this culture. Instead I was thinking about the heavy gloves I'd worn to keep her teeth away from my skin, and I was thinking about how gentle ground squirrels seem most of the time, but how she had scratched and bit when I grabbed her. I understood her fighting back, and certainly respected it. But I didn't, once again, yet take that understanding to the next level, that her fighting against being put in a cage was her telling me she didn't want to be caged, and that for that reason alone I should let her be. I didn't, in short, empathize with her. I know we've all been told that children naturally feel a connection to others, and I'm sure that's true, but I know that by the time I was seven, eight, nine, and ten this connection had at the very least been deeply frayed, and it took years of seeing others suffer as a consequence of my actions—or more precisely seeing the external trappings of their suffering, but not actually seeing their suffering at all—before it even occurred to me what I was doing. At that point I began the slow process of reweaving the braided connection between me and others.

The squirrel trying to bite me made me think of animals acting in ways you wouldn't normally expect, and that made me think of Old Yeller. That made me suddenly curious about how rabies works and sent me to the encyclopedia, which I brought onto the deck. Rabies, I learned, was a virus passed from creature to creature by saliva (this latter I knew from the book and movie). Creature A has rabies, and bites creature B, or less frequently, slobbers on creature B. The important thing is that viruses in creature A's saliva enter creature B. The viruses move quickly into B's nerves, and from there they inhabit B's spinal column and brain. Creature B will not show symptoms for a few weeks or even a few months. But once the viruses reach the brain, they reproduce rapidly, and soon inhabit the salivary glands. By now creature B will show signs of illness. In humans—and we've no reason to believe anything else for nonhumans—these include head-aches, fever, irritability, restlessness, and anxiety. Within days these symp-toms progress to cerebral dysfunction, anxiety, confusion, and agitation, leading to delirium, abnormal behavior, hallucinations, and insomnia. All of this is accompanied by muscle pains, salivation, and vomiting. At that point symptoms diverge into two distinct classes. In what's called "dumb rabies," creature B retreats steadily and quietly downhill, with some paral-ysis, to death. In what's called "furious rabies"—and this is what Old Yeller had—the creature begins to experience extreme excitement and is hit by painful muscle spasms, sometimes triggered by swallowing saliva or water. Because of this the creature drools and learns to fear water—thus the frequent references to rabid creatures being hydrophobic. The creature will also become extremely sensitive to air blown on the face. But there's more. During that final furious phase, the creature may, without provoca-tion, vigorously and viciously bite at anything: sticks, stones, grass, other animals. This stage lasts only a few days before the creature enters a coma and dies. Once infected, death from the disease is almost invariable.

I remember at that point putting down the encyclopedia, leaning against the deck railing, and staring at the light blue sky above the brown and gray and smoky blue and white of the distant Rocky Mountains, and I remember thinking about volition, free will. Of course I didn't use that language—I was precocious, but volition would certainly not yet have been part of my everyday vocabulary—and I couldn't have clearly articulated any of this, but I got it. I understood—or rather asked, which is almost always more important than understanding anyway—"Who's in charge? Who is actually doing the biting? Is it Old Yeller, or is it the virus?"

The virus knows that if it is to survive the death of its host, it needs to find a new host, which means it needs to get Old Yeller to slobber on or bite someone. Thus the painful spasms on swallowing and the excessive salivation, which combine to lead to the drooling. Thus the furious biting.

In some ways central to this discussion is the question of whether you perceive the world as full of intelligence, and so do not hesitate at the possibility of viruses knowing, viruses choosing; or whether you believe viruses act entirely unthinkingly, mechanistically, and so at most you'll allow viruses not to know, but to "know" that they need to find a new host. But in some ways that question doesn't matter at all, because in either case the viruses cause Old Yeller to change his personality, his behavior toward those he loves. Or perhaps loved.

The central point of R.D. Laing's extraordinary book *The Politics of Experience* was that most of us act in ways that make internal sense: we act according to how we experience the world. If, for example, I experience the world as full of wildly varied and exciting intelligences with whom I can enter into relationships I will act one way. If I experience the world as unthinking, mechanistic, and composed of objects for me to use, I will act another.

Clearly the virus changes its host's experience, at the very least by causing pain and hallucinations.

Now here's the question that struck me so hard on that hot summer afternoon: as Old Yeller snarls and snaps at those he so recently protected, what is he thinking? If I could ask in a language he could understand, and if he could answer in a language that I, too, could understand, what would he say? Is he terrified at this awful pain, and is he, because of that pain, lashing out at everyone around him? Is he confused? Is he asking where this pain comes from?

Or does he have his behavior fully rationalized? Has he—or the virus—created belief systems to support this behavior? Is he suddenly furious at the thousand insults large and small he has received from those who call themselves his masters? Certainly throughout the movie the humans—especially his "owner" Travis—have treated him as despicably as we would expect within this culture (where do you think I learned to mistreat animals?). Does he perceive himself as suddenly seeing things clearly, and as hating these others and all they stand for?

Or is he delusional, snapping not at Travis standing in front of him, but instead protecting him as he did before and biting at the rabid wolf who gave him the disease? Is he seeing phantoms dancing before him, just out of reach, so each time he lunges, it is at someone who is not there at all?

Or maybe Old Yeller fights with every bit of his emotional strength to not lash out at the humans who are his whole world, these humans for whom he has already many times offered his life. Maybe he feels like he has picked up some sort of addiction, a compulsion, and he just can't help himself.

Or maybe the virus has insinuated itself into his brain in such a way that Old Yeller now perceives the virus as God.

He hears its commands, and knows he must obey. Maybe this God tells him that he must convert these others to this one true religion, and that in doing so both he and they will achieve everlasting peace and joy—and a release from the torment of this world. Maybe he perceives himself as thus giving these others a gift.

We act according to the way we experience the world. The virus changed Old Yeller's experience of the world. When Old Yeller acts—or when any of us act—who's in charge? Who actually makes the decisions? Why does Old Yeller act as he does? Why do any of us act as we do?

⊕   ⊕   ⊕

I say, "I think the problem is God."

Allison opens her eyes wide, says, "Not . . ."

"No," I say, smiling, "not seeing god, not with you. The problem is God with a capital G."

Allison says, "Do you mean a belief in some distant sky God . . ."

"No . . ."

". . . the belief that God isn't of the earth?"

"No."

"That our bodies are shameful and that the earth isn't our real home?"

"No."

"No?"

"Can I say something?"

"Yes."

"I don't think the problem is a virus."

"No?"

"I think viruses get a bad rap. Viruses are necessary, natural. Some are even beneficial to us as individuals: we couldn't survive without them. We've got long relationships with them. And I think we can say that almost all, if not all, of them are beneficial to their landbases. Even the most predatory of them provide necessary checks, just like any other predator, or for that matter, just like almost any animal. You get too much vegetation, well, some bunnies have to come eat it. Too many bunnies, some lynx have to come eat them. If lynxes aren't around, then maybe a virus will come along to keep the bunnies in check. And the vegetation says, 'Thank you very much.' So do the bunnies. So does the landbase. In fact, the vegetation exists in part for the bunnies, who exist in part for the lynx, who exist in part for the viruses, who exist in part for the plants. We all exist for each other. The point is that viruses aren't malevolent. Whatever is killing the planet is."

Allison nods.

"Which brings us," I say, "to God. Let's pretend that God really exists, and He's just like the Judeo-Christians say. Well, what do we know about this God?"

"That He's one mean motherfucker?"

"He hates women," I say. "He hates sex. He's a God of rape. He's a God of war. He's a God of conquest."

"He's a projection of the patriarchal mindset," she responds. "A bunch of abusers—male abusers—figured out that if they simply went around raping women and children, it wouldn't take long for them to get called out. And maybe some of these abusers even had consciences, and felt bad about what they did. So in order to shut up their consciences and in order to get their victims to stop fighting back, they created this elaborate story of a God who gives them the right to rape and conquer and do all sorts of nasty stuff, who not only gives them the right, but the mandate: who tells them to commit atrocities, who tells them that if they don't they're not good servants of this God, and who tells their victims that they better not fight back, that if they do they will incur the wrath of God and be sent to hell, and tells them that if they are good enough victims, well, the meek shall inherit what's left of the earth."

"No," I say.

"What do you mean, no?"

"No. It means no."

Allison shoots me a look, then says, "This is all Post-Christian Feminism 101."

"But what if God is real?"

"As in . . ."

"Real."

"I don't know what you mean."

"What if the stories in the Bible are true? Oh, not all of them. God didn't create the world. He—and this is so typical of a patriarchal male—just took credit for it. But the smaller miracles, those are true. And the smiting. Lots of smiting."

"So you're saying—"

"Your muse really exists. My muse really exists. So why should we get so skeptical when it comes to the capital G God? Why are the spirits we experience real and the Big Guy is just a projection, a mass hallucination on the part of hundreds of millions of Christians, nothing more than an excuse to commit atrocities on the part of the powerful and a solace for the victimized?"

"Because your muse is good. My muse is good. They haven't told anyone to go forth and conquer. They aren't responsible for the murder of hundreds of millions of human beings. They aren't responsible for the mindset that's killing the planet."

"Why do all of these spirits have to be good? Why do they have to wish us well?"

Allison blinks hard, twice.

I say, "The central questions become: Why does He hate us so much? And, Why does He want to destroy the earth?"

Lives Less Valuable
A Novel

PM Press, 2009

Lives Less Valuable *begins with the understanding that every time we activists figure out ways to use the rules of those in power to stop destructive activities, those in power change the rules. This book is about the process of trying to figure out what we can do to actually stop the destructive activities. I knew it was going to be blasphemy to modern students of literature, and especially modern teachers of literature, because literature is supposed to be apolitical. But that rule is complete bullshit, and with the world being murdered it's beyond bullshit; it's inexcusable.*

The dream is always the same. It begins with the slightest feeling of unease, as from a misplaced sound or a sudden silence: the too-quick stopping of birdsong or the scolding of squirrels. Then from Malia a moment of hesitation, that inevitable aversion to the warning she knows she must heed, that resistance to acknowledging an unavoidable reality. Each time in the dream she pays attention not to the sound nor to the silence, but to the red-tinted lettuce leaves in her garden, and to her weeding. She pays attention to her niece Robin, and notices sunlight glinting off the twelve-year-old's dirty-blonde hair. She looks at the ground and notices the stems and leaves from yesterday's weeding lying shriveled in the brown dirt.

And then again she hears a sound from the forest across the pasture. Finally, always too late, she realizes that something really is wrong. Finally, always too late, she says, quietly yet firmly, "Robin, inside."

Always the response: "When I finish this row."

"Now."

"Just a minute."

A moment's inattention. In the battle between composure and panic, so often indecision wins out, spurred by a strange desire to appear calm when everything inside wants out, and everything outside is falling apart. The desire to remain asleep, comfortable, warm, hidden safely from what you know. A belief that if only you can remain steadfast in the dailiness of your activities, your world will never collapse. And so again Malia pushes aside the sounds, stoops to pick up a basket at her feet. She tells herself not to run, not to let even herself know anything is wrong.

She straightens, and hears another sound, then more silence. At last she understands, and in so understanding realizes the unforgivable stupidity of having ignored the warnings for so long. She starts to shout, "Run, Robin! Run!"

But the words never come. They are always too late. There is a shot, or silence, and an explosion of blood, red on the dirty-blonde back of Robin's head.

Always in the dream the basket falls, slowly, and Malia runs, slowly, for the house. Gunshots. So slow she can almost see the bullets. More shots, like fireflies in the distant forest. Closer, Robin lies in the brown dirt, the back of her head gone, her skull open, jagged like a broken glass.

The doorframe splinters from gunfire. Bullets whine above her head.

Into the house. And then the voices. Always the voices. Her parents, Dujuan, Dennis, Simon, Ray-Ray, and now Robin. "Run," they say, "Run." More gunshots. Men approaching. Room to room she runs in this dream, each room smaller than the last, until she squeezes into rooms the size of coffins, rooms the size of desk drawers, rooms the size of matchboxes. She hides from the men, hears the gunshots behind her, and always the voice of Robin, "Run, Malia, run."

The dreams. A moment's inattention. A single moment.

🖐 🖐 🖐

Dear Anthony,

I hardly know where to begin. Would "I miss you" be appropriate? After all these years, finally I write. After everything that's happened, somehow it seems unfair for me to suddenly reappear in your life, especially when our contact will necessarily be one way. I can write to you, but you, for obvious reasons, can't write back.

I hope you remember our relationship as fondly as I do, focusing not so much on its ending—which at the time seemed unbearably tempestuous to me, but now seems little more than a summer breeze—as on the time that made up its heart. Our relationship. It wasn't my longest, but it remains my dearest, and by a long stretch my most passionate.

I hope that after all this time you can still decipher my handwriting. For that matter I hope you're still living at the same place. I went to the library and looked you up on the Internet. Your address was the same. I'm glad for that, because that way I can picture you there, and I can picture us.

I can see you right now. You just walked to the corner to get the mail. It's hot, and already the tall grasses are turning yellow and brown. Leggy sweet clovers cascade with blossoms, and the vetch has just started to add its purple to the riot. It's dry. You kick up traces of dust with each step, and gravel rolls beneath your feet. As you walk, you don't look at the first neighbor on the left, because you never much cared for him. He never liked you either (or me, if you remember), so today when he sees you coming he busies himself a shade too quickly under his hood, fiddling with the carburetor so the two of you don't have to acknowledge each other. I remember these things. I remember so much about our time together. Little things, like this.

I guess the kids in the next house down don't play foursquare anymore, unless something has gone very wrong developmentally. Most likely they've graduated to basketball and football. Or maybe by now they've graduated altogether, and don't live there anymore.

The dogs are with you of course. Two. They were puppies then, and now they must be very old. Surely they're walking more sedately than before, maybe arthritically. I hope they've not died. One way or another there's been too much death these last few years. Theirs would add too much to the weight.

You reach the mailbox. A strange envelope. A typed address, and no return. You check the stamp: yes, first class, so it's not junk mail. The postmark. You stop and stand in the middle of the street, wondering who the hell you know in Odessa, Texas.

Well, no one now. I'm mailing this on my way out of town. I'm sure you understand why I can't say where. I'll let you know when I'm ready to leave the next place. Several months ago I moved here, on the run from the latest—and worst—of the deaths. I needed some relief. The first day I asked a woman at a restaurant, "What do people do for fun in Odessa?" She said, "They move away." I've saved a little money, so it's time for me to go.

You don't know how long I've wanted to write you, or come visit you. My family is all dead now. All of them. I don't have anyone anymore.

And I really don't have you. I did once, and I feel stupid for giving you away. I know that's not how I saw it at the time, nor maybe how you see it now, yet that's how I see it. But even that isn't so simple. If we'd stayed together I don't know if I would have followed this path, and despite it all, I'm not sure any other path would have been appropriate.

I don't know why I'm writing. It's stupid and dangerous. Yes I do. I need to talk. God, you don't know how I need to talk, and despite our problems we always knew how to listen to each other. But once again it's not so simple. It wasn't just our listening that was so beautiful about our conversations; it was our back and forth. Do you remember that night at the top of the stairs in the public library, interpreting each other's dreams, then describing the sexual play we each had in store for the other when we got home, only to learn to our horror that the stairs formed an echo chamber for the stacks? Knowing that everyone in the library had heard the details of your dream about the hermaphroditic tadpoles and the sixty-foot clam went a long way toward explaining the looks we got on the way out, though not quite so far, I'm sure, as the by-then-general knowledge that I was no longer wearing panties. And there was that time you got the book on the White Rose Society, and we stayed up all night talking about German resistance to Hitler. Do you remember? What was the name of that girl who was beheaded with her brother for distributing anti-Nazi literature? Sophia, I think. Isn't it too much that the Nazis beheaded a woman whose name means wisdom? I remember how beautiful she looked in that black and white photo. Those conversations are why I'm writing to you now, not just because we listen to each other, but because we hear, we understand, we mostly agree, and as happened so many nights, we anticipate each other.

I'm tired, and I want to come home. I can't, so this is as close as I can get.

If you are still friends with Charlie and the gang, please give them all a hug for me, especially Charlie. Of course do not tell them it's from me. I wish I could deliver it in person. And I wish I could give you a hug. I miss you.

I love you. I always have.

Malia

💻 💻 💻

Perhaps the story begins, as so many stories do, with water. Perhaps it begins with a stream, and perhaps it begins with a little girl spending summer days as long as lifetimes playing near this stream, getting wet, getting muddy, and when she gets tired, sitting on the banks to listen in on conversations between trees and frogs, grasses and water. Always water. Perhaps it begins with evenings overflowing with the sounds of crickets and early mornings heavy with fog. Perhaps it begins with this little girl watching water condense in tiny drops on leaves, then watching these drops join others to drip off the ends and into the stream.

Or perhaps it begins much later, still with water. Perhaps it begins with a river.

The river was not always this way. Once the river was full of fish: shad, river herring, sea lamprey, sturgeon, eel, trout, striped bass, salmon. The Atlantic salmon, long as an arm, swam seemingly with one goal in mind, to come home, where they would spawn. The fish—so many they kept you awake at night with the flapping of their tails against the water, so many that people were afraid to launch their boats for fear the fish would capsize them by their numbers alone—hurled themselves up waterfalls, and failing to make the top, hurled themselves again and again until through force of will they made it, battered, bleeding, exhausted, home. Now the salmon are gone. So are the bass, the eel, the sturgeon, the lamprey, the river herring, the shad. A few trout hang on, but not so well.

Once, you could drink the water. Once, there were no signs posted telling people not to eat the fish, no signs telling them not to swim in the river. That, too, has changed.

Perhaps this story starts with Malia sitting by this river. She comes to this spot often, because just right now, just right here, in the early evening sun, feeling against her skin the warmth the stones have stored through the afternoon, she can almost forget. Here she can pretend there is no city, no poison, no cancer, no dying children. Here she can pretend the fish still swim, only deep, where she can't see them.

She watches a dozen swallows dance over the surface, twisting and climbing and diving so suddenly that her breath comes in catches of surprise. In front of her, thin stems of willows quiver in the current. Some move slowly, in rhythm with the river's waves.

Others resonate with a different frequency, responding to a pulse she can't see.

Quick movement makes her look again to the swallows, and she follows one as he beat his wings, coasts, then flutters out of sight behind a pine downed in last winter's flooding.

Life, she thinks. It's not so fragile as sometimes we fear. We all want so much to live. The downed pine's branches point upward, and the light green of this year's growth shows the tree hasn't given up. Its torn roots still clutch at the soil, and its branches still reach toward the sky. It still produces cones, the next generation's attempt to carry on.

Once, an amusement park covered the far bank. That was long ago. Almost no sign of it remains, at least at this distance. Malia wonders how long it will take for the same to be true of the city as a whole. She hopes within her lifetime.

A pair of mallards wing their way from her right to her left, and she barely hears the whistling of their wingtips above the roll and whisper of the river.

She's been coming here for years, ever since she went to work for the Council Against Toxics—or CAT as they sometimes call it, or more often just the Council—but each time she comes, it's harder to go back.

It's getting late, though, and she has work to do, and she can't stay here forever, not this time.

Still she sits. A gnat lands on her hand. She looks closely, careful not to breathe. So tiny, the gnat could be crushed even by an accidental exhale. It opens and closes its wings slowly, and she reconsiders her position on the fragility of life. Life is supple and tenuous, she thinks, evanescent and tangible.

The gnat leaves, and she inhales deeply of a sweetness that takes her home. Childhood. Backyard. Picnics. Her parents. A locust tree. Climbing. A treehouse filled with the scent of locust.

The smell reminds her of her niece, Robin. Eight years old. Conceived under a locust tree out back at her parents' farm. Malia's sister Helene had brought a boyfriend up for the weekend, and they slipped away in the middle of the afternoon. Robin. She was named after the locust tree, *Robinia pseudacacia*, so that no matter where she went, she could take the tree with her. Helene died, and Malia and her parents raised Robin as their own. Malia has no children—she's never wanted to bring a child into an industrialized, overpopulated world—and so loves Robin all the more, fiercely, like a daughter.

Time passes, and still she sits. Even in the growing dark of early evening she can see a small school of minnows in the shallows at her feet, and to the side a scuttling crawfish. It's all so unfair, she thinks, so damned unfair.

It is this recognition, or rather remembrance, of the fundamental unfairness of what is being done to the river and to the people who live nearby that finally gives her the will to stand, stretch, and begin the long walk back to the office.

⊛   ⊛   ⊛

Or perhaps the story begins with someone else, with a young man named Dujuan sitting with his mother in a doctor's office, listening to the doctor—the white doctor—talk about Djuan's little sister. "Sometimes," the doctor is saying, "in advanced cases of leukemia, parts of the blood necessary for clotting are lost. Bleeding occurs more easily." The doctor tells them that his little sister, his mother's youngest child, had bled into her brain. Dujuan's mother grasps Dujuan's hand so tight her fingertips turn from brown to burgundy as the doctor describes Shameka's skull filling with blood, her brain being forced through the only open space, near the spinal column. "She suffered no pain," the doctor says, "because she was fully unconscious." And then he says, "All things considered, not a bad way to go."

In this moment, sitting across from the doctor in the doctor's office, Dujuan wants to kill him. Dujuan sees himself stand, sees himself pull out his knife, sees himself lean across the doctor's desk, sees himself cut the doctor's throat. Perhaps first, he thinks, he should knock the man unconscious, so he will feel no pain. Then he could say to the doctor's family—the white doctor's family—"All things considered, not a bad way to go."

But he doesn't move. He sits there and looks at his mother's face, brown, beautiful, tired. He continues to hold her hand. He holds onto the outrage as well, directed not so much, after that initial rush, at the doctor, who is the messenger, as at the death, and especially at how it happened.

🔹 🔹 🔹

He'd been there when she died, actually seen the life go out of her body, out of her. He hated that image of her body taut, every muscle straining as if to tear her apart, her eyes rolled back in her head, and then the convulsions, the rhythmic flailing of her arms and the arching of her back. A primitive groaning had emerged from her throat.

Dujuan had yelled when her body went rigid. That was when the doctors and nurses came. They'd pulled him from the room so they could work on her. But before they took him out, he saw her one last time, her body seizing. This was not his sister. Not any longer. She was gone.

🔹 🔹 🔹

The worst part for Dujuan, except, of course, for the death itself, was that she knew. All along Shameka had known she was going to die. Soon.

She had only cried about it once. Dujuan remembers a night about two months before she died. He'd gotten up about one in the morning, and walking by the room Shameka had shared with their sister, he'd heard her crying. He had stopped, stood, listened. He had wanted to go in and hold her, but hadn't known what to say, what to do. So he hadn't gone inside.

Since that night Dujuan has known that if he could do one thing differently it would be that night, and if he could change one thing about her death it would have been to make it so she didn't know, and didn't have to be afraid. He would have made it so she went out suddenly, like a light switch. *Nobody deserves to die afraid*, he thinks. *Nobody*.

🔹 🔹 🔹

Dear Anthony,

Immediately after the first murder, things stayed fine. Nobody talked. Our lives continued. Dennis, my coworker, was more brittle and spoke to me a little less, but until the whole thing broke and he killed himself I never was sure how much he remembered, not only because he was drunk that night, but also because he always had a stronger capacity for denial than most of us.

I've seen Dujuan and Ray-Ray a fair amount in the intervening years. They told me that drugs undid us. They told me about Simon, about the drugs, and about Simon talking to people he shouldn't have.

That brings me back to the question I've been asking about beginnings, only this time slightly differently. If Simon hadn't used drugs, would we have gotten away with it? If Simon had used drugs but hadn't talked, would Robin still be alive? If Simon had talked to someone less eager to collect the reward, would Simon still be spending the rest of his life in prison?

There are probably thousands of lessons here, but I've learned two hard ones. The first is to know whom I'm working with. If I were to choose three people from a thousand on whom my life would depend, in retrospect I'd put Simon somewhere between one thousand and, oh, one thousand. The second, and of course I'm violating this with these letters, is that if I'm going to fight back against the full power of the state, I need to keep my mouth shut about it.

I'll write more later.

Love,

Malia

⊛   ⊛   ⊛

Malia sits at her desk, working. She is in her office at the Council. The Council occupies a suite of five rooms plus a central lobby on the first floor of an old building near the river.

Her walls are bare except for two posters. The first is of Che Guevara, with the quote, "Let me say, at the risk of seeming ridiculous, that the true revolutionary is guided by great feelings of love." The second is a blown-up photograph of a grizzly sow standing on all fours in a meadow.

She hears the door to her office open, and by the time she spins in her chair to face the entry, Dennis is halfway across the room. He reaches for one of the room's other chairs, a sturdy padded roller with thick wooden arms, and drops into it. He leans forward, eyes bright, and says, "Guess what?"

"What?"

"It's on."

She hesitates. "What?"

He looks around, then eyes her closely. "We got bugs?"

"What?"

"Bugs. We got any?"

Malia finally understands. "We're clean."

"Good, because we just scored a huge victory."

"What happened? Did you blow up Vexcorp headquarters?"

"No, really. I just got the call . . ." He stops abruptly, then says, "Don't say that, there might be bugs."

"I swept last week," she says. "Besides, what the hell do we do that's gonna scare the Feds?"

"Vexcorp's my worry."

"Same difference," she says. "What's the story?"

Dennis, in his late thirties, began working as CAT's attorney a few years after Malia began there. About six feet, with dark hair, he's handsome in a clean-cut, energetic sort of way. Too handsome, Malia often thinks, or at least too attentive to his looks. Or perhaps just too theatrical. Malia has frequently seen him pause outside windows at restaurants for one final check before smoothing his hair and making his entrance.

Dennis says, "Guess."

"Shit, Dennis. Come on."

He pauses a moment, then says, "*60 Minutes* is going to do a segment."

"No!"

"This story is so sexy," he says. "It's got everything. Poisoned kids—" He interrupts himself, says, "—Poisoned poor kids—" He interrupts himself again, "—poisoned poor inner city kids, bought politicians, cancer rates through the roof, the fucking river's probably gonna catch fire like the Cuyahoga back in—"

She talks over him, asks, "How much time did they give you?"

Dennis keeps talking: "God, wouldn't it be great if it caught fire when the film crew was here? I can just see it. And we can juxtapose their denials with that handwritten cost-benefit analysis—"

She interrupts again: "—I need to take a look at that—"

Again he continues, "—showing it was cheaper to . . ." He trails off, says, "What? They said about fifteen minutes."

"Is that enough?"

His voice quickens again. "Enough? We're talking *60 Fucking Minutes*. The roving eye of American attention is gonna fall on Vexcorp, and the company is gonna feel the heat." He pauses, then says more slowly, "Fifteen minutes is enough. Besides, any more and people would get bored."

He stands and begins to pace. He says, "If enough people just know what's going on . . ."

"They do know, Dennis."

"But now it'll be undeniable. Pictures. Right there. TV's stamp of approval."

This is the hope, Malia thinks, that allows every activist to go forward: the belief that if only people could be given the right information, they would do the right thing. Malia long ago concluded this hope is essentially false. She says, "People either don't care or they don't know what to do."

"We'll tell them what to do."

"Write their fucking Congressman?" This is a sore point with Malia. The solutions presented by environmentalists, including her office-mates, including herself, are never sufficient to the problems. Getting poisoned by toxic effluents? Write to the head of the company requesting it change its practices. Democracy not functioning? Write your Senator begging him to not follow the money. She continues, "That guy doesn't take a dump without Vexcorp's OK."

Dennis stops pacing.

Malia stands. She asks, "Or maybe they should write letters to the editor. Of the corporate newspaper. Or maybe they should call the local TV channel. I read Vexcorp's board is interlocked with Viacom."

"Don't start."

"I also read Viacom holds four million shares of Vexcorp."

Dennis glares. "Just don't fucking start already."

More weary than angry, she says, "Viacom owns CBS."

He begins to pace again. "Can't you let me have five minutes before you start? Things are bad enough without your hardline bullshit." He stops, turns to face her, says, "People get turned off by your doom and gloom. They want happy. That's what sells."

"Like poisoned kids and a dead river?"

Patient, as though talking to a child, he says, "That's the great thing about this *60 Minutes* gig. I have an angle."

They stare at each other.

Finally, "Aren't you going to ask?"

"Of course, I'm just waiting for you to tell me."

"Jobs," Dennis says.

They stare at each other again. Malia is not going to ask what he means. She is not going to ask what he means. She is not going to ask what he means.

Having built up the suspense, he volunteers, "Vexcorp's going to clean up the river—"

"Shit," Malia says under her breath.

He continues, "—in one of those public/private partnerships. I heard Cash is going to introduce a bill—"

She interrupts, "Great name for a Senator."

He laughs. "There is a God."

"And She's got a sense of humor."

Dennis says, "Anyway, a source told me he's going to announce the bill's introduction tonight at a fundraiser."

She stares into space, then looks at him sharply, says, "Let's go."

Silence.

"To the fundraiser."

"Get serious."

"I am."

"I don't think you understand. It's a fundraiser. It costs fifteen hundred dollars just to sit down."

"Then we'll stand. We can make up some papers to look like a subpoena, or we can make a citizen's arrest for treason."

"Are you crazy?"

"We'll figure out something on the way."

Dennis takes a deep breath, then continues as though she hadn't spoken, "The unemployment rate in this slum is above fifty percent—"

Malia interrupts softly: "—Dennis—"

"—and this is going to create over 400 jobs."

"Let's go."

He says, "The Feds are going to provide the money."

"To Vexcorp, of course."

Dennis asks, sincerely, "What do you care who gets the subsidy, so long as the cancer rate goes down and the river gets clean?"

"You've been doing this too long to believe this bullshit."

"That's where *60 Minutes* comes in. We publicize the hell out of it, and Vexcorp can't back out."

"They won't back out. They'll just take the money and do nothing."

He shakes his head. "When are you going to learn you can't always fight them head on?"

She looks away, then back to him before she says, "You know, we've never even tried that."

He stares, blinks twice, slowly.

She knows she lost him. She continues, "We push paper around. They're not scared of us."

"You bet they are. Why do you think they plant bugs?"

"They're bored, paranoid. How would I know?"

"They plant bugs because we're effective," he says. "Just last year do you remember the legislation—"

"—Eviscerated in committee, and turned into an industry initiative by Cash . . ."

"It almost worked."

"That's one reason we always lose," she says. "We consider our losses near-wins."

"We got the message out. Right now that's what matters."

"No. Stopping the poisoning is what matters." She doesn't want to be having this conversation. She may as well be talking to an answering machine. Press one to hear *We must never be emotional.* Press two to hear *We must present only reasonable demands.* Press three to hear *Our tactics must fall within bounds declared acceptable by those on the other side.* Press four to hear *We must not call the other side "the other side" because the language is too divisive.*

She wonders if Dennis, too, feels unheard. Perhaps he does. She sees clearly that in order for him to keep at the work, Dennis has to focus on one particular task. When he maintains that focus, he's effective at achieving his goals. Whether the goals themselves accomplish anything is an open question. Malia sometimes wonders, too, if what she perceives as seeing the system clearly for what it is—a maze with no exits—in actuality is a way to allow her, too, to keep working. Perhaps she is as wedded to lost causes as Dennis is to superficial productivity.

Dennis shakes his head in a way that signals the topic closed. "This is a win-win situation. Vexcorp wants money. They get it. We want a cleanup, and we get that."

"No justice."

"Fuck justice. I want a clean river, and I want the kids to stop dying of cancer."

He has a point, if it will work. If.

They're both silent. Finally he says, "You have a special way of puncturing people's balloons, you know that?"

Dennis is right. This isn't the time to have this conversation. Not on the heels of his good news. And Dennis probably isn't the person to have it with. She says, "I'm sorry. Really. What you've done is great. I don't think anybody else could have done it."

He looks at her.

She thinks, *Time to mend some fences.* She says, "I sure as hell couldn't have pulled it off."

He isn't having it: "I don't know."

"No, absolutely. You're a genius. Brilliant."

He smiles a little, says, "Well . . ."

She asks, "When will they be here?"

"In eight weeks."

Neither speaks for a few moments, before Dennis says, "Say, you want to grab a bite to eat to celebrate?" Another pause, very short, before, "Like old times."

They had dated briefly right after Dennis began working for the Council. For Malia, the relationship had been at that boundary between the forgettable and regrettable: neither bad enough to regret nor good enough to remember or mourn. It seems to her that Dennis felt their relationship more important, in ways she was reasonably sure he couldn't articulate.

She says, "Thanks, not tonight." Then she gestures toward the document on her desk. "Friday's the deadline to appeal this EIS, and I've still got a half-dozen arguments to tear apart. Besides, I should crash the fundraiser."

"Want some help?"

"With the fundraiser?"

He points at the document.

"No thanks," she says. "How about a raincheck on dinner till next week, and also, can I borrow your Vexcorp files? That's where you've got the copy of that cost/benefit analysis, right?"

"It's on my desk. I'll get it."

Dennis leaves, then returns with a bulging file folder. "You can hang on to it for now. But don't lose it. I'll need that stuff when *60 Minutes* shows."

He turns to go, then turns back and says, "Are you going to be all right here?"

"If I get into trouble I'll yell and the wiretappers will rescue me."

"Don't joke. You gonna be all right walking to the bus?"

"I do it all the time."

"I worry all the time."

"Thanks. I'm a big girl."

Dennis leaves. She turns back to the document on her desk, rubs her eyes with the palms of her hands, and gets to work.

⊞ ⊞ ⊞

Dujuan sits at his mother's kitchen table, his father's .38 snubnose in front of him. The feelings are coming back. They began to return even before he got out of the car. He can't make them stay away. Sorrow, rage, emptiness, confusion, and most of all an indescribable weight. No longer can he carry his mother, nor his brother and sister. No longer can he carry the memory of Shameka.

He can't run away. Where would he go? How would that help? The feelings would follow close behind. Nothing helps for long. Drugs are useless, because he comes back down. Alcohol is no better because he eventually sobers up. After sex he still has to deal with another person. Sleep doesn't work because he always wakes up, and when the dreams follow him, even there he gets no rest. He had hoped that violence—not just violence used to achieve an end, but violence to which he can give himself up completely—would make the feelings go away, but it did no better than anything else.

He needs to talk to Montrell, to Boo. He doesn't know what to do to make himself feel better, or failing that, to make himself feel nothing at all. Boo would know. Their father would have known. Where is he when Dujuan needs him? Dujuan needs someone, and he knows he can't turn to his mother: he doesn't believe she would know what he's talking about, and in any case she has enough trouble just keeping Shane and Ketheia fed. Shameka's death hit her as hard as it did him.

And she doesn't know how hard he'd been hit, because he couldn't let that show. Had he shown it to her, she would, he was certain, have felt the need to take care of him, something she couldn't do. Not now. Maybe not ever.

He looks at the gun on the table, then watches his hands fumble open the box of ammo and pull out one bullet.

"Chickenshit," he says out loud. He's too damn chickenshit to take responsibility for even this decision.

He breaks out the cylinder.

He would never have done this here if his mom and the kids were home. But they're gone for the week to his grandma's for her birthday. This way his mother will never see the mess. Ray-Ray will check on him tomorrow or the next day, and walk in like he always does. Then he will take care of things. Like he always does.

The table and floor will be cleaned up by the time she gets back.

He inserts the bullet, then sees his hands reach for two more: one in six isn't good enough odds for him.

The sharp snap of the gun fitting back together—which would normally have been barely audible—echoes through the room and through his head. He hears it all down his spine and into the hard wooden chair on which he sits. He hears it in his feet and back up his legs. As he spins the cylinder every click of its ratchet makes its way into his bones.

He sees a hand draw the gun closer to his head. He doesn't know whose hand it is. Shameka's? His father's? Boo's? The man's from the street tonight? He sees tattered skin above one fingernail, then a freckle on the ring finger, and recognizes the hand as his own. He sees the finger squeeze, the hammer pull back. He has not yet reached the point of no return. The finger keeps squeezing. He wants to put the gun down, but can't make the hand do it. He closes his eyes to not see the flash.

It seems there is no one moment when the hammer stops going back and begins to come forward. There is a single smooth movement until the trigger stops resisting his finger and the spring-loaded hammer strikes, with a force weak enough that it could have been stopped by a finger and strong enough to blow apart someone's world.

There is a sense—perhaps even the deepest sense—in which it doesn't matter whether the firing pin strikes a cap or an empty cylinder, because someone is going to die this night, this moment. The only question is whether Dujuan's body will die as well.

⊛ ⊛ ⊛

Later, on the way home, what Dujuan remembers most is how good it felt to finally stop feeling, once the violence began. Until then he'd been edgy, holding down an anger that rose and rose inside of him. He'd snapped at Ray-Ray, and especially Simon, and he'd complained about the cold, but his anger wasn't directed at them. Nor was it directed—specifically—at the man who eventually stopped to see if they needed help, whom they robbed and whom Dujuan beat—shrugging off Simon's and even Ray-Ray's attempts to stop him—more severely than he'd ever beaten anyone. Dujuan knew, even as he heard again and again the thud of his booted foot against the man's ribs, and heard the grunting of the man's involuntary exhalations—the man long since having lost volition—and even as he felt the solidity and *rightness* of the impacts traveling back up his own leg, that he felt no unique anger toward this man as an individual. He didn't know this man, had never seen him before, and would never see him again. He didn't care to know him. He didn't care about this other's pain. What he cared about was how *good*—yet at the same time painful—it felt to feel the texture of the air at the moment the man realized he was in trouble, and to draw out that moment, feeling the other man's fear and tasting his questions, so tangible Dujuan could pluck them out of the air above the man's head: Will I live? Will this hurt? How much will this hurt? Will I humiliate myself in the pain? He cared about the crack in the man's voice, but only because it revealed a crack in the wall that in Dujuan's mind separated the two men. Dujuan accepted his own rage, his own violence, as part of who he was, and as a necessary response to his surroundings. And he somehow knew, as certainly as he knew his sister was dead, that who this man was and what he represented—though Dujuan didn't know what that might be—were based on violence against Dujuan and all he held dear. Dujuan could not have said how this was, but he knew it to be true. And so he beat the man, and continued to beat him.

He stopped when the air turned sour and he knew they had to leave. He hefted the man's shuddering body from in front of their car, dropped the hood, and dashed to the driver's side. He got in and started it up. They had barely pulled away when a police car passed them—presumably a random patrol, or perhaps officers called by the man on a cell phone before he stopped to help—coming the opposite direction.

Dujuan looked in the mirror, and the last thing he saw before he turned down a side road were the headlights of the cop car shining on the man he had just beaten. He did not feel a thing, and for that he was glad.

❀ ❀ ❀

They continue to drive. Ray-Ray still thinks about violence. The man in the alley isn't the only person he has killed. Ray-Ray killed his cousin Ricky, too. But that doesn't count: that killing was more gift than murder, more an expression of familial responsibility than violence.

Ray-Ray thinks back to last summer, in the back bedroom of his Aunt Claire's apartment.

As he does almost every afternoon—and toward the end it becomes several times each day—Ray-Ray enters without knocking, and makes his way down the hall. He passes the living room and notices that today Claire isn't watching television; she sits silently on the couch. For a moment, he considers going in to talk with her, but instead he walks into the kitchen and turns on the tap. He waits till the water is hot, then partially fills a glass. He pulls a spoon from the drawer. The silence in the apartment disturbs him, and he thinks again about asking Claire what's wrong. But he guesses he knows, and if so, it's better, he decides, to just not talk about it. He walks out of the kitchen and to the rear of the apartment. The door to the back bedroom is slightly ajar, as it is each time he comes.

Ray-Ray slips inside. He doesn't turn on the light. While he waits for his eyes to adjust, he listens for the soft sound of breathing that will tell him Ricky made it through another six hours.

Ricky grew up in this apartment, in this room, then left to live on his own, and at thirty-four came home to die. He has cancer, as his father did before him, and also his uncle. Cancer has become something of a tradition in this family, a sorrowful birthright that comes to them by way of where they live.

Ray-Ray hears Ricky say, quietly, "Still here, bro."

Ray-Ray has a hard time hearing the words, and can just barely make out a dark form on the bed.

"Your stuff," Ricky says, before taking another breath, "It don't work . . ."

Ray-Ray doesn't say anything.

Ricky continues, "It hurts."

The ticking of a clock in the hallway. Ricky's breathing, shallow and uneven, and Ray-Ray's own, too loud in the quiet of the room. Finally, Ray-Ray says, "We got to—"

Ricky cuts him off, "No hospitals."

Chemo. Surgery. He's been through it all. Nothing has helped. Nothing except the shit Ray-Ray hooked him on, and which he now gives him several times a day. It was enough of a struggle to get Ricky to go to the hospital in the first place: his father died in one, hands tied to bed railings to keep him from pulling out the feeding tubes.

Ricky says, "Too late."

"No—"

"I talked to Mom . . ."

"No."

"You said . . ."

Silence, until Ray-Ray says, "I know what I said."

More silence. The clock, the breathing, the sound of his blood pounding in his ears. It would be easy for Ray-Ray to walk out right now and never come back, to pretend the pain in the room is nothing to him. It would be easy, too, to try to talk Ricky out of the decision, as Ray-Ray and Claire had talked him out of it before. But the first isn't an option: it's not how family acts. As for the other, it's pure selfishness: to put off Ricky's death for another day or two or three would be doing no favors to Ricky. But Ray-Ray hates to do this. He hates to be the one.

"Cover your eyes," Ray-Ray says, and turns on the light. He looks at Ricky and hates what he sees. Hollow cheeks. Protruding forehead with deep hollows at the temples. Stick-like fingers covering sunken eye sockets. Ray-Ray looks away.

He puts the glass on the nightstand, then reaches into a drawer beneath, which he resupplied only a few days ago. He pulls out the plastic that holds the tar heroin, and unwraps it. He nearly retches at the bitter, vinegary stench, as happens each time he opens the package. He wouldn't do this for someone else. The heroin is dark, and tacky to the touch. He uses a pocket knife to scrape a dose into the spoon. He triples the dose just to be sure.

Between breaths, Ricky asks, "You know how much?"

"Yes," Ray-Ray lies, and triples it again. He considers giving him the whole damn chunk. No need to save it, since Ray-Ray doesn't use, Claire sure as hell doesn't, and there's no way he's going to feed any of Simon's habits.

Neither speaks as Ray-Ray rewraps the chunk, then pours a little hot water into the spoon and stirs it to dissolve the tar. Then Ray-Ray draws the liquid through a piece of cotton as a filter into the syringe. But there's too much junk for the gear. He'll have to slam him a couple of times.

Ricky extends his arm—a useless gesture, since the veins are gone— and Ray-Ray says, "There's gonna be a little prick here, Ricky."

Ricky says, "'Sides you?"

The same joke every day, and this is the last time. Silence. Ray-Ray begins to sweat. He says, "Are you sure?"

Ricky nods.

Ray-Ray asks, "Want me to get your mom?"

"She knows," Ricky says. "We talked."

"She don't want to hold your hand?"

"Fuck you," Ricky says. A long breath. "Don't make this hard."

It already is, Ray-Ray thinks. And suddenly he understands Claire's absence. It would be one thing to be present at your son's death, and quite another to be there for his killing.

Ray-Ray finds a vein in his cousin's neck—as is true of his arms, the veins in his legs have long-since collapsed—and injects the heroin. Afterwards, Ray-Ray draws up more of the junk and injects Ricky again. He does it a third time. If he's going to do this, he's going to do it right.

He looks away to pick up the spoon and the tar, and out of the corner of his eyes he sees Ricky shudder, once, and then sigh. Trying not to look at the body, Ray-Ray cleans up the nightstand—later that night he'll throw all the paraphernalia and shit into the river to have it out of his life—and leaves the room. He goes to tell his Aunt Claire. The whole time—even holding her as her whole body shakes—he doesn't let himself feel. He doesn't let himself feel until much later, and what he feels then is not so much rage or even sorrow as it is an emptiness that swells up inside of him until it's bigger than his heart, bigger than all of him, bigger than the whole damn city and everyone in it.

Ray-Ray never talks about the specifics of Ricky's death—not to anyone—but when anybody asks if at least it had been peaceful, Ray-Ray always replies, "Not for me it wasn't. Not for me."

Dreams

Seven Stories Press, 2011

For years I wanted to write a book that would be a mosaic of loosely connected dreams that I'd had. But this book, like so many of my books, instead began with a question: "What are dreams?" Books on dream interpretation generally speak of dreams as being at best messages from our unconscious. But that's never how I've experienced them. I've experienced dreams as coming from other sides, from a dreamgiver. This has always felt true to me, and I later discovered that it is compatible with every indigenous cosmology.

Dreams became an exploration of other sides, and then, a second question: If there are beings on other sides, why has the dominant culture continued to win? Why have these beings allowed the planet to be murdered? Are they not paying attention? Do they not care? Can they not stand up to it? I've explored the psychological, sociological, and physical irredeemability of the dominant culture; Dreams shows how it is irredeemable cosmologically.

It's also a study of epistemology. How do we know that dreams are from the unconscious? How do we know that they are not from other sides? How do we know that science is right? What are the central assumptions of science? What is knowledge? Richard Dawkins has written, "Science bases its claim to truth on its spectacular ability to make matter and energy jump through hoops on command." This culture's very epistemology is based on domination. Dreams is an exploration of an epistemology that is not based on domination.

For many years I've wanted to write a book about the mutually beneficial and extremely complex relationships between dreaming and waking realities. But each time I'd finish a book and begin preparing for this one, some other book would suddenly demand my attention, and this one would get pushed aside. Then this year this book made clear that it was to be next. There was to be no more delay.

As so often happens with messages from the places where writing comes from, or where dreams come from, I chose to ignore it. I thought my reason for ignoring this message was, as we always seem to think our reasons for ignoring these messages are, inescapable. In this case my reason was that the real, physical world is being murdered, and I didn't want to waste my time writing about dreams. With all the world at stake, any book—any action, any thought, any day, any lifetime— that doesn't help us *succeed at stopping* this culture from killing the planet is inexcusable, unforgivable. And how could a book about dreams possibly help?

And then it hit me. Part of the reason that this culture is killing the planet is that it ignores, devalues, or demonizes messages from those places where writing comes from, where dreams come from, where so many other impulses and ideas and beings come from. It tries to create a rigid separation between what it calls the human, on one hand, and what it calls the natural, and especially what it calls the supernatural, on the other; it then favors what it calls the human at the extreme expense of everyone else.

The fundamental difference between civilized and indigenous ways of being is that for even the most open-minded of the civilized, listening to the natural world is a metaphor. For traditional indigenous peoples it is not a metaphor; it is how you relate to the real world.

I am not indigenous. Not in the slightest. I will never be indigenous. I am simply a living member of a living universe, and so are you. The experience of listening to and communicating with nonhumans—including other mammals, other animals, fungi, plants, bacteria, and others; and also beings this culture does not even consider to be living, such as rivers, rocks, mountains, stars, soil, and others; and also beings this culture does not even consider to exist, such as muses, dreamgivers, spirits, and others—is the birthright of every one of us. Our culturally imposed exile from these relationships—this culturally imposed echo chamber in which we find ourselves imprisoned—is one of the costs this culture inflicts upon us.

Because this exile is so unnatural, it is extremely difficult to maintain. It must be more or less constantly reinforced with messages that other intelligences do not exist, with messages of self-proclaimed superiority, with frenetically defiant messages of self-imposed alienation. To communicate with nonhuman others, to "hear voices," is, we are told again and again, to be insane.

⊞ ⊞ ⊞

Today I saw a standard intake questionnaire for a psychiatric clinic. One of the questions to help determine whether you suffer from paranoid schizophrenia was, "Do you hear voices when no one is around?" Presumably, nonhumans in your physical presence don't count as someone (and just as presumably, humans on television do). So me hearing the voice of my muse as she gives me the words I write would be one strike against me. Receiving help from trees when I get to especially difficult parts of books would be another. Perhaps strike three would be that the night before I got into a terrible car wreck in which my mother's neck was broken and she was made functionally blind I heard a voice telling me again and again to stay home that next day; I did not listen, which from a psychiatrist's perspective may have been sane. But in this case, me being "sane" cost my mother vision loss and decades of pain.

Here is what I know. I have, myself, consciously experienced communication with beings from the other side. My muse is a real being, not a reification of unconscious processes. So is the being who gives me dreams. There are others I know of, too. And beyond these I have no idea how many more of these beings I routinely communicate with, any more than I have any idea how many different beings I communicate with who live *inside* my body (such as bacteria and whipworms in my guts, white blood cells throughout my body, and so on).

This book is in part a hard-headed look at what it means to communicate with those on "the other side," however we may conceptualize "the other side," or more accurately, however "the other side" may actually be. It's also an attempt, as all of my work is, to break the stranglehold that scientific, materialist, linear thought has on how we perceive, think, experience, and act on (as opposed to *with*) the world, and then through us the stranglehold that scientific, materialist, linear thought has on the real, physical world itself.

At this point in this culture's unraveling of the world, we desperately need every bit of help we can get, from whatever sources we can find. I want to find out who lives over there on "the other side," to get to know them at least a little bit (insofar as they want to be known; I don't want to poke in where I'm unwelcome and reproduce the same old patriarchal, pornographic, scientific mind-set by attempting to force others to reveal themselves if they do not want to do so), and if they are interested (and I'm guessing some will be), to ask them for their help. For in many ways this book is a desperate attempt to find more allies to help stop this culture before it kills the planet.

In my two-volume set *Endgame*, I explicitly excluded the possibility of "help from the other side," not because I believe that we can't get help from the other side, but rather because, frankly, so many people in this culture are insane, and they are lazy, and they will use any excuse in the world (and any excuse outside of the world) to not act against this culture. If I even hinted at the possibility of help from the other side, far too many people would respond, "Well, then, the Great Mother is just biding her time, and when she's ready she'll save us all. So there's really no reason for me to take out that dam or oil refinery, is there? Besides, if I *do* take out that dam or oil refinery, I might get my hands (and my spirituality) dirty, and we certainly don't want that, do we?" Basically, most people in this culture are addicted to this culture, and addicted to their own slavery to this culture, and to have any chance of reaching an addict, one must not—and I cannot too strongly emphasize the word *must*—give them any outs at all.

I know that both the muse and the dreamgiver help me more than I can say. We have deep and abiding relationships. And I know that there are others over there, too. Some friendly. Some not. I also know that in times of profound trouble, many of the indigenous have called upon these others for help.

We need their help now. If it is the case that traditional indigenous peoples had constant intercourse with beings on these other sides, and with the other sides themselves, then these other sides are also parts of our homes, just as are trees and slugs and stones and soil. And it is long past time that we returned home. We may very well find friends and neighbors there, friends and neighbors who are ready to help us defend our—and their—homes.

❦ ❦ ❦

Let us begin with dreams. They are gifts from "the other side," whether you believe the other side is the unconscious or, as seems clear to me, the home of other beings like muses or dreamgivers. In either case, dreams have their origin outside our conscious selves, outside our realm of control. Sure, I can force myself to sleep and hope I get a dream, and I can even request that the dream have a certain shape, texture, feel, but my request remains a request and the dream remains its own.[1]

Even though many of my dreams are lucid, in that I'm usually aware that I'm dreaming and can often at least somewhat guide the dreams, it's also true that the disappearance of control is often a marker for when I've fallen asleep. As I'm daydreaming my way toward sleep, the moment the dream becomes more dream than fantasy is the moment sleep takes over. One moment I'm fantasizing about blowing up a dam on the Klamath River, carefully setting charges where they'll have the greatest effect, and the next a thousand spiders crawl from the concrete and begin to expand until they fill the world. That's the moment the dream begins. That's the moment I lose control.

❦ ❦ ❦

I owe my life to a dream. I was in college, majoring in mineral engineering physics, on a path toward gainful (or painful) employment, toward financial security, toward the life that most of us are told we're supposed to want. And I was miserable. Every day on my way to school, or summers on my way to work in physics, I fantasized about driving past the school or past the buildings where I worked, and never stopping, but just driving, not looking back. I never did that, though. I was a good little worker, wasting my life, twenty-one and already looking toward retirement. There seemed something desperately wrong with that, something as desperately wrong as the obscene phrase, "Thank God it's Friday." I looked around, and saw that nearly all of my fellow students were as unhappy as I was, but were attempting to rationalize their unhappiness—insofar as they even thought about it or allowed themselves to feel it—by thinking about everything they would buy with the money they'd make. Automobiles, vacations in the Bahamas, and big houses as anesthetics. It became increasingly clear to me that the more aware one became of one's own life and one's cultural surroundings, the less happy—or given what I now understand, using language I didn't have at the time, happy™—one was. It seemed I had a choice: I could numb myself out the way I saw all those around me numbing themselves out, or I could be miserable. Some choice.

One day a friend asked me the obvious question: "If increased awareness means less happiness, why bother?" Of course I had no answer.

My dreamgiver did. A few nights later I had a dream: I was driving. To my right I saw baby cranes—blue-green, all legs, beak, and wings—standing in a field. They took off and crashed, took off and crashed. I stopped the car and got out. "That looks like it hurts. Why do you do it?" One of the cranes looked me square in the eye. "We may not fly very well yet, but at least we aren't walking."

I awoke, happy. From that moment, there has been no turning back.

❦   ❦   ❦

Part of the reason we are told that other intelligences, and conversations with other intelligences, cannot happen is because the events are willfully unrepeatable (that is, unrepeatable because the actors in the events have volition, as opposed to unrepeatable because the events are random), and therefore not predictable, and therefore not controllable. This culture is based on the assumption that all of the world (except humans, sometimes) is without volition, is mechanistic, and is therefore predictable (most often absolutely, because of this lack of volition, or at the very least probabilistically, because of randomness). Therefore, the existence of the willfully unpredictable destroys a foundational assumption of this culture. The existence of the willfully unpredictable also invalidates this culture's ontology, epistemology, and philosophy, and reveals them for what they are: lies upon which to base this omnicidal system of exploitation, theft, and murder. It's much easier to exploit, steal from, or murder someone you pretend has no meaningful existence (especially if you have an entire culture's ontology, epistemology, and philosophy to back you up); indeed, it becomes your right, even your duty. The existence of the willfully unpredictable reveals this culture's governmental and economic systems for what they are, as well: means to rationalize and enforce systems of exploitation, theft, and murder (for example, try to stop Monsanto's exploitation, theft, and murder, and see how you are treated by governments across the world).

But willfully unpredictable nonhumans exist. Sometimes some of them allow some of us who are willing to look to see them, and sometimes they don't.

⊛ ⊛ ⊛

If the indigenous had (or have) access to these other sides, and to those who might be allies on these other sides, why has the dominant culture consistently been able to dispossess and destroy the indigenous? (And why limit it to humans asking for help? Why have these potential allies not helped passenger pigeons, Eskimo curlews, Falkland Islands wolves? Why have all of these wild humans and nonhumans not been able to call on unseen allies to help, as Tecumseh and so many others have so desperately desired, to push the civilized back wherever they came from?) And a related question: if the earth really is intelligent—which I fully believe it is—why hasn't it killed us off?

Perhaps the answer is that the scientists—and more broadly, the members of this culture—are right, and essentially every other human culture that has ever existed is wrong. There is no plan. Everything is random. The existence of life on earth is random. Natural selection consists of random genetic mutations that either take hold or do not. As Richard Dawkins, the extraordinarily influential and popular scientific philosopher—he's got more Google hits than Mick Jagger, for crying out loud, even though he's a freakin' scientific philosopher—put it, we exist in "a universe of electrons and selfish genes, blind physical forces and genetic replication."[2] Humans are the only meaningful intelligence on earth, and possibly in the universe. The world consists of objects to be exploited, not other beings to enter into relationships with. There is no magic. No meaning inheres in the world; the only meaning is what we project. Says Dawkins again, "You won't find any rhyme or reason in it [the universe], nor any justice. The universe that we observe has precisely the properties we should expect if there is, at bottom, no design, no purpose, no evil, no good, nothing but blind pitiless indifference."[3] The only mysteries are those we've not yet cracked. Because nonhumans have no meaningful intelligence, they have nothing to say, to each other or to us. Thus interspecies communication is bunk, no matter who the nonhumans are: animals, plants, rivers, rocks, stars, muses, allies on the other side, and so on. Anyone who thinks otherwise is superstitious, that is, delusional, maybe primitive, maybe crazy, maybe childish, maybe just plain stupid. If this culture is right and every other culture is wrong, then there are no muses, no fates. There are no messages from stars. Astrology is crap. Prayer is crap. Lucky socks are crap. Premonitions are crap, and intuition is either just unconsciously paying really close attention to something, or it's crap. Nothing more. Heaven is crap. Hell is crap (or maybe just being forced to read Richard Dawkins). All notions of reincarnation or an afterlife are crap. Spirituality is crap. Dreams are purely psychological. Love is nothing but a series of chemical reactions in the brain. The same is true for awe. The same is true for loneliness (calling loneliness a purely chemical response certainly lets this alienating culture off the hook, and makes us all just feel worse; first you tell us no one else exists and no meaning exists, and then when we feel lonely you tell us it's just chemicals in our brains. Maybe now you can slide us some *soma*, and the chemicals in our brains will cause us to think we're not so lonely, which in this rubric means we won't *be* quite so lonely). The same is true, embarrassingly enough, for thought. Further, if this scientific, materialist, instrumentalist perspective is right and every other culture is wrong, then the universe is a gigantic clock—a machine; a very predictable and therefore controllable machine—and God (insofar as we can use God as a metaphor, since neither God nor gods exist) is nothing but a blind watchmaker,[4] or more accurately to this perspective, God is Himself a giant clock.

Power in this case, then, is like meaning; there is no inherent power in the world (or outside of it)—just as no power inheres in a toaster or automobile until you put it to use—and the only power that exists is that which you project onto and over others (or that which others project onto and over you). Power exists only in how you use raw materials.

And science is a potent tool for that. That's the *point* of science. Dawkins—and remember that he is a preeminent contemporary scientific philosopher, with more Google hits than Mick Fucking Jagger—writes, "Science boosts its claim to truth by its spectacular ability to make matter and energy jump through hoops on command, and to predict what will happen and when."5 If you use raw materials more effectively than anyone else, well, then, more power to you. This means, of course, that might makes right—or rather, right, too, is like meaning and doesn't inhere anyway. If nonhumans are not in any real sense beings and are here for us to use (and not here for their own sakes, with lives as meaningful to them as yours is to you or mine is to me), then using (or destroying) them raises no significant moral questions. Right is what you decide it is, or more accurately, it's irrelevant (except insofar as you can use the concept of *right* as an opiate to allow you to live with yourself and/or keep those you exploit from killing you). Right is whatever you want it to be, which means it's really nothing at all. This malleable notion of right means that you can fairly easily talk yourself into feeling good about exploiting the shit out of everyone and everything else.

If all of this sounds sociopathological, that's because it is. Let me put this another way. Last night I was at my mom's house, watching a documentary on David Parker Ray, a serial killer from Truth or Consequences, New Mexico, who is suspected of killing up to sixty women. He kidnapped women—with the help of his daughter and her friend, and then later with the help of his girlfriend—and held them as sex slaves, or more accurately rape slaves. He turned an entire tractor-trailer into a well-stocked torture chamber, where he videotaped what he did to them. One of the main characters of the documentary was an FBI profiler. She compared Ray's attitudes toward his victims to those most people have toward tissue paper: once you use it, are you concerned about what happens to it? Of course not, she said. And that was how Ray perceived—or rather didn't perceive—his victims: simply something to use and throw away.

When the profiler said this, my first thought was *passenger pigeons*. Then *chinook salmon*. Then *oceans*. Then *cows in factory farms*. How deeply do members of this culture mourn passenger pigeons? Salmon? Oceans? How much do they consider the suffering of victims of factory farming? How about the victims of vivisection? Three days ago the California legislature passed a law (as an Urgency Measure) that codifies animal testing as a human right.6 This culture as a whole, and most of its members, considers the victims of this way of life no more than David Ray Parker considered his victims. And if the scientific, materialist, instrumentalist perspective is right, this is only natural. As Dawkins says, "Blindness to suffering is an inherent consequence of natural selection."[7]

If this scientific, materialist, instrumentalist perspective is correct, if the world (and the universe) is here for you to use, and entering into meaningful relationships with nonhumans and/or the unseen is insane, impossible, "anthropomorphic," and a waste of time and energy, then you will certainly have a huge competitive advantage over all of these superstitious, childish, primitive, insane peoples who, it ends up, are wasting their time "communicating" and "communing" with dumb animals and with what they believe are spirits or beings from "other sides." While medicine men or witch doctors speak mumbo jumbo with the "spirits," and while war chiefs put on special clothing they stupidly believe makes them invulnerable to your bullets, you're mustering your limitless armies, preparing your even more limitless rifles, and loading grapeshot and canister into your also limitless cannon. Who would you rather put your money (and the survival—and let's not forget growth!—of your culture) on: army after army of well-trained soldiers equipped with the most modern killing technology; or a bunch of very brave yet pathetically under-equipped American Indians who may, through loss after loss, be starting to lose faith in the spirits who heretofore they believed guided them? Given those choices, give me the big guns. Bullets to the brain somehow always seem to trump spiritual sophistication. Maybe that's because, if the scientific, materialist, instrumentalist perspective is right, spiritual sophistication is just a fancy way of saying delusional and primitive. Maybe, if the dominant culture is right, the American Indians were calling for help from those who simply did not exist.

If the scientific, materialist, instrumentalist perspective is right, then the earth hasn't fought back, and won't fight back, simply because the earth has no volition and therefore can't choose to do anything. The earth (and by extension all its inhabitants except humans, by which we really mean civilized humans, by which we really mean rich, white, male civilized humans) is an object. All "beings" are here for us to use, and if we're going to ask why the earth hasn't killed us we may as well ask why a tool box doesn't kill us when we take out tools and use them, or why a woodpile doesn't kill us when we take out wood to burn, or why a refrigerator doesn't kill us when we take out food to eat. It's a silly question. Sure, we might run into problems someday when the refrigerator is empty, but we're surely smart enough to just find another refrigerator. Like the bumper sticker (invariably on a huge muddy pickup driven by a smug ass-hole) says: "Earth first. We'll log the other planets later."

If the scientific, materialist, instrumentalist perspective is right, and all nonhumans on the planet (and the planet itself) are just objects to be used, that means we, just like the indigenous, will never be able to summon help from others, be they Kamchatka brown bears, deadly viruses, oceans, fungi, forests, muses, fates, demons, angels, spirits, or ancestors. None of these exist. We can ask, but no one will hear us, and certainly no one will respond. We are, as this culture tells us in so many ways, all alone.

If we are all alone, and we care about the planet, our actions become clear: we must do everything necessary to decisively and finally bring down civilization before it kills any more of the planet. Because if the scientific, materialist, instrumentalist perspective is true, this culture will continue its routine and necessary destructiveness until it collapses or is stopped. The only real responses the civilized have to this destructiveness are the same ones they always have: primarily to call on everyone to rely on the generosity, graciousness, and skill of the civilized (and to kill or otherwise severely punish those who do not heed this call). The modern name for this generosity, graciousness, and skill regarding the natural world (and the most overtly exploited humans) is "sustainable development." But of course "sustainable development" will for many reasons fail to materially help the natural world (and the most overtly exploited humans). It is an oxymoron, since "development" is a euphemism in this case for indus-trialization, which is by definition unsustainable;[8] in fact, industrializa-tion is utterly, irrevocably, and functionally antithetical to sustainability. This absurdly obvious oxymoron remains in common usage primarily for three reasons: (1) pushing this particular lie well serves those in power; (2) a lot of people are too busy, too emotionally drained and defeated, too fearful, too fully metabolized into the system, too incapable of thinking for themselves, too financially well-rewarded by the system, too dishonest, too greedy, too insane, too defensive of and about this culture, and/or too stupid to see the phrase for what it so obviously is (and of course different people can have multiple reasons for their inability to perceive the absur-dity of "sustainable development"; George W. Bush, for example, would fall into at least ten of the above categories; and President Barack Obama would fall into at least nine); and (3) "sustainable development" is nothing more nor less than the twenty-first century version of *the white man's burden.*

In Rudyard Kipling's late-nineteenth-century poem "The White Man's Burden," he attempted to show just how damn difficult it is to be a white man in a world where you are constantly—and with great reluctance and heavy sighs—having to civilize benighted savages. This is a profound obligation carried by white men. How did these savages somehow survive on their own—lazy and wasteful as they are—for tens of thousands of years? Left unsaid in Kipling's poem—as is often left unsaid in public discourse about these topics—is any inconvenient discussion of genocide, ecocide, enslavement, or mass organized theft of resources. Left unsaid is that the point of empire is to conquer, subdue, enslave, steal, and murder. Of course.

Fast forward a hundred or so years, and it's still damn difficult to be a white man in a world where you are now constantly—and with great reluctance and heavy sighs—having to civilize (I mean, develop) benighted savages (I mean, underdeveloped nations). Only now the burden is even heavier, since these white men must now attempt to fulfill their obligations to rule over the entire planet, to "sustainably" manage forests and oceans (how did these forests and oceans ever survive for millions of years without scientific management?), to be "good stewards" of land and air and water that can all evidently fare no better without our assistance than could the savages of a hundred years ago, that need our help just to survive. Now left unsaid in all this talk of "sustainable development"—as is often left unsaid in public discourse about these topics—is any inconvenient discussion of genocide, ecocide, enslavement, or mass organized theft of resources. Left unsaid is that the point of empire—the point of industrial civilization, the point of civilization—is still to conquer, subdue, enslave, steal, and murder. Of course.

If the scientific, materialist, instrumentalist worldview is right, and we really are all alone in a universe bereft of nonhuman intelligences or beings, but if we for some strange reason care about the continuation of life on this planet (if perhaps we are not terminally narcissistic and psychopathic), we're still where we started. We either need to fight by ourselves or find allies to fight alongside us. But if the allies aren't there, we better roll up our sleeves and get to fighting.

⊕   ⊕   ⊕

Let's return to a line Richard Dawkins wrote: "Science boosts its claim to truth by its spectacular ability to make matter and energy jump through hoops on command, and to predict what will happen and when."[9]
Does anyone else see the fundamental flaw in logic here? Let's say I have a gun. Let's say I point this gun at your head. Let's say I command you to jump through hoops. Let's say you do it. I do, after all, have a gun pointed at your head. Now, with this gun pointed at your head, I tell you to jump through those hoops again. And then I predict that this is precisely what you will do. You do it. Whaddya know, I'm a fucking genius; I commanded you to jump through hoops, and I correctly predicted that you'd do it.

Richard Dawkins was with this sentence incredibly intellectually dishonest—and sneaky as hell—and the only reason he hasn't been called out on it is that he has a whole culture of sociopaths for company. He has conflated the power to command with truth. He has, and this should come as no surprise to anyone paying any attention to the trajectory of this culture, conflated domination with truth. But neither the power to command nor domination are the same as truth. The power to command is the power to command, domination is domination, and truth is truth.

Richard Dawkins could put a gun to my head. He could even kill me. But that wouldn't mean that he is telling the truth. This culture is dominating the planet. This culture's domination of the planet is killing it. That does not mean this culture is telling the truth, or is even capable of understanding it.

⊕   ⊕   ⊕

Sam Harris wrote that he doesn't trust "the wisdom of Nature," and that he believes that to trust "Nature is a stultifying and dangerous mythology."[10] Now, wait a goddamn minute. Do you want to talk about wisdom? Do you want to talk about a wisdom that shouldn't be trusted? Do you want to talk about a "stultifying and dangerous mythology"? Like any True Believer, I'm sure Harris would claim that *his* religion—science—is not a mythology, although it clearly is. A myth is "a popular belief or tradition . . . embodying the ideals and institutions of a society or a segment of a society," and at this point the scientific, materialist, instrumentalist, mechanistic perspective is certainly that. Nonetheless, I'm sure scientists would argue that science is not a mythology, but rather an objective description of reality *as reality is*. Of course, fundamentalist Baptists argue the same for their religion. So would anyone who refuses or fails to question the assumptions of his or her faith.

None of which is to say that a mythology—the set of stories a particular group lives by—cannot also accurately represent physical reality. I am merely pointing out that science, too, is based on faith, no matter how fevered the denials of dogmatic scientific philosophers, and no matter how much they may claim that science merely describes physical reality, and makes no ethical or moral claims.

The notion that science makes no ethical or moral claims is absurd, and I'm surprised that otherwise intelligent people so often accept this. First, the *precepts* of science—including the notion that the universe is mechanistic, and including the emphasis placed on repeatability (which follows from and reinforces the notion that the universe is mechanistic, or not a willful decision-maker, or not filled with willful decision makers)—carry with them extraordinary moral weight, in that they lead to certain behaviors that carry with them moral consequences. For example, if you perceive the world as alive and filled with willful subjects with whom you can enter into relationships, you will treat those others differently than if you perceive them as dead, or as parts of a machine, or as resources. Second, even something as simple as counting carries with it moral weight. I have written elsewhere[11] that arithmetic presumes that the items to be counted—the digits—are identical. Before you dismiss this as so much hairsplitting, consider that Treblinka and other Nazi death camps had quotas to fill. Guards held contests among the inmates in which winners lived, and a preset number of losers didn't. But they're just so many numbers, right? Not if you lose. It's easier to kill a number than an individual, whether we're talking about so many tons of fish, so many board feet of timber, or so many boxcars of *untermenschen*. I don't have anything against counting, but valuing the quantifiable over the nonquantifiable (or at least what you are incapable of quantifying) carries with it moral weight. And pretending that none of this is true or accurate carries with it moral weight.

What is at stake is our understanding of the nature of reality, our relationship to reality, and the stories by which we live our lives. The stories by which we live our lives—the myths we live by—deeply influence how we behave. And further, if the stories we tell are destructive or evil—as in a primary mythology that has as its stated purpose extending one's "dominion" over others (whether this dominion comes from God or Science), that attempts to "bind ['Nature'] to your service and make her your slave" or to "make matter and energy jump through hoops on command"—it should surprise no one when our behavior is destructive. And as we've seen, the stories of science are extremely dangerous: is it possible to be more dangerous than to be facilitating the murder of the entire planet?

So in whose wisdom do a huge percentage of people in this culture trust? Harris demonstrates one of science's most dangerous stories in his line, "Science, in the broadest sense, includes all reasonable claims to knowledge about ourselves and the world."[12] It's an absolutely extraordinary statement. Remember, he called a trust in "Nature" a "stultifying and dangerous mythology." But certainly any mythology that makes this claim is pretty much by definition stultifying (from the Latin meaning "stupid-making." Can any mythology be more stupid than that which is both leading to and facilitating the murder of the planet?). It is similar to the Catholic Church's claim that "outside the Church there is no salvation." This was an attempt to convince people that the Catholic Church had (or has) a monopoly on salvation, which was (and is) an attempt to claim a monopoly on a particular sort of power. Science's attempt to convince people that science has a monopoly on knowledge is also an attempt to claim a monopoly on a particular sort of power. After all, as Francis Bacon said, "knowledge is power."

Harris has also said, "It is time we realized that to presume knowledge where one has only pious hope is a species of evil. Wherever conviction grows in inverse proportion to its justification, we have lost the very basis for human cooperation. Where we have reason for what we believe, we have no need of faith; where we have no reasons, we have lost both our connection to the world and to one another. People who harbor strong convictions without evidence belong at the margins of our societies, not in our halls of power."[13]

For once, I agree with him (or at least I agree with the parts that aren't pretty-sounding-but-empty rhetoric). And I challenge Harris or any other scientific fundamentalist to provide evidence that science, and a scientific, materialist, instrumentalist, mechanistic perspective that has as its aim extending human "dominion" and causing the world to "jump through hoops on command," is not gravely harming the world, is not driving species after species extinct, is not destroying natural community after natural community, is not diminishing life on this planet. I challenge him and those like him to show how the belief that science and the scientific, materialist, instrumentalist, mechanistic perspective can do anything other than gravely harm the world is anything other than a "pious hope." I challenge him and those like him to show how their convictions that science is not a "species of evil" can still grow even in the face of the horrors caused by science and scientists and the whole scientific, materialist, instrumentalist, mechanistic perspective.

It is true that where we have evidence for what we believe, we have *less* need of faith. And so I would challenge scientists to show their reasons for believing that the mythology of science—based as it is from its genesis to its current endgame on increasing power and control—is not destructive of life on this planet. Where is the evidence? We can talk till the end of the world about the importance of science in helping us understand global warming (and believe me, we *will* talk about the importance of science in helping us understand global warming until the very end of the world, and we will talk and talk and talk, and do anything to distract us from actually stopping this culture before it causes the end of the world), but carbon dioxide emissions continue to rise, and without science and the scientific, materialist, instrumentalist, mechanistic perspective, there'd be no industrial civilization heating up the globe. It's ridiculous to sing the praises of science for helping to understand or even (almost undoubtedly unsuccessfully) helping to solve a problem that either would not exist or would not be so severe were it not for science and the scientific mindset in the first place. This applies not only to global warming, but to any of the other myriad problems caused by science and the scientific, materialist, instrumentalist, mechanistic perspective.

Harris says, "People who harbor strong convictions without evidence belong at the margins of our societies, not in our halls of power." Yet that is

precisely where the firm believers in the mythology of science reside, this
mythology based on the ridiculous conceit that the world is mechanistic,
or dead, this mythology based on the equally ridiculous conceit that only
humans are *really* sentient, this mythology based on nonhuman others
not existing for their own sakes, this mythology based on torturing the
planet into compliance, this mythology based on the notion that "we will
find better and better ways to exploit our resources and maintain our way
of living while still protecting our forests and oceans and the rest of our
environment," this notion for which there is no evidence whatsoever, this
notion for which there is an entire world of evidence to the contrary.

⊛ ⊛ ⊛

Science is imbued with arrogance, especially the supreme arrogance of
believing that everything can be known; or that science is about control—
about, as Dawkins said, "making matter and energy jump through hoops
on command"; or, to put all this another way, that having done away with
the distant monotheistic sky God, humans are now trying, through science, to take God's place, to know as much as God, to become as powerful
as God, to become God.

What evidence is there for this? Well, we can talk about the words of
J. Robert Oppenheimer, "father of the atomic bomb," who after the first
Trinity test, famously said, "Now I am become Death, the destroyer of
worlds."[14] Or we can talk about the billions of dollars spent to create artificial life or artificial intelligence while this culture destroys real natural life
and denies the existence of real natural nonhuman intelligence. We can
talk about the widespread belief among many of those who belong to the
cult of the scientific, materialist, instrumentalist, mechanistic perspective
that someday science and technology will "solve" not only such existen-
tial "problems" as pain, aging, physical limitations, and death; but also
the very real problems *caused* by science and technology, such as toxic or
radioactive waste, global warming, biodiversity crash, human overpopulation and overconsumption, and so on. We can talk about the fact that
this widespread belief is an article of faith that fails to stand up to even
the most superficial scrutiny, and we can talk about the fact that all the
evidence in the (dying) world fails to shake the faith of the True-Believing

members of the cult of the scientific, materialist, instrumentalist, mechanistic perspective. We can talk about this culture's frenzied insistence that there be no limits on growth, "knowledge," exploitation, power, wealth. We can talk about the fact that this culture is killing the planet. We can talk about the fact that although methane burps have started, and although it is widely understood that anthropogenic global warming is caused in great measure by the burning of coal, oil, and gas, none of the mainstream proposals to curb global warming seriously propose stopping the burning of coal, oil, and gas. These proposals take industrial capitalism as a given and the real world as that which must conform to industrial capitalism (just as an omnipotent God could make all others conform to His wishes). If we define insanity as being out of touch with reality, this is by definition insane. The real world is *the real world*. This culture is not the real world. The stock market is not the real world. The US government is not the real world. Laboratories are not the real world. The real world is sockeye salmon, black terns, Ethiopian wolves, Mekong giant catfish, Sicilian fir, the Columbia River, the Amazon Basin, polar ice caps, the Pacific Ocean. And one of the many things these cult members do not allow themselves to understand is that without a real world you do not have a social structure, even a social structure in which you can make believe that you can force matter and energy to jump through hoops on command. No planet, no you, no matter how megalomaniacal you may be. Fantasies aside, you ain't God. Believers in the Cult of the Industrial God, believers in the Cult of the Scientific God, believers in the Cult of the Mechanistic God, are staking the life of the planet on their entirely unsupported and unsupportable faith that, through science, humans, or rather Humans—*Homo sapiens sapiens*: the wisest of the wise—will be omniscient enough to be able to find solutions to the crises caused by the burning of coal, oil, and gas without stopping the burning of coal, oil, and gas, and will be powerful enough, omnipotent enough, to be able to make these plans work, in violation of the straightforward "scientific laws" of cause and effect which they say govern the universe. Doesn't that sound like God, or rather a cult who pretends they're God? How is the continued belief in these plans, in the face of the violation of those "scientific laws," any less ridiculous than fundamentalist Christians believing that Jehovah stopped the earth from spinning so that Joshua could win the battle of Gibeon?

You've heard of the governmental agency called the God Squad, right? The official title is the Endangered Species Committee, and the purpose of the committee is to explicitly determine whether or not to condemn a specific species to extinction because it is in the way of economic activities. Of course most of the time the God Squad doesn't need to become involved; hundreds of species are driven extinct every day by this God culture.

Still not convinced? Let's try this. We can talk about the faith—supported by no evidence whatsoever, and contradicted by a (dying) world of evidence—that members of this culture can through science and technology know enough, can become omniscient enough, to manage forests without killing them, to manage rivers without killing them, to manage oceans without killing them. Can play God. It is an utterly fanatical religious belief that through science, humans can know "more or less everything there is to know" about these others, or at least enough to make them "jump through hoops on command, and to predict what will happen and when." Well, I've got a prediction for you, you arrogant, murderous motherfuckers: if you try to control forests, rivers, oceans, and so on, you'll kill them. And truth be told, I've got another prediction for you, and listen to this one well: I and people like me, allied with nonhuman people, and allied with those on the other sides, are going to stop you before you can fully manifest your desire to become Death, destroyer of worlds.

⊛ ⊛ ⊛

Wait, you say, how do you know these others are real subjects? How do you know you aren't projecting? You've told us a bunch of ways you think are invalid for determining the subjectivity of another. Aren't you going to tell us a way that is valid?

I don't think that's the right question. Some better questions might include: How do we survive? What is the relationship between one's definition of truth and survival? What is true?

Having said I don't think it's the right question, I'm going to take a stab at answering it anyway. Q: How do we know whether it is more accurate to perceive others as having subjective experience, or to perceive them as non subjects, that is, objects to be exploited, matter and energy which we can force to jump through hoops on command? A: Results.

Exactly, I can hear Richard Dawkins say. Results. That's what we get when we make them jump through hoops.

To which I respond, that's not what I mean by results. I mean something far more sophisticated, far more important, and far more real. I mean the ability to live sustainably. I mean survival.

I live on Tolowa land. The Tolowa lived here for at least 12,500 years without killing the place. This culture has been here 180 years, and the land is dying. Why?

Richard Dawkins says you can tell that a worldview is true if that worldview can help you make matter and energy jump through hoops on command, and predict what will happen and when. I say that you can tell a worldview is true if you and those around you can live with that truth for 12,500 years.

Those sorts of results—the ability to live in a place and to make that place better, on its own terms, because of your presence—are the most important results there can be, and they are the greatest and most sophisticated and most important measures of a worldview's truth.

Ⓢ Ⓢ Ⓢ

I received a note asking that I sign on to the following "Call to Conscious Evolution" in the form of a petition: "Climate change, economic disparity, educational inequities, geopolitical tensions—these mounting concerns are symptoms of a world that is out of balance. Together we can shift consciousness by co-creating a new way of being together.

"The Call to Conscious Evolution was born following a gathering of global visionaries. It's a movement that fully supports that the future is not what happens to us, but rather what WE create.

"Together, we can co-create a new narrative of conscious evolution by building a global community and creating a culture of peace; restoring ecological balance to nourish all life, and mitigate the effects of climate change; engaging in social and political transformation by calling for a more conscious democracy; promoting health and healing by acknowledging the profound mind-body-spirit connection; supporting research and education that optimize human capacities; encouraging integrity in business and conscious media.

"In this great time of uncertainty, join us in elevating consciousness to create a better world. One governed by meaning and purpose. Accept nothing less.

"Every voice counts—your voice counts [in this meaningless Internet petition].

"THE PLEDGE: I join with Deepak Chopra, Marianne Williamson, Jean Houston, and other evolutionary leaders to pledge to make my conscious evolution an important part of my life, and in so doing help make the world a better place."

This is their fucking solution to global warming? To sign a petition (petitioning whom? No one: a petition sent into the Internet's ether, and they're not even burning sacred Internet incense) pledging to make "conscious evolution" an important part of their lives? Would it be possible to be more narcissistic? Would it be possible to come up with a less meaningful solution?

My next wave of response was, as you can probably guess, one of anger. This is all so wrong.

I need to be clear. I'm in favor of good intentions. But it's offensive to me that people try to meditate carbon dioxide out of the atmosphere, or pretend that working on their precious little "conscious evolution" will help stop global warming, *when we aren't doing the work that is needed in the physical world*. This is analogous to standing on railroad tracks trying to meditate a train to a stop, focusing our intention on the train instead of focusing our intention on stepping off the tracks, or better yet, pulling up the tracks and blowing up the bridges. To stop a train, you dismantle the infrastructure that allows the train to run. To curtail global warming, you dismantle the infrastructure that causes global warming.

We all know what we must do to curtail global warming. We must dismantle every oil refinery, every pipeline, every oil and natural gas well. We must dismantle the infrastructure that is killing the planet. Global warming is caused by the burning of oil and gas. Those who profit from the burning of oil and gas will not voluntarily stop profiting from the burning of oil and gas, even at the expense of life on the planet. This entire economic and social system is based on the burning of oil and gas. Those who benefit from this economic and social system will not voluntarily stop burning oil and gas, even at the expense of life on the planet. We need to deprive them of their means to kill life on this planet. Meditating won't do that.

The Derrick Jensen Reader

Meditating or conscious evolution wouldn't have stopped Hitler. It wouldn't have stopped Stalin. It wouldn't have stopped Ted Bundy. It wouldn't have stopped apartheid, US chattel slavery, the horrors of the Khmer Rouge, of Manifest Destiny. It won't stop the horrors of industrial capitalism. Of industrial civilization. It won't stop global warming.

I got a note recently from a very smart Anishinaabe/Nehayow woman saying she loved how I bash hope—which I define as a longing for a future condition over which we have no agency—but also saying that I was forgetting the rightful role of prayer. However, she didn't take the ridiculous New Age direction of saying we need to pray our way out of difficulties. She said, "Of course we hope for things when we have no control over them. And, indeed that is why we ought to not hope for the survival of the planet because we do have control. Stop hoping, start acting. But, what we don't have control over is the response of the planet: the other two-leggeds, the four-leggeds, the swimmers, and the flyers, the grandfathers, the soil, the trees. . . . You get the idea. Pretty much the rest of creation. And this is where I believe we must remain hopeful. This hope is the hope you see in prayer. So that, after I blow up the dam (to use your example), I will pray that the rest of creation accepts my offering. And I have hope that with enough of these actions, creation will eventually accept these offerings. If I do not have hope that creation will listen, than I have almost no motivation to continue. I might as well help accelerate our downward spiral (which may be a technique given some of our prophecies . . . but that's another story for another time)."

In other words, once the destructiveness of this culture is brought to a stop, it will be up to trees and oceans and soils and grasses to do their part by helping to heal the grievous harm this culture has caused. That is where prayer helps. But if we don't first do our part, then our so called prayers—our "intention," our "conscious evolution"—are in all truth blasphemy. We have to do the real work, and then we may do the real praying.

⊛ ⊛ ⊛

I think we underestimate the permeability of our psyches. Just as with every breath I take in, I breathe in trees, breathe in soils, breathe in rain—after all, that's what scents *are*—I think the same thing happens psychically. I pass a tree, who does not end at the tip of a needle or a broken branch, but stretches far in all directions, sensing, thinking, speaking, waiting, acting. This tree reaches in to me, like my muse, like my dreamgiver, like you, like the scent, taste, sight, sound, feel of you reaches inside of me to ride along with and within me, to feel with me, to feel what I feel, to feel me and who I am. For a time the tree rides along with me, and then it leaves. Except it doesn't, since parts remain behind, as happens when you are with me, and I am with you. I, too, enter the tree, feel with the tree, communicate with the tree, if only I remember. The same happens from fungus to stone, river to sand, sand to shell, shell to star, star to me, star to you, everyone to everyone, speaking, listening, hitchhiking, possessing, surrounding and being surrounded by, permeating and being permeated by.

That's life.

❦ ❦ ❦

Some possessions are mutually beneficial. My possession by my muse helps her, and it helps me. Trees give off hormones in spring to help fish grow, and (decaying) fish feed trees and help them in ways I literally cannot imagine, much less know. The sights and sounds and smells and tastes and textures of redwoods make me happy, and my hope is that mine may do the same for them.

Many possessions that at first seem harmful, and that *are* harmful to the one possessed, may in fact be helpful on the far more important level of the community. But not all possessions are helpful or even benign, to individuals or to communities. I think there exist those others who, for their own reasons, do not wish us—individually, collectively, or life in general—well. Given that this culture is killing the planet, and given the manifest insanity of this culture's ubiquitous destructiveness, and given the manifest insanity of this culture's response to this culture's ubiquitous destructiveness—we are, after all, talking about life on this planet—it is long past time we at least consider learning about who some of these might be.

❦ ❦ ❦

I asked the frogs for a dream, asked them to tell me what it's like to be them, asked them to tell me what I can do, asked them to tell me anything they wanted me to know.

I had one of the worst dreams of my life. To be a frog right now is more horrifically, painfully, terrifyingly difficult than I can reasonably imagine.

I awoke a bit before dawn. In the first part of the dream that I remember I was on a train (and I've put some tentative interpretations in endnotes). The train consisted of one long car.[15] A big man walked up and down the length of the car, raping whomever he chose in that moment. He and his rape were relentless, incessant. He was never not raping, and no one escaped his rape. He raped women. He raped men. He raped children. He killed women. He killed men. He killed children. He raped corpses.[16] The train was moving at breakneck speed. But several of us[17] conspired to get off. We jumped off the back of the train. Some of us died when we jumped off. Some of us did not die. Those of us who survived kept running, on or alongside the tracks.[18] No matter how fast the train went, we could not slow down. We were so tired. We saw the big man push the top halves of people's bodies out through the train's windows as he raped the bottom halves of their bodies. We could often no longer tell if these bodies still contained life. But no matter, he kept raping, and we kept running.

Being outside the train did not protect us. He stood in a doorway, wearing a blue military coat and no pants, his penis hugely, throbbingly erect, shouting at us that we needed to run faster, and that he would make sure that each and every one of us had to "suck my dick." Constantly, ceaselessly, he shouted this as we ran alongside the speeding train.

The train started to pull away from us. It was traveling so fast. We wanted to rest, wanted to get away from this ceaseless rapist, and knew that if we kept near him he would kill us and if we stopped running we would die.[19] We ran faster and faster. More and more of us died. And we saw through the windows that the rapes never ceased. And we saw that he would stick his penis out through windows, out through doors, and make some of us who were running, who were barely hanging on to life, service him.[20]

Far ahead was a tunnel. Somehow we knew that here would be our chance to catch him, to kill him. Up to this point none of us had considered fighting back. It simply never occurred to us. We were too busy surviving the rapes, too busy running alongside the train to hell, too busy trying to stay alive.[21]

The tunnel provided promise.[22] I do not know how or why. I just know it did. But the train was pulling away from us. And if it got too far ahead we would die. We were so tired of running.

We saw another train coming along far behind us. We did not know if this train was different.[23] We knew it could be no worse. But it did not matter, because if we did not keep up with *this* train, we would die.

The tunnel got closer and closer. The train disappeared into it.

We saw that before the entrance to the tunnel were a pair of long and winding concrete bridges. These bridges, mirror images of each other, wound left and right, up and down. We each had to choose which bridge we would take. We no longer made any decisions communally, as we had done previously. We were too tired from running, too traumatized from the rapes, from the murders. We no longer had the reserves to do anything but keep running as fast as we could, with no rest, to keep up with this awful speeding train of rape, this train moving so fast to its own and everyone else's doom, to keep running or die.

The bridges were surrounded by masses of high-power electric lines. I approached the bridges, running as fast as I could. I was vaguely aware that some chose one bridge, some chose the other. I no longer had the luxury to pay attention to who chose which, or to the fates of those who chose the other. I was nothing but exhaustion, still running.

As I ran along the concrete bridge, I found myself running faster than I ever thought possible, then faster still. And when I got to the mouth of the tunnel, someone said to me, "Touch these wires, and it will all be over. You can rest. There will be one moment of blinding pain, and then oblivion."[24] I did not want oblivion. It was not yet my time to die. But I was so tired. I wanted to rest. I saw or heard or imagined so many sparks, so many pops, as others chose oblivion over the torment that had been thrust upon us.

I did not touch the wires. I needed to continue. I entered the tunnel.

I emerged driving a car. A woman was with me in this car. The streets were crowded with cars, so crowded it was hard to move. I knew that she and I had to get out of the city or we would die. We tried to maneuver the car, but the street was too crowded. So we abandoned the car, and pushed our way into a small but crowded restaurant.

The woman was still with me. We needed to stay together[25] But there were so many people that we got separated in the crush of humanity.[26] We tried to stay in contact, but she was slowly pulled away. I tried to swim through the crowd but could not reach her.

And then I felt a pain in my leg. I looked down. I saw someone chewing on my leg.[27] I kicked him. Someone else reached for me. I punched her. But now there was someone else, and someone else besides that. I realized I was surrounded, not merely by humans, which would because of their sheer numbers have been bad enough, but also by those who were once human, but who were now undead, who were now zombies, who now tore apart and consumed every living being. And I was fighting for my life.

That was the dream that the frogs gave me. That is what they wanted me to know. That is what they wanted me to tell you.

❦ ❦ ❦

Do you have a fate? Have you ever experienced yourself as having been "fated" to do anything? Meet anyone? Were you ever "fated" to be in a certain relationship? If you answered yes to any of these, how have you explained this fate to yourself or others? Have you tried to explain it?

In my late teens and early twenties I read everything I could about being fated; what it means; who gives us these fates; who helps us to accomplish them; whether it is possible to have a fate and blow it; the role of free will in fate.

In retrospect this was absolutely age appropriate; I was figuring out who I was in relation to and with myself, local communities, society at large, the world, the universe, other sides. I was preparing for the work of being an adult.

Two quotes in particular captivated me. The first was by Seneca: "*Ducunt fata volentem, nolentem trahunt* (The fates guide him who will, him who won't they drag)."

The second was by Thomas Mann, commenting on Schopenhauer, "Precisely as in a dream it is our own will that unconsciously appears as inexorable objective destiny, everything in it proceeding out of ourselves and each of us being the secret theater manager of our own dreams, so also in reality the great dream that a single essence, the will itself, dreams with us all, our fate, may be the product of our inmost selves, of our wills, and we are actually ourselves bringing about what seems to be happening to us."[28]

These two seemed to me at the time (and still do seem) to carry two widely disparate presumptions about the nature of fate. In the former, you have a fate, and you can go along willingly with this fate, or you can be dragged. This seemed (and seems) to argue for some external beings or forces leading us (and when necessary dragging us) in some direction it has been determined we are supposed to go. The forces could be God or gods. They could be ancestors. They could be "the universe." They could be just about anyone or anything.

The second seems to argue that it is our own will who was guiding us all along: the guiding actually comes from ourselves.[29]

It should not surprise us that the former was written thousands years ago, while the latter was written in the twentieth century. This follows the movement of the entire culture, of systematically cutting ourselves off from the universe (and other sides) and simultaneously denying that the external world has meaning, and that other sides exist. This takes us back to what I argued earlier, about how members of this culture can sometimes allow that some dreams may have some meaning because in this perspective dreams come from us. So our own will, like our own dreams, can have meaning, and our own will can guide us, but because this culture has made a fetish, a god, and a lifestyle of control, and is based on the terrifying and fear-based notion that we must always be in control of everything (most especially ourselves, except when we can't control ourselves because nasty selfish genes cause us to do nasty selfish things we would never have done on our own) then *of course* we can (and *must*) never be guided by some external fate.[30]

Please note I used the word "guided," not "controlled. There is a line from the book *Sacred Possessions*: "The external stimuli, like the stars, influence but do not compel."[31]

There is much I don't like about Greek civilization. It was racist, sexist, classist, a democracy every bit as self-congratulatory and false as the modern United States, a conquering empire as self-justifyingly aggressive and as based on theft and murder as any modern empire, including the United States. It was based on destruction of its own and other land-bases. In other words, it was a typical civilization.

That said, I do like parts of the Greek notion of *eudaimonia*, which is what some of the ancient Greek philosophers called the point of life. It's commonly translated as "happiness," but I believe a more accurate translation would be "fittingness": how well your actions match your gifts, match who you are. My understanding of it is that after we die, we spend a hundred lifetimes being treated how we treated others here on earth—and I hope the scientists, capitalists, pornographers, rapists, vivisectors, politicians, and other zombies enjoy their time in hell—after which we go back into the pool of those to be reborn. When our turn comes we decide who will be our parents and what will be our gifts, our purpose.[32] Just before hopping back to this side we drink something that causes us to forget. And here we are. It becomes our task in this world to remember our gifts, our purpose, and to realize them, with the help of guiding spirits, or daimons. Thus *eudaimonia*, which literally means "having a good guardian spirit."

I think again of that line, "The external stimuli, like the stars, influence but do not compel."

And I think also about what so many indigenous peoples have said to me, that humans are to be guided by "original instructions" on how to live, including how to live in concert with one's nonhuman neighbors, who presumably receive their own original instructions, too, all given them by their Creator. It is a person's responsibility to live according to these original instructions. Many who do not live right lives do so not because humans are evil, selfish, and destructive, nor because of a Christian original sin, but rather because they have forgotten or actively ignored their original instructions, and they continue to forget about and actively ignore those who continue to give them their instructions.

In fact, this notion of original instructions could not be further in its spiritual, behavioral, and social implications from the implications of original sin, or original sin's intellectual derivative, selfish genes. Original instructions presume we come into this world carrying with us advice on how to live properly, how to fit in, how to do what is right; and even more crucially, we come into this world having been given a personal and social framework for looking for that advice, for finding it in our daily lives, in dreams, in our relationships with others, and in these others' actions. These instructions are available to individuals through direct unmediated experience, and although one is often encouraged to ask assistance from one's elders and one's community in understanding these instructions and in comprehending messages one may get from dreamgivers, muses, others on other sides, and others on this side, one is also not only encouraged but expected to take increasing responsibility—from the root meaning "to give in return"— for one's own understanding and interpretations, and how this understanding translates into action and behavior toward one's community, which includes nonhumans and other sides. This is called "wisdom." The default in this cosmology is sufficiency; and becoming a selfish, destructive, narcissistic asshole takes effort. On the other hand, original sin presumes that we enter this world damned, insufficient, and only through the intercession of the one and only Jesus Christ (as always, with this culture, there can only be one way, one door; whether it is Jesus Christ, capitalism, science, or anything else about this culture, it's this culture or death), and normally with the assistance of experts of the cloth, or at least through the mediated Word of God that is the Bible, are we able to finally exit this world and go somewhere in a galaxy far, far away where there is no death, which means where the fundamental responsibility of life, which is death (my death is one of the offerings I must make to have this beautiful experience of having been alive), is denied (which I guess means the machines have won, because there will be no life). The default in this cosmology is insufficiency, is "sin." The cosmology of the selfish gene is, as I said before, even worse: Christianity without the redemption. It presumes that all of life is based on behavior that under almost any definition would be deemed sociopathological, and that sufficiency, indeed knowledge itself, is based on domination and control. The default in this cosmology is sociopathy. Becoming a (by definition) selfish, destructive, narcissistic asshole takes no effort at all, as we see. In fact it's in our very genes.

We've been circling the subject of tragic inevitability, preordination, narcissism, what we can and cannot control, and this culture's strange relationship with cause and effect that we discussed earlier when speaking of proof. When I attack science, and when I attack tragedy, I am in no way attacking the observation (when we can) of cause and effect, nor attacking the perception (or attempted perception) and certainly not the existence of consequences. If I poison the creek near my house, those who live in the creek will die. If I poison the oceans, those who live in the oceans will die. If I burn carbon and change the atmosphere, there will be consequences.

If I take prednisone, as I did in my twenties for Crohn's disease, on the advice of doctors, I may destroy bones in my ankles. I understand *how* prednisone destroyed my bones no more than I understand *how* my muse, my dreamgiver, the gambling god, or others communicate with me, but I no longer take prednisone. I certainly understand the existence of cause and effect, and I also understand the existence of consequences. I am attacking, among other things, the mind-numbing stupidity and absurdity of fetishizing the insistence that a failure to perceive a mechanistic cause and-effect relationship *means* that such a relationship does not exist (as in my reputation-threatening proposal that someone may be attempting to encourage me to write this book by helping me to win bets when I write on it and to lose them when I don't; why is this culture so afraid of the numinous, or even just the inexplicable?) and at the same time utterly refusing to see cause-and-effect relationships that would, if perceived, lead one, if one were even remotely sentient, and were not a sociopath or a zombie, to act decisively to stop the actions that are leading to unspeakably dreadful consequences. Let me put it simply: many people in this culture guffaw at the mere possibility of connections for which (or whom) we can discern no evidence for a cause-and-effect relationship; and they simultaneously fail to discern perfectly obvious cause-and-effect relationships when the perception of these relationships might activate their conscience, and the halting of these relationships might threaten the comforts and elegancies on which they have come to rely.

🐚 🐚 🐚

Traditional divination practices have often been based on the notion that those on other sides are able to influence physical objects such as tea leaves or tarot cards or knucklebones (and how interesting is it that the talus— the primary bone transplanted into my body during the writing of this book—was one of the bones[33] used as early dice for divination and game play). The point is that I begin the process, whether by stirring tea, shuffling cards, tossing dice, starting a fire, or what have you, and then those on other sides send signals through these media, or through dreams, or through whatever other means we and they have to communicate, and then I attempt to interpret these signals.

This takes us right back to the question, if there are those on other sides, and if they are able to influence the shape tea leaves assume in the bottom of a cup, why haven't they influenced the shapes of viruses to wipe out this nasty culture before it kills the planet? (In related news: just tonight liberal US President Barack Obama announced a three-year spending freeze on many domestic programs, all the while increasing spending on war—conquest and murder—dramatically.)

I don't know. Hundreds of pages and a year later, I still don't know.

But here is another attempt at an answer. Maybe those on other sides need someone to roll the die, to cast lots, to brew the tea. Maybe these acts are done together, us on this side and the others on other sides. Maybe we need each other for these acts. Just like my muse needs me to write the words, and I need her to give them to me, perhaps divination requires one with a physical body on this side to help the other to act here. Perhaps the same is true not only for casting lots, but for many other actions as well. Perhaps there are those on other sides who want desperately to stop this culture, but they need our help. Perhaps they need our help as much as we need theirs. They cannot do it alone.

And perhaps part of the reason this culture has time and again won is that it has had the help of some on other sides who for whatever reasons do not wish us well, and perhaps those who do not wish us well have required the help of those on this side who are more interested in destruction than in life (remember that freeze on domestic spending and increased spending on conquest).

Perhaps, for better and for worse, we work together with those on other sides to create the future, or the lack of one.

⊛   ⊛   ⊛

Richard Dawkins says that science boosts its claims to truth on its "spectacular" ability to make matter and energy jump through hoops on command, and to predict what will happen and when.

But I'll tell you the strongest evidence I can see that the scientific worldview may be "objectively" true: that no one has yet stepped in to stop this psychopathic culture. If anyone or any group is capable of stepping in, why have they not done it?

⊛ ⊛ ⊛

Here's another thought. Perhaps no one has stepped in to stop the dominant culture from killing the planet because the actions of the dominant culture on this planet are one battle in a larger conflict or war. Certainly many mythologies have been based on the concept of cosmic wars between good and evil, although because of the co-optation of the words "good" and "evil" by monotheism (and indeed by capitalism and patriotism, both of which declare the acquisition of power to be good), we may wish to describe these conflicts as between those who serve life and those who wish for whatever reasons (please note I'm still not asking what those reasons are) to destroy life. And certainly one can win a war while still losing some battles; perhaps those who serve life are busy elsewhere. Or perhaps they are losing on many fronts. I obviously have no knowledge of this, and I can think of no way I can gain this knowledge except by asking for a dream. So I ask.

⊛ ⊛ ⊛

What if the Aztecs were right? What if their gods did indeed require blood sacrifices in order to be sustained? And what if every action requires a sacrifice? What if we owe spiritual debts to other sides? What if every invention carries with it a spiritual debt that must be paid? What if a knife is an expensive piece of equipment, and a gun even moreso, and an airplane even moreso? What if an automobile costs more than money—what if it creates a spiritual debt?

Let's ask that question about the Aztecs again: what if their gods did indeed require blood sacrifices in order to be sustained?

Now let's ask it one more time, with two slight changes of emphasis that change everything: what if *their* gods did indeed require *blood* sacrifices in order to be sustained?

And now let's ask a few more questions: What if other gods require blood sacrifices as well? What if in return for these sacrifices they will give you power? And what if they do not themselves have bodies, do not themselves have hands and fingers and thumbs with which to make things, and what if they need some human followers—some slaves, some addicts, some domesticates, some zombies—to do their bidding?

And what if these other gods do not wish us well?

⊛ ⊛ ⊛

Do you really think the Aztecs thought their mass sacrifices were anything other than how the world works? Don't you think the Aztecs had an epistemology and cosmology that described the universe in ways that made sense to them, in ways that led them to think that they were just *living*?

⊛ ⊛ ⊛

What do you call the murder of the oceans? Once, whales were so thick in the oceans they were a hazard to shipping. Once, fish were so thick you could drop in a bucket and remove it filled with fish. Once, seabirds were so thick you could not see the sun. What do you call these murders?

What are they if not sacrifices?

Most any sane person already recognizes them as sacrifices: the planet is being sacrificed on the altar of economics, or on the altar of technological progress, or on the altar of civilization.

But what if this is not merely a figure of speech, but is in all spiritual truth what is happening?

What do you call the murder of the passenger pigeons? Prior to the arrival of this culture, there were more passenger pigeons in North America than all other birds combined. Members of this culture slaughtered them, and then slaughtered them, and then slaughtered them, until there were none left. Prior to the arrival of this culture, there were penguins in the northern hemisphere: they were called great auks, and they were plentiful beyond modern belief. Members of this culture slaughtered them, and then slaughtered them, and then slaughtered them, until there were none left. Prior to the arrival of this culture, the plains (and woods) were full of bison. This culture slaughtered them, and then slaughtered them, and then slaughtered them, until there were almost none left.

What if these are real blood sacrifices, and what if there are gods leading this culture, and these sacrifices are absolutely necessary in order for them to be sustained? What if these sacrifices are the source of this culture's "luck"?

Name one indigenous culture which has not been harmed by the dominant culture. You cannot do it, because it cannot be done. Every indigenous culture has been harmed. Most have been driven from their land, most have had their ways of life destroyed, most have had their languages extinguished.

Why?

Five hundred thousand children die every year as a direct result of so-called debt repayment from so-called colonies to the institutions at the center of empire. Are these not sacrifices? Certainly they are sacrifices to capitalism. Who is to say they are not sacrifices to some awful god of capitalism?

Sixty thousand people die every day from the effects of pollution. On whose altar are those lives sacrificed?

One out of every four women in this culture is raped within her lifetime. An additional one out of about five has to fend off a rape attempt. Whom does this terror serve? Who feeds off of this mass terror? To what gods are the unbroken psyches of these women offered?

There were forty-some attempts made to assassinate Hitler. These attacks often failed for nearly unbelievable reasons. Fog saved Hitler's life. Bombs unaccountably failed to explode, even after the detonators fired. A fly landing on someone's hand caused an attempt to be aborted. Soviets could have trapped Hitler deep inside Russia had their tanks not run out of gas directly at the edge of the runway Hitler used to escape. In the famous July 20, 1944, plot, Stauffenberg was only able to activate one detonator instead of the two they had planned because someone interrupted him in the bathroom as he was setting the detonators, and would not leave him alone. The single detonator may have been sufficient, had the briefcase carrying the bomb not been pushed far beneath a massive oak table by someone trying to look more carefully at the maps on top. The list goes on.

How did Hitler get so lucky?

Maybe the tens of millions of victims of World War II and the Holocaust can buy a fair amount of luck.

I don't believe the Aztecs considered their sacrifices as strange or disgusting. Indeed, they were simply a part of everyday life. The sacrifices were routine. They were simply part of how things were. And how are things now?

Here is my final possible reason the dominant culture has been able to dispossess and destroy the indigenous, and why the earth has not killed us off. It is the possibility in which I most believe. This has all come to pass because the dominant culture and its members have been willing to sacrifice so much to their gods. Call these gods God, capitalism, science, technological progress, machines, civilization, or what have you. Or find and call out these gods by the names behind these rationalizations. Find and call out these others who do not wish us well.

⊛ ⊛ ⊛

What if many members of this culture really have become zombies? What if they serve some other master or masters? What if this culture has consistently gotten so fucking lucky in part because of the sacrifices it makes—and heavy sacrifices they are—to some others on some other sides, some others who do not wish us well?

⊛ ⊛ ⊛

I am fully aware that many people, and indeed entire cultures, upon facing their own powerlessness in the face of seemingly unstoppable horrors, sometimes fall back on mystical visions of unseen forces coming from out of nowhere—perhaps literally from out of nowhere—to save the day, to drive the conquerors back to where they came, or even to drive them all the way to hell, to restore the birds to the sky, the bison to the plains, the fish to the water, the people to the land. I am fully aware that, faced with the seemingly unstoppable horror that *is* this culture, many of these visions have made people believe they now wore magical clothing that would stop the bullets of the conquerors.

These mystical visions have clearly not yet come true. The conquerors have not been driven back to hell, but rather have continued to bring hell with them. The birds have not been restored to the sky, nor the bison to the plains, nor the fish to the water, nor the people to the land. The soil hasn't even been restored to the land.

I am fully aware that with writing this book I have been this entire time in danger of falling into similar traps. When all work on this side fails to stop the members of the cult of the scientific, materialist, instrumentalist, mechanistic, managerial perspective, fails to stop the sociopaths, fails to stop the zombies from murdering all whom I love, from murdering the planet, I am in danger of, in desperation and sorrow, harnessing my hopes for a future onto some miraculous assistance from beings this culture staunchly and feverishly inculcates us into believing do not exist.

But I feel comfortable that this is not what I'm doing right now. I'm not suggesting we call on those on other sides in lieu of fighting back. Quite the opposite; I'm suggesting we try to learn to listen better so that we may gain whatever assistance they may be able to give us, as we fight the zombies on this side with every tool at our disposal.

Many freedom fighters have resisted oppression with the assistance of those on other sides. Harriet Tubman is certainly one such example. When she was young, she suffered a head injury when a white man threw a weight at a fleeing slave and struck her instead. This injury led to what we would now probably call seizures, where she would fall seemingly unconscious of her surroundings, but would remain aware of all that happened around her. After her injury, she began to experience visions and dreams which she took as signs from God (she was a devout Christian). These visions remained with her throughout her life, and she relied on this "consulting with God," as she called it, to help her find her way and to save her from danger as she brought slaves to freedom. Evidently her consultations were successful, as she was able to famously remark, "I was conductor of the Underground Railroad for eight years, and I can say what most conductors can't say—I never ran my train off the track and I never lost a passenger." Thomas Garrett said of Tubman, "I never met with any person of any color who had more confidence in the voice of God, as spoken direct to her soul."

Another resistor who listened to those on other sides was Lozen, the sister of the Chihenne-Chiricahua Apache chief Victorio. As well as a skilled warrior, she was a prophet, and those on other sides were able to help her learn the movements of the white invaders. One of her methods of divination was to stand with her arms stretched wide while chanting a prayer to Ussen, a primary deity of the Apaches. The prayer went, "Upon this earth / on which we live / Ussen has Power / This Power is mine / For locating the enemy. / I search for that Enemy / Which only Ussen the Great / Can show to me." She would slowly turn around, and the feelings in her arms would tell her not only where the enemy was, but how many there were.[34]

Those are the alliances I am looking for.

※ ※ ※

This is the point in this book where I'm supposed to give you recipes to follow on how you, too, can connect to those on other sides, how you, too, can find your own muse, can listen to trees who will help you write, can listen to God so He can help you avoid slave catchers, can listen to Ussen who will help you find and defeat your enemy.

I have to disappoint, for a number of reasons. The first is that I don't know how you should connect to those on other sides. Maybe one way would be to have a white man throw a weight at a fleeing slave and instead hit you in the head. Another might be to have you raised in a culture with a several-thousand-year tradition of regular intercourse between sides, with long-term relationships already established, and subtle (and obvious) means of passing information (and possibly oneself) from this side to other sides and back.

I don't know how these communications happen, really any more than I know how I breathe; or how the dog comes when I call, or doesn't; or where my hand is when I'm scratching my back; or how or why sometimes dreams come and sometimes they don't.

I just know that I have made listening to these communications a central part of my life, and I have made making time to listen to these communications a central part of my life. And I have, as with any other friendships, proudly claimed these others who I have relationships with. And I have, as with others who give me assistance, gladly given credit to these others who have helped me.

I also know I have never let anyone convince me, against the evidence of my own experience, that these others are not real.

※ ※ ※

I ask for a dream. It is so repetitive that at first I do not even recognize it as a dream. In this dream time and again I awaken from a dream. Each time I awaken I am a different person. And each time I awaken I am not alone in my body. Each time I awaken there are others with me, around me, above me, below me, without and within me. Each time I awaken there are these others I can see, and with whom I can speak. Each time I awaken there are these others who accompany me and each of us. These others are here, and they are willing to help.

This is the dream I have time after time after time, until finally I recognize it is a dream, and until finally I recognize its obvious meaning, and until finally I recognize it is not even a dream. It just is.

Forget Shorter Showers
Why Personal Change Does Not Equal Political Change

Orion Magazine, **July/August 2009**

One of the big con jobs that has been foisted on us the last thirty years is that the way to make social change is through personal change. It's completely absurd, and it serves those in power. Personal change is completely trivial compared to institutionalized social change. We don't need to take shorter showers; we need to stop big industry and big agriculture from stealing water.

Would any sane person think dumpster diving would have stopped Hitler, or that composting would have ended slavery or brought about the eight-hour workday, or that chopping wood and carrying water would have gotten people out of Tsarist prisons, or that dancing naked around a fire would have helped put in place the Voting Rights Act of 1957 or the Civil Rights Act of 1964? Then why now, with all the world at stake, do so many people retreat into these entirely personal "solutions"?

Part of the problem is that we've been victims of a campaign of systematic misdirection. Consumer culture and the capitalist mindset have taught us to substitute acts of personal consumption (or enlightenment) for organized political resistance. *An Inconvenient Truth* helped raise consciousness about global warming. But did you notice that all of the solutions presented had to do with personal consumption—changing light bulbs, inflating tires, driving half as much—and had nothing to do with shifting power away from corporations, or stopping the growth economy that is destroying the planet? Even if every person in the United States did everything the movie suggested, U.S. carbon emissions would fall by only 22 percent. Scientific consensus is that emissions must be reduced by at least 75 percent worldwide.

Or let's talk water. We so often hear that the world is running out of water. People are dying from lack of water. Rivers are dewatered from lack of water. Because of this we need to take shorter showers. See the disconnect? Because I take showers, I'm responsible for drawing down aquifers? Well, no. More than 90 percent of the water used by humans is used by agriculture and industry. The remaining 10 percent is split between municipalities and actual living breathing individual humans. Collectively, municipal golf courses use as much water as municipal human beings. People (both human people and fish people) aren't dying because the world is running out of water. They're dying because the water is being stolen.

Or let's talk energy. Kirkpatrick Sale summarized it well: "For the past 15 years the story has been the same every year: individual consumption—residential, by private car, and so on—is never more than about a quarter of all consumption; the vast majority is commercial, industrial, corporate, by agribusiness and government [he forgot military]. So, even if we all took up cycling and wood stoves it would have a negligible impact on energy use, global warming and atmospheric pollution."

Or let's talk waste. In 2005, per-capita municipal waste production (basically everything that's put out at the curb) in the U.S. was about 1,660 pounds. Let's say you're a die-hard simple-living activist, and you reduce this to zero. You recycle everything. You bring cloth bags shopping. You fix your toaster. Your toes poke out of old tennis shoes. You're not done yet, though. Since municipal waste includes not just residential waste, but also waste from government offices and businesses, you march to those offices, waste reduction pamphlets in hand, and convince them to cut down on their waste enough to eliminate your share of it. Uh, I've got some bad news. Municipal waste accounts for only 3 percent of total waste production in the United States.

I want to be clear. I'm not saying we shouldn't live simply. I live reasonably simply myself, but I don't pretend that not buying much (or not driving much, or not having kids) is a powerful political act, or that it's deeply revolutionary. It's not. Personal change doesn't equal social change.

So how, then, and especially with all the world at stake, have we come to accept these utterly insufficient responses? I think part of it is that we're in a double bind. A double bind is where you're given multiple options, but no matter what option you choose, you lose, and withdrawal is not an option. At this point, it should be pretty easy to recognize that every action involving the industrial economy is destructive (and we shouldn't pretend that solar photovoltaics, for example, exempt us from this: they still require mining and transportation infrastructures at every point in the production processes; the same can be said for every other so-called green technology). So if we choose option one—if we avidly participate in the industrial economy—we may in the short term think we win because we may accumulate wealth, the marker of "success" in this culture. But we lose, because in doing so we give up our empathy, our animal humanity. And we really lose because industrial civilization is killing the planet, which means everyone loses. If we choose the "alternative" option of living more simply, thus causing less harm, but still not stopping the industrial economy from killing the planet, we may in the short term think we win because we get to feel pure, and we didn't even have to give up all of our empathy (just enough to justify not stopping the horrors), but once again we really lose because industrial civilization is still killing the planet, which means everyone still loses. The third option, acting decisively to stop the industrial economy, is very scary for a number of reasons, including but not restricted to the fact that we'd lose some of the luxuries (like electricity) to which we've grown accustomed, and the fact that those in power might try to kill us if we seriously impede their ability to exploit the world—none of which alters the fact that it's a better option than a dead planet. Any option is a better option than a dead planet.

Besides being ineffective at causing the sorts of changes necessary to stop this culture from killing the planet, there are at least four other problems with perceiving simple living as a political act (as opposed to living simply because that's what you want to do). The first is that it's predicated on the flawed notion that humans inevitably harm their landbase. Simple living as a political act consists solely of harm reduction, ignoring the fact that humans can help the Earth as well as harm it. We can rehabilitate streams, we can get rid of noxious invasives, we can remove dams, we can disrupt a political system tilted toward the rich as well as an extractive economic system, we can destroy the industrial economy that is destroying the real, physical world.

The second problem—and this is another big one—is that it incorrectly assigns blame to the individual (and most especially to individuals who are particularly powerless) instead of to those who actually wield power in this system and to the system itself. Kirkpatrick Sale again: "The whole individualist what-you-can-do-to-save-the-earth guilt trip is a myth. We, as individuals, are not creating the crises, and we can't solve them."

The third problem is that it accepts capitalism's redefinition of us from citizens to consumers. By accepting this redefinition, we reduce our potential forms of resistance to consuming and not consuming. Citizens have a much wider range of available resistance tactics, including voting, not voting, running for office, pamphleting, boycotting, organizing, lobbying, protesting, and, when a government becomes destructive of life, liberty, and the pursuit of happiness, we have the right to alter or abolish it.

The fourth problem is that the endpoint of the logic behind simple living as a political act is suicide. If every act within an industrial economy is destructive, and if we want to stop this destruction, and if we are unwilling (or unable) to question (much less destroy) the intellectual, moral, economic, and physical infrastructures that cause every act within an industrial economy to be destructive, then we can easily come to believe that we will cause the least destruction possible if we are dead.

The good news is that there are other options. We can follow the examples of brave activists who lived through the difficult times I mentioned—Nazi Germany, Tsarist Russia, apartheid South Africa, antebellum United States—who did far more than manifest a form of moral purity; they actively opposed the injustices that surrounded them. We can follow the example of those who remembered that the role of an activist is not to navigate systems of oppressive power with as much integrity as possible, but rather to confront and take down those systems.

The Man Box

This essay was written at the request of Eve Ensler and James Lecesne for their upcoming collaborative theater piece, with the working title "Breaking Out of the Man Box."

The man box is full of proof. Except that there is no man box, the man box can never be filled, and real men don't need proof.

Let's start with Abraham and Isaac. You know the story. God tells Abraham to slit his child's throat. Abraham ties up his son, raises the knife, and at the last moment God says it was a test. End of story. Lesson? Abraham shows, by his willingness to violate his child, the proof of his worth. And Isaac learns that his father was willing to kill him rather than act against the cult of masculinity, against the rules of the man box.

There are many rules of the man box, even though there is no man box, and there are no rules. Why call it a box when it's the way things are? And why call it a rule when it's who you are?

Rule 1: There is no man box.

Rule 2: There is no box *but* the man box, and thou shalt have no other boxes before it.

Rule 3: That's the way things are.

Rule 4: That's who you are.

So I'm in a restaurant, and I overhear one guy say to another that he's in pain. The other responds, "Suck it up. When are you going to quit being such a woman?"

Yes, I understand that men are taught to not feel. Yes, I understand that the cult of masculinity is all about not feeling. I understand that must be hard. But honestly, I don't give a shit about understanding the emotional state of members of the cult of masculinity, except insofar as that understanding might help stop them. It's a bit late in the game to be worried about the feelings of perpetrators.

426	The Derrick Jensen Reader

The ones I care about are their victims, because the man box isn't about putting men in a box, it's about putting everyone else in a box, the box of other, of less than, of trophies, the box of the violable, the box of targets, the box of victims, the box of the violated, the box of proof of the men's own manhood.

Have you ever done the math on how many women who are alive right now have been raped? There are almost seven billion people on the planet, so there are about 3.5 billion women. About one in four women is raped in her lifetime, and another one in five fend off rape attempts. So more than 800 million women living today will be raped in their lifetimes. Let's say half of those have not yet been raped. So 400 million women living now have been raped.

And another now.

And another now.

This also means, among many other things, that unless a few men are excruciatingly busy, there are a lot of rapists out there, a lot of members of the cult of masculinity, a lot of men who adhere to the rules of the man box.

But you already knew that.

But of course there is no man box, and there can be no man box, because if there were a man box, that would mean there's something outside the man box, and there's nothing outside the man box because there *can be* nothing outside the man box, and there *can be* nothing outside the man box because there *must be* nothing outside the man box.

Because if there were, well, there isn't, and can't be, and mustn't be.

Because if there were, that would mean members of the cult of masculinity aren't as omnipotent—as completely potent—as they *must be*. And also because if there were, why would any victims put up with this shit?

So there must not be a man box, because everything is part of the man box.

That is, everything is violable. And everything must be violated.

Rule 5, which is actually Rule 1, which is actually the only rule there is: I exist only insofar as I violate you.

But of course rule 5 does not exist. Nor does rule 1.

The other day I saw an astronomer saying why he thought it was important to explore Mars and other planets: "It will," he said, "answer that most important question of all: Are we all alone?"

I have an even more important question: is he fucking crazy?

No, just a member of the cult of masculinity.

Did you know that 200 years ago there were flocks of passenger pigeons so large they darkened the sky for days at a time? And flocks of Eskimo curlews so thick that ten, fifteen, twenty birds would fall to a single shot? There were so many whales in the North Atlantic they were a hazard to shipping, and there were runs of salmon so thick they would keep you awake all night with the slapping of their tails against the water. And he asks if we are alone?

Only if you're a member of the cult of masculinity, in which case you are of course alone, with other members of your cult, because you have declared yourself to be the only one who matters, the one who does *to* as opposed to everyone else, to whom it is done.

Did you know that this culture is driving two hundred species extinct each and every day? Did you know that stolid scientists are saying the oceans could be devoid of fish in fifty years?

And do you know why?

And did you know that the world used to be filled with thousands of vibrant human cultures? And that human cultures are being driven extinct at an even faster relative rate than nonhuman species?

And do you know why?

The man box is full of women. It is full of passenger pigeons. It is full of whales. It is full of indigenous humans. The man box is full of the entire world.

But the man box isn't full, because the man box—which does not exist—can never be full.

The psychiatrist R.D. Laing famously asked, "How do you plug a void plugging a void?"

That's the question, isn't it?

But of course it isn't the question because men don't have a void, and if they did have a void they certainly wouldn't plug it with a void.

Someone once told me that any hatred—or maybe any void—felt long enough no longer feels like hatred, but rather like religion, or economics, or science, or tradition, or just the way things are.

With all the world at stake I need to speak plainly. The problem is that within this patriarchy, identity itself is based on violation. Violation becomes not merely an action but an identity: who you are, and how you and society define who you are. Within this patriarchy men's masculinity defines itself by identifying others—any and all others—as inferior (which is why those stupid fucking scientists can ask "Are we all alone?" as they destroy the extraordinary life on this planet), and as being therefore violable, and then violating them. For men under this patriarchy, these acts of violating others are how we become who we are. They validate who we are. They then reaffirm who we are, as through these repeated acts of violation we come to perceive each new violation as reinforcement not only of our superiority over this other we violated but as simply the way things are.

So without this identification of others as inferior, without this violation, we are not. We are a void. And so we must fill this void, fill it with validations of our superiority, fill it with violations. Thus the rapes. Thus the violation of every boundary set up by every indigenous culture. Thus the extinctions. Thus the insane belief in an economic system based on infinite growth despite the fact that we live on a finite planet. Thus the refusal to accept any limits on technological progress—more properly termed technological escalation, as it really involves an escalation of the wielders' ability to control and violate at a distance—or on scientific "knowledge." Thus the sending of probes to penetrate the deepest folds of the ocean floor. Thus the bombing of the moon.

What makes this problem even worse is that because there are always those who have yet to be violated, and because this violation isn't really solving the needs it purports to meet—it's a void plugging a void—this drive to violate is insatiable. This culture will continue to violate, until there is nothing left to violate, nothing left.

So what is at stake in this whole discussion is life on this planet. This cult of masculinity must not merely be left, and must not merely be exposed. It must be destroyed, or it will continue to violate its way to the end of all that is alive.

But before we can leave this cult we must understand that it is *not* all that is. That there is a cult of masculinity, and there is a man box, and you can leave them both. Burn this into your heart: this imperative to violate is not natural. It is cultural. And we must resist every effort by the abusers, by the violators, to "naturalize" this drive to violate. For this is what abusers, violators, must do. They must attempt to convince themselves and everyone else that their way is the only way, that there is no other way. They must convince themselves and everyone else that not only is there nowhere outside the cult of masculinity, nowhere outside the man box, but indeed there exists neither a cult of masculinity nor a man box. There is only this one way of life, which is not *just* a way of life because it encompasses all that is or ever was or ever will be. It is everything.

They say.

But they are lying, to themselves and to you. Even if they have an entire culture to back them up, they are still lying.

We must never forget that. There is a cult of masculinity, and there is a man box, and we can leave them. We can not only leave them. We can destroy them. We must. With all the world at stake, we must.

An Interview with Derrick Jensen

by Lierre Keith

Conducted July 16, 2011, in Crescent City, CA

Lierre Keith: Gore Vidal said that first novels are thinly disguised autobiographies written for purposes of revenge. Your first book, *A Language Older Than Words*, was autobiographical. Though it wasn't written for purposes of revenge, it was written to expose something that was very primary in your life, which makes it a very personal book. What compelled you to write it?

Derrick Jensen: Originally, *A Language Older Than Words* was supposed to be a happy face discussion of interspecies communication, where I would compile various people's stories of how they communicate with their dog or cat, or with a tree or river. What I found was that a lot of people were routinely having conversations with nonhumans, but they couldn't talk about it publicly because of fear of ridicule. So I wanted to put enough of these stories together that people could begin to have a public conversation about it. I was compiling these stories and having a good time doing so, but I was having trouble writing the book. I couldn't figure out why until I realized that at this stage in planetary murder, it would be entirely inappropriate to write a book that in any way tries to put a happy face on current human and nonhuman relations. It would be insulting and a lie.

I also realized that to write a book that purports to show that nonhumans can think and communicate would be grossly insulting. It would be like writing a book that shows that Jews really aren't subhuman or that blondes really can think. It would maintain the same old bigotry and chauvinism that characterizes this culture anyway. It would hold up humans as a standard by which everyone else must be judged. What broke the book open for me was the realization that I shouldn't be asking, "Can nonhumans communicate, or can they not?" Instead I was really interested in asking, "Why is it that some people are able to listen, and some people are not? Why is it that some people are *willing* to listen, and some people are not?" And from there the book really opened out to an understanding of how before you can exploit someone you have to silence them, you have to deafen yourself to them, you have to systematically pretend they cannot speak. The Portuguese explorers said of the Africans, "When they speak, they fart with their tongues and their mouths." So because the Portuguese stated that the Africans could not communicate, it was acceptable, in the minds of the Portuguese, for the Portuguese to enslave them. They were rationalizing their exploitative behavior. And we see the same thing on every level with the silencing of women, the silencing of children, the silencing of other cultures, other races, other species. All those to be exploited have been systematically silenced. I became interested—interested is too weak a word—I became consumed by exploring the ways that this culture silences these others. And silences our own experiences as well.

LK: What makes the book so compelling to so many people is that you are exposing your own vulnerable experiences. That really drives the power of the book. Was talking about your own history as compelling for the writer as it is for the reader?

DJ: I don't believe I'm the first person who ever said the personal is the political. And I learned a lot from many very good feminist writers who intimately interweave their personal stories with the philosophy. Another reason I include so many personal things in my writing is that one of the rules of writing is "show, don't tell." It's important to make an intellectual argument, and at the same time, for the writing to really affect people, there has to be an emotional component, too. And then I have my own experiences of having been silenced and of silencing myself, so if I'm going to write about silencing, it would never have occurred to me to not include a discussion of that silencing.

Writing *A Language Older Than Words* was extremely difficult for me because of the intense examination that I had to do of my own abusive childhood. I'd been doing that examination for years, but that didn't alter the fact that, as is true for many healing crises—both psychological and physical healing crises—the process itself was extraordinarily difficult. Prior to writing that book I used to have nightmares pretty constantly and awaken fifteen, twenty times a night with night terrors. That happened routinely my whole life, because of my childhood. During the writing of that book, the nightmares and night terrors got so bad that I wrote most of it sleep deprived: I usually couldn't get to sleep until after dawn. Then I'd just have very fitful, drifting sleep, full of nightmares until 9:00 or 10:00AM and then I would get up for the day. There were a lot of times I just wanted to quit writing the book because that sleep deprivation was so extreme. But I'm glad I stuck with it, for a number of reasons. One is of course now the book and the analysis in it is out. And personally, also, it helps a lot. After I finished the book, I discovered the truth of what so many people from Robert J. Lifton to Judith Herman have said about recovering from trauma: it doesn't really help to pretend it didn't happen, to pretend you can go on with your life without dealing with it. And those who survive best are generally those who are somehow able to take their trauma and turn it into a gift for the community, to metabolize the pain, to make meaning of it and to then use this prior wounding as a gift to the world. Lifton calls this a "survivor mission."

That said, I didn't write the book in any way for personal healing. I'm not interested in writing that is done for personal healing. That's why God invented journals. If you're going to actually attempt to communicate, then you need to have a larger message. The personal healing was a very nice benefit from writing this book, a nice side-effect, but I wrote the book to explore and expose the ubiquitous lies that characterize this culture and that allow it to continue. And not merely to explore and expose these lies, but to affect real change in the real world.

LK: I think that that's one of the reasons that your book is so compelling. I've read a lot of survivor narratives, but yours is the only one I've ever read that made me think: this is somebody I need to be friends with. "Breaking the silence" has become kind of a cliché, and it's really important for survivors to do that, but your book went far beyond that because it wasn't just your personal pain and your personal telling of your story. There's a way bigger meaning than just "I was hurt."

DJ: Well, a couple things about that. One is, once again, this is why God invented therapists, so that you can tell the story to therapists. And the stories should be told to therapists or to journals or to friends or to extended family. It's absolutely crucial, but there's absolutely no reason for me to read about it. I fully recognize that I am writing about painful subjects, and so one of the things that I strive very hard to do is to give readers gifts that are commensurate with the pain that is caused by the reading of the book. If you're going to inflict pain on the reader by talking about extremely painful subjects, then you have to give them gifts in return. Otherwise, it's just a painful experience.

LK: Reading some of those survivor narratives feels like lancing a wound. And that can be an important stage for when you are first coming to grips with your own story. But in *A Language Older Than Words*, you took pain and you transformed it into something else. You made it luminous. And in a way, that is bigger than personal redemption. You are looking at the whole scope of the last 10,000 years of history and the terrible situation that we find ourselves in now as animals on a dying planet.

DJ: On a planet being murdered.

LK: Yes, right. It's the micro and the macro. Your personal story is reflected everywhere.

DJ: While survivor narratives certainly have their importance, *A Language Older Than Words* was never intended as a survivor narrative. It's intended as an exploration of the ways in which this culture silences those who are exploited, and the ways it rationalizes exploitation. And my own narrative was one of the supports for that. And not even a central support. A little bit of trauma can go a long way in a book. There are very few pages in that book that are explicit depictions of the abuse. Maybe three or four pages out of the total of three hundred.

LK: I am reminded of, actually, a line from *The Lion, The Witch, and the Wardrobe* where C. S. Lewis is describing the battle at the end. He's writing for children, and what he says is very simple, "And if I told you how horrible it was, your parents wouldn't let you read this." That is so much more evocative than him actually going through descriptions of chopping and slicing and people dying and all the rest of it. That line has haunted me. In some ways, *A Language Older Than Words* did much the same thing. Like you say, there are only three or four pages of the horrors that you went through as a child, but it was enough to make an impact. It was really all the reader needed.

DJ: Well, thank you. Something that's interesting about the book is that there's a place where it says "Everything I'm saying is a lie. None of this really happened." The very last thing I wrote was that section, which means the book could have worked without the graphic descriptions of abuse.

LK: It's right in the beginning.

DJ: I put that in because my publisher was afraid that my father might sue, so he wanted me to put in a disclaimer. And I told him to get lost. I said I'm not going to say that this didn't happen. And then I realized that I could actually use that as an exploration of how silencing works, where I tell you what happened in the very same sentence that I am denying that any of it occurred. My father *didn't* leap across the table to beat my brother, he *didn't* chase my brother around the house, he *didn't*—

LK: —give your brother epilepsy—

DJ: —he didn't do any of that stuff. None of this happened. Which mirrors the experience of abuse across the whole culture. The world is not really being killed. Men don't routinely rape women. The level of denial that goes on. It's like R. D. Laing's three great rules of a dysfunctional family. Rule A is don't. Rule A1 is rule A does not exist. And Rule A2 is never discuss the existence or non-existence of Rules A, A1, or A2. That section is really a description of that process of amnesia that is central to all abusive behavior. Judith Herman begins her wonderful book *Trauma and Recovery* by saying that there are some things that happen that are too terrible for words, and that's the meaning of the word unspeakable. Milan Kundera said that the struggle against oppression is the struggle of memory against forgetting.

That's what's being manifested in that section, which was written basically as an attempt to protect my publisher from a lawsuit.

LK: One of the things that really struck me excerpting your work was how if you take any small section, it would be really hard to know which book it came from. You have this long trajectory, one meta-theme. I wondered if you felt that in your writing career, you've sort of been writing one book in pauses. Or whether you feel like there's been a change in the themes that you're addressing.

DJ: I feel like my work is one long consecutive story. I remember a conversation with a creative writing instructor I had in my early twenties. We agreed that you really don't judge an author by a book, but rather by their body of work. So I'm more concerned with the whole body of work than I am with any individual book.

Also, I've been really heavily influenced by music. Beethoven has been a huge influence on me with the ways he works with themes. Think about the beginning of his Fifth Symphony. He takes that and then he puts it upside down, he puts it backward, he puts it forward, and he keeps coming back to it and putting permutations on it. A writing teacher, John Keeble, helped me understand how we as writers can do the same sorts of things with our writing as composers do with music. Through that I began to see writing in terms of how the writer can play with the theme, just like Beethoven turning it upside down, playing it backward and forward. So in a musical sense I certainly have themes in my work. And also, certainly there are topics to which I keep on coming back—I mean, this culture's killing the planet.

One goal of physics seems to be some sort of a unified field theory. One of the things that I've tried to do is to come up with a unified field theory of this culture's destructiveness. I don't really write about environmental destruction. What I'm writing about are the connections between that destruction and other forms of this culture's destructiveness, the motivations, the similarities and differences, and then of course exploring resistance to that.

Books are living. As I gain understanding, my books change. If I want to revisit some idea that I wrote about six books ago, how do I do it in such a way that it's going to still be interesting for the people who read that older book, and, at the same time, will introduce new people to the material? I wish that my work could be one big, long book that people could read. Of course that's totally unrealistic. The point is that, yeah, there are things that come up in one book and then I want to explore again in another book, they're not completely independent. Each book stands alone, but there still are references back and forth. I'm presuming that's probably true for a lot of other writers. It certainly happens in fiction. Authors have some characters that repeat, and they have themes. I mean, how many times has Stephen King scared us? Or how many mysteries did Agatha Christie write? How many times did Hercule Poroit figure out the mystery in the end?

LK: Did you always want to be a writer?

DJ: I wanted to be a writer when I was a kid.

LK: Did you like stories? Why did you want to be a writer?

DJ: I liked reading a lot. I read a lot when I was a child. And I would write stories that were essentially plagiarism. I felt bad about this until I found out that a lot of writers do it. I really liked the *Narnia* series, so in second grade I wrote stories about kids who went through a closet instead of a wardrobe into this other world called Palmia instead of Narnia, and the stories were essentially the same as the *Narnia* characters' with very minor changes. Another thing I used to do, when I was seven and eight, was to begin typing a book. I would have the book next to me and then I would copy it. I wasn't claiming it. I was just typing it. I don't know why I did that particular thing, but . . . yes, I wanted to be a writer when I was a kid. I wrote plays and I wrote adaptations—

LK: Do you still have any?

DJ: Probably somewhere. In fourth grade, I wrote a play about somebody who was wounded at the battle of Fredricksburg and had to make his way back to a hospital, where they cut off his arm.

LK: Was it ever produced?

DJ: It was done in class, and I think it was a bit hyperdramatic for most of the students. Most of the plays that the other kids wrote were "me and my pet dog went for a walk" or something, so my plot was pretty over the top in comparison.

LK: A plot about having your arm amputated must have been fun for an eight year old.

DJ: It took place behind a screen. And we actually—

LK: —the drama, my God.

DJ: And we actually made a cardboard hatchet to raise above the screen and bring down and made the person . . .

LK: Scream?

DJ: Scream. Anyway, I still liked to write, and in ninth grade I took a typing class. Instead of typing from the textbook our class was to use, I made a very bad deal with the typing instructor, which was that I could write stories in class and he would count one page of a story the same as one page of typing from the book.

LK: Wow.

DJ: Then in high school, I was good at science, and so I did what people who are good at science do, which is to believe that you should get a good job. I still remember one of those aptitude tests that I took in high school. I was getting great grades in calculus. I was going to go that direction. But the aptitude test said that I should be a cross between an artist and a popular entertainer. I said to a friend of mine, "These tests are so stupid. Look how ridiculous they are, because I know I'm going to be a scientist. This is just crazy!" I went to the School of Mines on a scholarship. I went there *because* I got a scholarship, and I couldn't afford to go to college otherwise.

I hated it, and realized that what I really wanted to do was be a writer. Then, like Joseph Campbell said about his own life, I spent my twenties learning how to write, so that when I had something to write about later, I would know how to do it.

LK: How did you learn to write? What was the most important thing that you did to become a writer?

DJ: When I was in college, I was unhappy, but so was most everybody else there. I had this habit of asking people if they liked their jobs. About 90 percent would say no. I started asking, "So what does it mean when the vast majority of people spend the vast majority of their waking hours doing things they don't want to do?" And I would ask my fellow students, "Aren't you unhappy?" And they would say, "Yeah, I am. But when I get out of here, I'm going to buy myself a nice car." Some fellow students were actually looking forward to retirement.

LK: So age twenty, they were already looking forward to—

DJ: —to being sixty-five.

LK: Right.

DJ: I realized that that wasn't really the life for me. On the other hand, I worked all through college. I worked summers at NOAA, the National Oceanic and Atmospheric Administration, and one of my bosses there got sick one time and he showed up at work anyway at noon. He said it was because he couldn't stand to miss work, because he was so looking forward to doing the experiment. I remember thinking, I don't ever want to work a job that I wouldn't do if I were sick, that I don't want to do *anything* with my life that I wouldn't do if I were sick. What is it that would excite me so much that I would want to do it all the time? I have an environmentalist friend who says that a lot of environmentalists begin by wanting to protect a specific piece of ground, and end up questioning the foundations of Western Civilization. Because you start asking, why is this particular piece of land being destroyed? And if you keep asking questions, you end up questioning the foundations of this whole exploitative culture. If you ask, why do men rape women? You end up at the roots of patriarchy. And if you ask, why is there racism? Why does this happen? You end up in the same place. For me, one of the questions was, why is there a wage economy? Why do people work jobs they don't like? If you ask that question, it takes you to the wage economy. Okay, why is there a wage economy? That leads you to start questioning capitalism. Okay, why is there capitalism? Well, that leads you to ask, why is there systematic exploitation?

When I was in second grade, there was a subdivision built right next to where I lived. I wouldn't have called it a habitat at the time, but this home for meadowlarks, garter snakes, toads, crawdads, cottonwood trees, and grasshoppers was turned into houses. Even at seven years old, I remember thinking that this couldn't go on forever, because if you keep expanding, you eventually run out of space. From a very early age I understood that you can't have infinite growth on a finite planet. That was abundantly clear to me. And it was also clear to me that I was on the side of the meadowlarks and the grasshoppers and the toads. I was deeply saddened by what happened to them, and also deeply moved by their troubles.

LK: I think a lot of kids feel that. I wonder why some of us are able to hang on to it and why others forget. In your case, what do you think it was?

DJ: Well, I asked both Robert J. Lifton and Judith Herman why some people open out from trauma and some people shut down, and they both laughed and said, "I have no idea." And they are pretty much two of the world's experts on the effects of trauma, so if they are going to defer on that question, I think I might punt as well. And I have no idea why. You know, twenty books later I have no idea why some people are empathic and some people are not.

Back to how I learned to write. Well, I spent my twenties in some senses not doing much of anything. I started the beekeeping business and then the bees died. Then I got very sick and I couldn't do anything for awhile. But in other senses, I was getting grounded. One reason I'm so prolific today is because I spent so much time in my twenties vomiting up the effects of the whole process of schooling, teaching myself how to really think. Because critical thinking is not encouraged in schools. I was teaching myself to question, to question everything, and teaching myself to learn how to listen. Not even necessarily teaching myself, but instead sitting by a river and learning how to ask questions from the river. I don't know how the river taught me, but it did.

And then in my late twenties, I started to write. I had been practicing. Somebody had said to me, "You're not a real writer until you've written a million words." And nerd that I was, when I started to get serious about writing, I started keeping track of how many words I wrote. I was going to try to write a thousand words a day, because then I figured in a little under three years I would be a writer. And a thousand words a day, as you know, is a lot. There's no way I made that. But if I wrote five hundred words in a day, I'd still be a writer in less than six years. Then the system got more complicated, because obviously it's easier to edit a draft then to write it. Where did that fit with my five hundred or a thousand words a day? Here's what I came up with. If I wrote five hundred words of a first draft, that would count as five hundred words. Then each word of my second draft would count as half a word, so if I edited five hundred words it would count as two hundred and fifty. And then each word in the third draft would count as a third of a word, and so on. It's pretty silly.

My time of getting grounded really ended one day in my late twenties when I was talking to a friend who said to me one of the most important things I've ever heard. He said, "You have gifts, Derrick. And if you don't use those gifts in the service of the community, then you're not worth shit. In order to succeed at anything, you have to make it the most important thing in your life." Some thirty-three years later I still remember where I was when I heard that. I didn't have a telephone at that point—not because of philosophical reasons, but because I had no money—and so I was standing at a payphone outside a tiny grocery store in Spirit Lake, Idaho.

That night I dedicated my life to writing.

Fast forward ten years. I read in a book on creativity that the magic number between when you dedicate your life to intensive study of an art and when you release your first truly creative work is supposed to be about ten years. It was true for Mendelssohn, Mozart, Beethoven, a lot of the big composers. And the same was true for me. I dedicated my life to writing in 1987, and in 1997 I wrote *A Language Older Than Words*. I wrote a couple of books before that, but that was my first book where I found my voice, or where my voice found me.

So in 1987, I started writing seriously. I started practicing a lot. I did things like copying good or great writers' work. You know, painters do that all the time. Writers don't do it nearly enough. I would take writing that I liked and hand copy it word for word, which is interesting now that I think about it because that's essentially what I was doing as a kid. Except now I wouldn't type it, because that's too fast. I would hand write it slowly, and I learned a lot . . .

LK: Your synapses form in a whole different way.

DJ: Absolutely. And then after that, not only would I copy it, but I would then modify it and make it my own. For example, if the author wrote a description, I tried writing the same description with a few changes: if the author described the secretary for the detective's office, I described the postal clerk in the same way.

I was doing all this writing, and the writing was really hard, and my writing was pretty horrible. It's just tremendously painful work to write like that, simply forcing out the words. Then in 1987 I became good friends with a family whose daughter was married to an abuser. In fact, when I first met her, her husband was in jail for raping her. And we talked a lot. Meanwhile, I was going to copy a story by James Herriot, the *All Creatures Great and Small* writer. I'd read one of his stories about a man who had only one friend in the world, and that was his dog. Every day he would go to a bar, and the dog would sit next to him on a barstool. Then one day the dog died and the man killed himself. It's something like a ten-page story. By the first sentence, I knew what was going to happen, and by the end of the end of the story I was bawling. I thought, "Wow, that's a really good trick. I need to learn how to do that." So, I decided to write a story about a guy whose only friend is his dog and the dog dies, and, since I was a coward, I was going to have the protagonist move out of town instead of kill himself. I tried and tried and tried to write this story, and couldn't, in great measure because it would have been a terrible story.

Meanwhile, this friend of mine was thinking of going back to her husband. One night I talked to her mom until 4:00 in the morning, asking each other, "How can we keep her from getting back together with her husband? What can we do, what can we say to her?" And then I went home and went to bed, and I woke up the next morning with the plot of a story in my head. Here is the plot: there's a woman who is in an abusive marriage. She becomes friends with a guy who has a dog. They talk about her abusive relationship. Eventually her husband finds out about their friendship and kills the dog. The guy moves out of town. The woman hits bottom, and that's the catalyst for her to finally love herself enough to give her husband the boot.

This was going to be about a ten-page story, which I vowed to finish before I went to sleep. When I got to about page seven, I realized it was going to be about a fifteen-page story, which I couldn't finish that day. Soon I got to page twelve and realized it was going to be about twenty-five pages. I got to page twenty and realized it was going to be about fifty pages. Two hundred pages later I had my first novel, which never got published. The point here is that I wrote it as a two-hundred-page attempt to help my friend understand why she should leave her husband. That was the moment when I became a writer because I wasn't writing for practice, and I wasn't writing from my head. Instead I was writing because I had something that I desperately needed to communicate.

Some writer asked Amy Tan, "Does writing ever get any easier?" And she said, "No, but it gets better." That's not my experience at all. My experience is that writing has gotten much easier. If it were still as hard to write now as it was when I was twenty-five, I would have quit a long time ago. The thing that's hard is finding the place where it's easy. But if you can find that, if you can tap into that place where the writing comes easily, it flows. Years ago I read an interview with the writer Charles Johnson, in which he said that the writing he wants to read is that which the writer writes as though there is a gun to the writer's head, and the writer will be killed as soon as the writer writes the last sentence of the last paragraph of the last page. Now, if you write with that sense of honesty and urgency, as though this is the last thing you can ever say, as opposed to writing for publication, writing for a grant, writing for this or that other motive, that if you write with that intensity, then you're not going to lie, and you're not going to mess around, and every word is going to be sharp. Like I said, that was the moment I became a writer, because I wasn't writing to become a writer. I was writing to try to save my friend from that terrible situation.

LK: And now you're writing to save the planet from that terrible situation.

DJ: That's exactly it. And that's one reason I'm so prolific; if writing makes any difference whatsoever, I need to write as much and as fast as I can. If I have this gift, I need to use it in the service of my community, or I'm not worth shit, you know? I have a tremendous responsibility. If I believe that words are worth anything, I can't dilly-dally around. All of my writing does come from that sense of urgency and the sense of wanting to communicate something.

I've always loved the line by Ursula LeGuin that writing is like sex: it's a lot better with two people. What I take that to mean is that journals are fine and important, but real writing is an attempt to communicate something of importance. There are a bunch of implications of Ursula LeGuin's line. One of them is that there is a responsibility to one's reader, also, to not be self-indulgent, to attempt to communicate, to be clear and to not be selfish. Years and years ago, I was at a storytelling festival and a very experienced storyteller corrected an incorrect word I had used. I'm horrified to admit this, but I said to her, "It's just a word." And she said, "No, you mugged me because you are responsible for every moment that every person who is reading your book or listening to your talk is giving to you. They're giving you these tremendous gifts of their time, and if you lie to them, if you use a word incorrectly, if you don't give them gifts for every moment, for every page, you're mugging them. You're mugging them with words just as surely as if you stole their wallet. You're stealing their time." That's always been very crucial to me, to be giving gifts to the reader.

So how, then, does someone learn how to write? It's an accretion of all these lessons. I went back to get an MFA because I had learned as much as I could by reading Doestoevsky. I wanted to be able to have a teacher of whom I could ask questions. I hated school a lot, and the only really positive experience I had is when I got my MFA, which I enjoyed very much. That's probably because I went to school not to get a degree, not to pass time, not because I had to, but instead I went specifically and solely, explicitly and completely, to learn how to become a better writer. And I was completely ruthless about it. If something helped me become a better writer, I was very focused on it. If it did not help me to become a better writer, I ignored it or got it out of my life.

LK: Is that how you are in your life now as a writer?

DJ: In terms of being ruthless about things . . . yes. Writing is the most important thing to me. I mean, obviously, the real world, the living world, is more important to me than writing. But in terms of things I do in my life, everything else is pretty secondary. I know some writers who have had regular jobs, I've known some writers who have had families, and I don't know how they do it.

LK: I don't either.

DJ: I also just know that the best way that I can contribute to a living planet is by writing. I'm not saying my contribution is more important than someone else's. I'm just saying that this is my gift. I was working a few years ago with an attorney who has done great work protecting trees and protecting forests. She's argued before the State Supreme Court of California. She's great. She literally wrote the book on fighting the California Department of Forestry. A local neighborhood group hired her to protect some forests here, and to lower her fees so it would be manageable, I had to do a lot of the basic writing for her. As we were writing documents for court, I asked, "Oh my god, do you like doing this stuff?" And she loved it! I said, "That's really great because I wouldn't like this at all." On the other hand, I really love the sort of writing that I do. She and I are both contributing in ways that suit us.

I don't talk about my romantic relationship life at all in my books, in part because I share so much of the other parts of my life that I need to keep some things just for me. Nonetheless, I'm going to tell one story. I was dating somebody about ten years ago, and that relationship was kind of a disaster for any number of reasons. My mom didn't like her, and the woman wrote a letter to my mom asking her why. My mom wrote back, "I don't know you. The only thing I care about really is that Derrick is happy, healthy, and writing. And since he met you he's been unhappy, he's been sick, and he hasn't been writing. So, actually, no, I don't like you." And she was right, those are the things I care about in my life. But the order of importance would probably be am I writing, am I healthy, am I happy.

A very important thing happened to me when I was in eighth grade. I went out for the football team and a friend of mine who was a couple years older said, "Even if you get really tired, keep trying as hard as you can, because someday it's going to be over and you're going to look back and wish that you would have tried harder. Don't look back with regrets." That stuck with me.

That's part one of that story. It continues. My experiences in athletics were terrible in junior high. The coaches were horrible, discouraging, and mean, in general, and specifically to me. They were so awful that it intimidated me and kept me from going out for any sports in high school even though I was quite athletic. Then in college, I was in a handball class where there were too many people for the number of courts, so every day somebody would be randomly assigned to run laps on the track. I've always hated purposeless running, and one day when I was assigned to run I noticed there was a high jump pit set up on the infield. I'd loved high jumping ever since I was a kid, so much that I made my own pit and bar and standards when I was in fourth grade. So I set up the standards and I started doing some jumping. The handball coach came out and saw me. I thought he was going to yell at me for not running, but instead he said, "You know, you're jumping higher than anybody we have on the team. Why don't you come out for track?" My confidence was so terrible that I told him, "No." And, what do you know, the next day we have handball class I was again assigned to run. I went out and high jumped, and he came out and asked me to join the team again. I said, "No." It happened again the next day I was assigned to run. The coach kept this up until I finally agreed to come out for the team. That was my sophomore year, and by the time I was a senior, I had broken the school record, tied the conference record, and won our collegiate conference championship. And then I graduated.

I will never know how good I could have been as a high jumper had I jumped seriously for more than two and a half years. That is a lesson I learned very well. I don't want to die with any books in me. Hell, I don't want to die with any life left in me. I don't want to get to the end of my life and say, "I could have tried harder. I could have saved more land. I could have written more books. I could have done things better." I want to live such that I use the gifts that the universe has given me to their fullest.

LK: Your books are a lot of things, but one thing that they definitely are is arrows. You've put the arrows in your bow, you've pulled back the string, and you've aimed. How will you know when the arrows have hit the target?

DJ: I'll know when there are more migratory songbirds every year than the year before, and when there are more salmon every year than the year before, and when there are more newts every year than the year before, and more moths and native slugs and redwood trees and Port Orford cedars. And when the oceans are recovering instead of being murdered. That's when I'll know. When there is less dioxin in every mother's breast milk. One of the most quoted lines that I've written is, "Every day when I wake up I ask myself whether I should write or blow up a dam." What that line is really getting at is the difference between symbolic and non-symbolic action. Writing is by definition symbolic. And symbolic actions are nice enough. I'm glad when I hear from someone that my work has changed his or her life. I'm glad when I help them understand that they're not crazy, that instead the culture is crazy. That's all great, but what really is important to me is that my work affects change in the real world.

I don't want to understate the importance of symbolic action, of helping people to understand this culture's destructiveness. That understanding is a necessary step. I can still remember in 1988 being in a public library in Spokane, Washington. The book *The Natural Alien* by Neil Evernden jumped off the shelf at me. I started reading it and a huge weight went off my shoulders. It was the first book I ever read that did not take the utilitarian perspective as a given. It was the first book I ever read that said nonhumans have more than utilitarian value. They have lives that are their own that are as precious to them as ours are to us. There's a part where he said, "What do you do if you make some impassioned defense of some creature and then when you're done making it the other person says 'Well, that's fine. But what good are they?'" His answer is to say "Well, what good are you?" Not to insult them, but to point out the stupidity of the utilitarian argument. Evernden then asks, "How much are you worth? If you crush your bones into bone meal, that's probably worth about two dollars. And your blood's worth about five dollars." He goes through it. You're worth something like seventeen dollars. But of course that's not your real value. Yet people think that way, of course about nonhumans, and about humans. Recently I was doing a talk and one guy asked what I thought about full-cost accounting, where you put a value on everything, and that will supposedly help to preserve it. So a tree is not merely worth money as two-by-fours, but instead is assigned a value for the CO_2 it sequesters, and so on.

I said, "Well, okay, that's great, but how much are you worth?"

And he said "Actually, the UN has developed numbers for that and in terms of wages and productivity I'm worth about four-and-a-half million dollars."

I said, "Okay, great, so why don't we make a deal? The federal government is going to pay your family five million dollars and kill you. Does this sound like a good deal to you?"

My point is that, yes, it's important to be able to analyze this culture, to deconstruct it, to decolonize our hearts and minds. Neil Evenden's book saved my sanity and saved my life. But none of that alters the fact that changing lives by themselves doesn't do a damn thing for the salmon.

When I say, "Every morning when I wake up I ask myself whether I should write or blow up a dam," I'm really asking whether writing actually affects change in the real physical world. Because we can do all the analysis we want, but at some point, there has to be action.

LK: In all your books, you mention this other being who influences you, your muse.

DJ: The muse is not a projected part of my unconscious. It's not my super-conscious. It's not any of that. The muse is an actual being. It's a she in my case. She's an actual being who doesn't have a body. She lives on other sides, somewhere else, I don't know where. Part of the reason that I was chosen by this muse is that I have opposable thumbs, and the muse can't physically write. She can give me the words, but she can't write them because she doesn't have a body. I know that in many ways this sounds crazy to members of this culture, but a disbelief in muses is only recently a part of even this culture. Only with the so-called enlightenment—isn't that a nice piece of propaganda?—did people begin to think they wrote their own words. Before that, it was pretty much accepted by everybody that muses create art. The modern view is much lonelier. And it also makes it much easier to destroy the world. It's far easier to destroy the world when you are the only one who matters.

We started by talking about *A Language Older Than Words* and how this culture silences other beings. The fundamental difference between Western and indigenous ways of being is that even the most open-minded Westerners generally perceive listening to the natural world as a metaphor, as opposed to the way the world really is. That's one of the central reasons this culture is killing the planet, because it perceives the world as consisting of objects to be exploited as opposed to other beings to enter into relationship with. In this culture, it is considered crazy by many to believe that a tree can speak to you. Or to believe that a muse speaks to you. But this has been commonly accepted by essentially every other culture besides this one. This culture is an extreme aberration in terms of not believing that others have something to say. That's an extraordinarily narcissistic perspective, and exposing this narcissistic perspective for what it is, is one of the things I've wanted to accomplish with my work, to knock the materialists off of their intellectual high ground, to show that their intellectual high ground is based on many untenable assumptions.

Part of my job as a writer is to articulate things I know in my heart to be true, but to which I've not yet put words, and in so doing, to help other people articulate things they know in their hearts to be true, but to which they've not yet put words. The single most common piece of praise that I've gotten is, "Thank you so much for saying the things I knew but I couldn't put words to." It's not my job to write something new or amazing, and I don't need to do cartwheels or backflips. What I need to do is attempt to tell the truth as honestly as I can. And in so doing it will help other people to tell their truths. And part of my job is to say these things and then have people laugh at me "Oh, he's crazy. He thinks that trees communicate." But that will give another person the courage to say it. And then another person and another person, until finally people don't feel strange saying that. Until people once again don't feel strange saying that they had a conversation with their muse. And when they don't feel strange saying, "We need to bring down civilization before it kills the planet."

There's another reason that I'm so prolific. Early in my career when I was writing *A Language Older Than Words*, I had an agent. The agency's address was 1 Madison Avenue. They had an entire floor of that building. So it's a huge, prestigious literary agency. I sent my agent the first seventy pages of the manuscript, and she told me that if I cut the social criticism and the family stuff, I'd have a book. I fired her. It was the same day that the members of the MRTA got massacred in the Japanese ambassador's house in Peru. I wrote to her and said, "If they're going to give their lives, then the least I can do is tell the truth. You're fired." I was in therapy for twelve years, talked about all the abuse in my childhood, and I never once cried. Except that day. The only day I cried in therapy was that day, because I thought I had destroyed my career even before it began. All my adult life I had wanted to be a writer. I had a Madison Avenue agent. I had a book I was working on. And I fired her.

The good news is that the muse has rewarded me greatly for that loyalty that I showed. That's one reason I have such a good relationship with the muse. In a tangible sense, I showed her early on that what she said to me was more important to me than financial success, more important to me than fame, more important to me than my career, that the important thing was having integrity with the work. In retrospect that was the best decision I've ever made in my life. Which doesn't mean that it wasn't difficult. I finished the book and sent it to another agent, who said I could keep the family stuff and the social criticism but that I would have to get rid of the non-linear organization. I could have one chapter about coyotes and one chapter about child abuse. She didn't sell the book either. I sent it off to a number of publishing houses. They all rejected it. In fact, here's another piece of trivia: the serial killer in *Songs of the Dead* is named after an editor who sent me an especially nasty rejection letter. He said that he really hated books like this one and that he always told young writers never to project their tiny epiphanies onto the natural world. And it made me so mad that I saved up his name to use it years later.

Anyway, the muse is real. People, presumably including nonhuman people, have them.

LK: How does she communicate with you?

DJ: That's a good question. When I was a teenager, I read the *Merlin Trilogy* by Mary Stewart. There was a line I've never forgotten: "The gods will not speak to those who have no time to listen." If I don't make time, communication doesn't happen. The muse is there, but if I don't ask for help, I don't get it. Another thing I've found is I have to try on my own to write something, and I have to try hard enough, but then when I can't, I just say, "I need some help. Can you please help me?" Usually she gives me help right away. It used to take a long time. Now it's pretty fast. We have a much better relationship. I've nurtured this relationship. I give her credit. How's that, a white man giving a female being credit for something? I always thank her. I treat the muse as I would treat any other being. I also don't insult her work. I don't say bad things about the words she gives me. I can certainly say some bad things about my processing of those words.

The muse is as terrified of this culture as I am, and she wants to stop it. That's one reason she works so hard. We're a team. We work on this together. I can't put out the books without her, and she can't put them out without me. We are, together, trying to save the planet, as best we can.

Bibliography

"Animal Planet: From India's famed camel fair to Indonesia's fierce Komodo dragons - all the world's a zoo." *San Francisco Chronicle*, November 28, 2004, F1, F4.

Agence France-Presse. "Wilderness almost non-existent on planet Earth: study." *Breitbart.com*, June 28, 2007. Accessed February 19, 2012. http://www.breitbart.com/ article.php?id=070628185002.qek4e5qz&show_article=1&catnum=0.

Americas Watch. *Rural Violence in Brazil.* New York: Human Rights Watch, 1991.

Associated Press. "Millions Of Children Enslaved Group Reports On Conditions Of Asian Child Labor, Slaves To Western Consumerism." *Spokesman Review*, (Spokane, Wash.) June 20, 1999. http://www.spokesman.com/stories/1995/sep/19/millions-of-children-enslaved-group-reports-on/.

Associated Press. "Pentagon Hopes to Identify People by the Way They Walk." *News-Star* (Shawnee, Okla.), May 20, 2003.

Associated Press. "UN chief calls for action on climate change." *San Francisco Chronicle*, November 8, 2007.

Anderson Valley Advertiser. November 1, 2000.

Bacon, Francis. *The Works.* Edited by James Spedding. New York: Garrett Press, 1968.

Bancroft, Lundy. *Why Does He Do That? Inside the Minds of Angry and Controlling Men.* New York: Berkley Books, 2002.

Baratay, Eric, and Hardouin-Fugier, Elisabeth. *Zoo: A History of Zoological Gardens in the West.* London: Reaktion Books, 2002.

Barsamian, David. "Expanding the Floor of the Cage, Part II: An Interview with Noam Chomsky." *Z Magazine*, April 1997.

Barry, Tom and Preusch, Deb. *Central America Fact Book.* New York: Grove Press, 1986.

Bentham, Jeremy. "Panopticon; Or The Inspection-House: Containing The Idea Of A New Principle Of Construction Applicable To Any Sort Of Establishment, In Which Persons Of Any Description Are To Be Kept Under Inspection; And In Particular To Penitentiary-Houses, Prisons, Houses Of Industry, Work-Houses, Poor-Houses, Lazarettos, Manufactories, Hospitals, Mad-Houses, And Schools: With A Plan Of Management." *The Panopticon Writings.* Ed. Miran Bozovic. London: Verso, 1995, p. 29-95. Accessed February 19, 2012. http://cartome.org/panopticon2.htm.

Bettelheim, Bruno. Introduction to *Auschwitz: A Doctor's Eyewitness Account*, by Miklos Nyiszli. New York: Frederick Fell, 1960.

Blaisdell, Bob, ed. *Great Speeches by Native Americans.* Mineola, NY: Dover, 2000.

Block, Lawrence. "Keller's Therapy." From *Enough Rope.* New York: HarperCollins, 1999.

Bock, Paula. "Oceans of Waste: Waves of junk are flowing into the food chain." *Seattle PI*, April 23, 2006.

Bogo, Jennifer. "Crying Rivers (Romania gold mine spills cyanide into rivers) (Brief Article)." *High Beam Encyclopedia*, May 1, 2000.

Bonewits, P. E. I. "The Laws of Magic." *The Deoxyribonucleic Hyperdimension*. Accessed February 19, 2012. http://deoxy.org/lawsofmagic.htm.

Boyer, Mike, and Peale, Cliff. "International Paper to add 400 jobs by mid-'97." *Cincinnati Post*, February 14, 1996, B6.

Brandon, William. *New Worlds for Old: Reports from the New World and their effect on the development of social thought in Europe, 1500-1800*. Athens: Ohio University Press, 1986.

Breining, Greg. "South China Tiger as Good as Extinct." *San Francisco Chronicle*, January 9, 2003, A1.

Brewer, E. Cobham. "Deluge." *Dictionary of Phrase and Fable*. Bartleby.com. Accessed February 19, 2012. http://www.bartleby.com/81/4834.html.

Brown, Dee. *Bury My Heart at Wounded Knee: An Indian History of the American West*. New York: Holt, Rinehart, and Winston, 1970.

Bryant, D., Nielsen, D., and Tangley, L. *The Last Frontier Forests: Ecosystems and Economies on the Edge*. Washington, DC: World Resources Institute, 1997.

US Department of Justice, Bureau of Justice Statistics. "Correctional Populations in the United States, 1997." November 2000. Accessed February 19, 2012. http://bjs.ojp.usdoj.gov/content/pub/pdf/cpus97.pdf.

Burke, Edmund. Ed. Charles W. Eliot. *Reflections on the Revolution in France*. In The Harvard Classics Edition, *On Taste, On the Sublime and the Beautiful, Reflections on the French Revolution*, and *A Letter to a Noble Lord*. New York: P. F. Collier & Son, 1937.

Byczkowski, John J. "Firm's Plans Still Not in Stone." *Cincinnati Enquirer*, September 5, 1994, D1.

Casey, Susan, and Gregg Segal. "Our oceans are turning into plastic. . . . Are we? " *Best Life Magazine*, May 11, 2007.

Cherfas, Jeremy. *Zoo 2000: A Look Beyond the Bars*. London: British Broadcasting Corp., 1984.

Chomsky, Noam. "The Victors: Part I." *Z Magazine*, September 1990.

Churchill, Ward, and Wall, James Vander. *Agents of Repression: The FBI's Secret Wars against the Black Panther Party and the American Indian Movement*. Boston: South End Press, 1988.

Churchill, Ward, and Wall, Jim Vander. *The COINTELPRO Papers: Documents from the FBI's Secret Wars Against Domestic Dissent*. Boston: South End Press, 1990.

Colorado School of Mines. "Acid Producing Potential of Mine Overburden." Accessed July 6, 2007. http://www.mines.edu/fs_home/jhoran/ch126/app.htm. No longer available online.

Colorado School of Mines. "Microbial Influences: Thiobacillus ferrooxidans." Acid Mine Drain Drainage Experiments at CSM. Accessed July 6, 2007. http://www.mines.edu/fs_home/jhoran/ch126/microbia.htm. No longer available online.

Conot, Robert E. *Justice At Nuremberg*. New York: Carrol & Graf, 1983.

Croke, Vicki. *The Modern Ark: The Story of Zoos: Past, Present and Future*. New York: Scribner, 1997.

Daly, Mary. *Beyond God the Father: Toward a Philosophy of Women's Liberation*. Boston: Beacon, 1985.

Dawkins, Richard. "God's Utility Function." In *River Out of Eden: A Darwinian View of Life*. New York: Basic Books, 1995, 85.

_____ . "What is True?" *The Devil's Chaplain*. New York: Houghton Mifflin, 2003.

De Rooy, Sylvia. "Before the Wilderness." *Wild Humboldt* 1, Spring/Summer 2002.

DeLong, J. Bradford. "The Corporations as a Command Economy." Accessed March 17, 2004. http://www.j-bradford-delong.net/Econ_Articles/Command_Corporations.html. No longer available online.

Densmore, Frances. *Teton Sioux Music*. Bulletin 61. Bureau of American Ethnology, Smithsonian Institution, 1918.

Descartes, Rene, and David Weismann, Ed. *Discourse on the method; and, Meditations on first philosophy*. New Haven: Yale University, 1996.

Dew, Thomas Roderick. "Abolition of Negro Slavery." *American Quarterly Review*, XII (1832), 189-265.

Diamond, Stanley. *In Search of the Primitive*. New Brunswick: Transaction Publishers, 1993.

Dimenstein, Gilberto. "Little Girls of the Night." *Report on the Americas*, North American Congress of Latin America, May/June 1994. Accessed February 19, 2012. http://pangaea.org/street_children/latin/brzpros.htm.

Dimitre, Tom."Salamander Extinction? " *Econews: Newsletter of the Northcoast Environmental Center*, March 2002, 10.

Douglass, Frederick. "The Significance of Emancipation in the West Indies." From *The Frederick Douglass Papers, Series One: Speeches, Debates, and Interviews*, Vol. 3. Ed. John W. Blassingame. New Haven: Yale University Press, 1985.

Draffan, George. "Directory of Transnational Corporations." Accessed February 19, 2012. http://www.endgame.org/dtc/directory.html.

Drake, Samuel G. *Biography and History of the Indians of North America, from Its First Discovery*. 11th ed. Boston: Benjamin B. Mussey & Co., 1841.

Drinnon, Richard. *Facing West: The Metaphysics of Indian-Hating & Empire-Building*. Norman, OK: University of Oklahoma Press, 1997.

Edwards, David. *Burning All Illusions*. Boston: South End Press, 1996.

_____ . *The Compassionate Revolution: Radical Politics and Buddhism*. Devon, UK: Green Books, 1998.

"The Effects of Strategic Bombing on the German War Economy." The United States Strategic Bombing Survey, Overall Economic Effects Division, October 31, 1945.

Evernden, Neil. *The Natural Alien: Humankind and Environment*. Toronto: University of Toronto, 1985.

Faust, Drew Gilpin. *The Ideology of Slavery*. Baton Rouge: Louisiana State University Press, 1981.

Foucault, Michel. *Discipline & Punish: The Birth of the Prison*. Translated by Alan Sheridan. New York: Vintage Books, 1979.

French, Marilyn. *Beyond Power: On Women, Men, and Morals*. New York: Ballentine, 1985.

Fromm, Erich. *The Anatomy of Human Destructiveness*. New York: Holt, Rinehart, and Winston, 1973.

Froyen, Richard T. *Macroeconomics: Theories and Policies*. New York: MacMillan, 1983.

Galeano, Eduardo. *Upside Down*. Translated by Mark Fried. New York: Metropolitan Books, 2000.

Gast, W.R. et al. *Blue Mountain Forest Health Report*. US Forest Service. Malheur, Umatilla, and Wallowa-Whitman National Forests, 1991.

Geisler, Charles C. "Ownership: An Overview." *Rural Sociology* 58 (1993) p. 532-546.

Glaspell, Kate Eldridge. "Incidents in the Life of a Pioneer." *North Dakota Historical Quarterly*, 1941, p. 187-188.

Glick, Brian. *War at Home: Covert Action Against Us Activists and What We Can Do About It*. Boston: South End Press, 1989.

Goettlich, Paul. "Plastic in the Sea, " *Mindfully.org*, October 5, 2005. Accessed February 19, 2012. http://www.mindfully.org/Plastic/Ocean/Sea-Plastic-LN-PG5oct05.htm.

Gordon, H.L. *The Feast of the Virgins and Other Poems*. Chicago: Laird and Lee, 1891.

Griffin, Susan. *Woman and Nature: The Roaring Inside Her*. New York: Harper & Row, 1978.

Gruen, Arno. *The Insanity of Normality: Realism as Sickness: Toward Understanding Human Destructiveness*. Translated by Hildegarde and Hunter Hannum. New York: Grove Weidenfeld, 1992.

Gugliotta, Guy. "Rats Turned Into Remote-Controlled Robots: Techniques Potential Uses Include Aid to Victims of Disaster or Neural Injuries." *Washington Post*, May 2, 2002, sec. A. Accessed February 19, 2012. http://www.washingtonpost.com/ac2/wp-dyn?pagename=article&node=&contentId=A18261-2002May1&notFound=true.

Hastings, Max. *Bomber Command*. New York: Touchstone, 1979.

Hancocks, David. *A Different Nature: The Paradoxical World of Zoos and Their Uncertain Future*. Berkeley: University of California Press, 2001.

Handler, Marisa. "Indigenous Tribe Takes on Big Oil: Ecuadoran Village Refuses Money, Blocks Attempts at Drilling on Ancestral Land." *San Francisco Chronicle*, August 13, 2004. Accessed February 19, 2012. http://www.sfgate.com/cgi-bin/article.cgi?file=/chronicle/archive/2004/08/13/MNGHB86B4V1.DTL.

Harris, Sam. "Mother Nature is Not Our Friend." *Richarddawkins.net*. Accessed February 19, 2012. http://richarddawkins.net/article,2096,Mother-Nature-is-Not-Our-Friend,Sam-Harris.

————. "Science Must Destroy Religion." *MachinesLikeUs.com*. Accessed February 19, 2012. http://machineslikeus.com/articles/ScienceMustDestroy.html.

————. *The End of Faith: Religion, Terror, and the Future of Reason*. New York: W.W. Norton, 2004.

Hartmark-Dounas, Laura. "Summitville, the Exxon Valdez of the Mining Industry, " *Sprol*, October 17, 2005. Accessed February 19, 2012. http://www.sprol.com/?p=268m.

"Hate Directory." Accessed December 2, 2000. http://www.bcpl.net/-rfrankli/hatedir.htm. No longer available online.

Haynes, Tim. "High-tech Robot Gives Glimpse of Future." *Boston Globe*, July 25, 2003.

Hearst, David. "Sci-fi War Put Under the Microscope." *The Guardian,* May 20, 2003. Accessed February 19, 2012. http://www.guardian.co.uk/uk_news/story/0,3604,959469,00.html.

Hechler, David. "Child Sex Tourism." Accessed December 3, 2000. ftp://members.aol.com/hechler/tourism.html. No longer available online.

Herman, Judith Lewis. *Trauma and Recovery: The Aftermath of Violence—from Domestic Abuse to Political Terror.* New York: Basic Books, 1992.

Higgins, Margot. "Extinction Debts Come Due Long after Deforestation." *ENN: Environmental News Network,* October 12, 1999.

Hoage, R.J., and Deiss, William A., foreword by Michael H. Robinson. *New Worlds, New Animals: From Menagerie to Zoological Park in the Nineteenth Century.* Baltimore: John Hopkins University Press, 1996.

Hoffmann, Peter. *The History of the German Resistance, 1933–1945.* Translated by Richard Barry. Cambridge, MA: The MIT Press, 1977.

Knights of the Ku Klux Klan National Party Headquarters. "Does the Klan Hate Negroes." http://www.kukluxklan.org/doesthe.htm. No longer available online.

Inglis, Brian. *Poverty and the Industrial Revolution.* London: Hodder and Stoughton, 1971.

"International Paper to cut 215." *Cincinnati Enquirer,* February 13, 1997, B16.

International Union for the Conservation of Nature, Species Survival Commission. 1996 IUCN Red List of Threatened Animals. Gland, Switzerland: IUCN, 1996.

Jefferson, Thomas. Ed. Andrew A. Lipscomb and Albert Ellery Bergh. *The Writings of Thomas Jefferson.* Vol.11. Washington, DC: Thomas Jefferson Memorial Association, 1903.

Jensen, Derrick. *Endgame, Volume 1: The Problem of Civilization,* and *Endgame, Volume 2: Resistance.* New York: Seven Stories Press, 2006.

_____. *A Language Older Than Words.* New York: Context Books, 2000.

_____. *The Culture of Make Believe.* White River Junction: Chelsea Green, 2004.

_____. *Walking on Water: Reading, Writing, and Revolution.* White River Junction: Chelsea Green, 2004.

_____. *Welcome to the Machine: Science, Surveillance, and the Culture of Control.* White River Junction: Chelsea Green, 2004.

Johnson, Allan. G. *The Gender Knot: Unraveling Our Patriarchal Legacy.* Philadelphia, PA: Temple University Press, 2005.

Keegan, John. *The Second World War.* 1st Amer. ed. New York: Viking Penguin, 1990.

Keye, William Wade. "Managing Forests, Protecting Watersheds." *San Francisco Chronicle,* December 1, 2002, D5.

Kirchshofer, Rosl, ed. *The World of Zoos: A Survey and Gazetteer.* New York: Viking, 1968.

Klare, Michael T. *Resource Wars: The New Landscape of Global Conflict.* New York: Metropolitan Books, 2001.

Klauk, Erin. "Environmental Impacts at Fort Belknap from Gold Mining." *Impacts of Resource Development on Native American Lands,* Carelton College. Accessed February 19, 2012. http://serc.carleton.edu/research_education/nativelands/ftbelknap/environmental.html.

Kupers, Terry. *Prison Madness.* San Francisco: Jossey-Bass, 1999.

La Boétie, Éttiene de. *Discours de la Servitude Volontaire*. Trans. Harry Kurz as *The Politics of Obedience: The Discourse of Voluntary Servitude*. Buffalo: Black Rose Books, 1997. Abridged and edited version available at Personal Empowerment Resources. Accessed February 19, 2012. http://www.mind-trek.com/.

LaChapelle, Dolores. *Sacred Land, Sacred Sex: Rapture of the Deep: Concerning Deep Ecology And Celebrating Life*. Durango: Kivaki Press, Durango 1988.

Laing, R. D. *The Politics of Experience*. New York: Ballantine Books, 1967.

Liddell Hart, B.H., ed. *The Rommel Papers*. Translated by Paul Findlay. New York: Harcourt, Brace, and Company, 1953.

Lorde, Audre. "The Master's Tools Will Never Dismantle the Master's House." *In Sister/Out-sider*. Trumansburg: The Crossing Press, 1984.

Lukas, Ellen. "Children For Sale: The Stockholm Congress Against the Commercial Exploitation of Children.*" Insight on the News*, December 1996.

Luoma, Jon R. *A Crowded Ark*. Boston: Houghton Mifflin , 1987.

Mann, Thomas. *Death in Venice, Tonio Kröger, and Other Writings*. Ed. Lubich, Frederick A. New York: Continuum Publishing Company, 1999.

Maslow, Abraham H. *The Farther Reaches of Human Nature*. New York: Viking, 1971.

Mason Jr., Herbert Molloy. *To Kill the Devil: The Attempts on the Life of Adolf Hitler*. New York: W.W. Norton & Company, 1978.

Masson, Jeffrey Moussaieff, and McCarthy, Susan. *When Elephants Weep: The Emotional Lives of Animals*. New York: Delecorte Press, 1995.

McBay, Aric. "Interview with Lierre Keith." Accessed October 30, 2008. http://www.inthewake.org/keith1.html. No longer available online.

McKie, Robin. "NASA aims to move Earth." *Guardian Unlimited*, June 10, 2001. Accessed February 19, 2012, http://observer.guardian.co.uk/international/story/0,,504486,00.html.

Merchant, Carolyn. *The Death of Nature: Women, Ecology and the Scientific Revolution*. San Francisco: Harper SanFrancisco, 1983.

Mies, Maria. *Patriarchy and Accumulation on a World Scale*. London: Zed Books, 1999.

Mineral Policy Center. "Cyanide Leach Mining Packet." Washington, DC: GPO, 2006.

Minnesota State Public Defender. "Sources: South Africa's incarceration rates compared to the U.S." Accessed February 19, 2012. http://www.pubdef.state.mn.us/homepages/statepd/south_africa.htm.

Montague, Peter. "Rachels' Democracy and Health News n n n932." Rachel's News, Environmental Research Foundation, November 8, 2007. Accessed February 19, 2012. http://www.precaution.org/lib/07/ht071108.htm.

Montgomery, David. *King of Fish: The Thousand-Year Run of Salmon*. Boulder, CO: Westview Press, 2003.

Mostert, Noel. *Frontiers: The Epic of South Africa's Creation and the Tragedy of the Xhosa People*. New York: Alfred A. Knopf, 1992.

Mowat, Farley. *Sea of Slaughter*. Toronto: Seal, 1989.

Mullan, Bob, and Marvin, Garry. *Zoo Culture: The Book About Watching People Watch Animals*, second edition. Urbana: University of Illinois Press, 1999.

Mumford, Lewis. *The Pentagon of Power*. New York: Harcourt Brace Jovanovich, 1970.

_____. "Authoritarian and Democratic Technics." *Technology and Culture* 5: 1 (1964).

_____. *The Myth of the Machine: Technics and Human Development*. New York: Harcourt Brace Jovanovich, 1966.

"NMFS Refuses to Protect Habitat for World's Most Imperiled Whale: Despite Six Years of Continuous Sightings in SE Bering Sea, NMFS Claims It Can't Determine Critical Habitat for Right Whale." Center for Biological Diversity, February 20, 2002. Accessed February 19, 2012. http://www.biologicaldiversity.org/swcbd/press/right2-20-02.html.

National Criminal Justice Commission. "Key Findings." Accessed December 2, 2000. http://www.igc.apc.org/ncia/KEY.HTML. No longer available online.

Oliveria, D. F. "Act Provocatively and You Provoke; Dumb Decisions; Sometimes Women Lead Men into Temptation." *The Spokesman-Review*, November 7, 1997.

Olmos, Mararite Fernández and Paravisini-Gebert, Lizabeth. *Sacred Possessions: Voudou, Santería, Obeah, and the Caribbean*. New Brunswick: Rutgers University Press, 2000.

Orwell, George. *1984*. New York: Signet Classics, 1962.

Perlin, John. *A Forest journey: The Role of Wood in the Development of Civilization*, Cambridge: Harvard University Press, 1991.

Ponting, Clive. *A Green History of the World: The Environment and the Collapse of Great Civilizations*. New York: Penguin Books, 1991.

Public Citizen. "Corporate Welfare Examples for 1999." Accessed February 19, 2012. www.citizen.org/congress/welfare/articles.cfm? ID=1053.

Beers, David, Ed. Quinn. *The Voyages and Colonizing Enterprises of Sir Humphrey Gilbert*. London: Hakluyt Society, 1940.

Rainforest Foundation-US. "Why? Save The Rest." Accessed July 25, 2003. www.rainforestfoundation.org/why.html. No longer available online.

Raven-Hart, Major R. *Before Van Riebeeck: Callers at South Africa from 1488 to 1652*. Cape Town: C. Struik, 1967.

_____. *Cape of Good Hope, 1652-1702: The First Fifty Years*, 2 vols. Cape Town: A.A. Balkema, 1971.

Raymond, Janice. "Legitimating Prostitution as Sex Work." *Coalition Against Trafficking in Women*, December 1998. Accessed February 19, 2012. http://sisyphe.org/spip.php?article689.

Remedy. "Mattole Activists Assaulted, Arrested after Serving Subpoena for Pepper Spray Trial." *Treesit Blog*, August 27, 2004.

Reuters. "Robot Shows Prime Minister How to Loosen Up." August 22, 2003.

"Reviving the World's Rivers: Dam Removal." Part 4, Technical Challenges. *International Rivers Network*. Accessed July 11, 2004. http://www.irn.org/revival/decom/brochure/rrpt5.html. No longer available online.

Robinson, Laurie Nicole. "The Globalization of Female Child Prostitution: A Call For Reintegration and Recovery Measures Via Article 39 of the United Nations Convention on the Rights of the Child." *Indiana Journal of Global Legal Studies* 5, no. 1 (Fall 1997).

Ross, Colin A. *Bluebird: Deliberate Creation of Multiple Personality by Psychiatrists*. Richardson, TX.: Manitou Communications, 2000.

Rothfels, Nigel. *Savages and Beasts: The Birth of the Modern Zoo*. Baltimore: Johns Hopkins, 2002.

Rousseau, Jean-Jacques. *On the Social Contract*. Mineola, NY: Dover Publications, 2003.

Russell, Diana E. H. *Sexual Exploitation: Rape, Child Sexual Abuse, and Sexual Harassment*. Beverly Hills: Sage, 1984.

Rybcynski, Witold. *Taming the Tiger: The Struggle to Control Technology*. New York: Viking Press, 1983.

Sauer, Carl Ortwin. *The Early Spanish Main*. Berkeley: University of California Press, 1966.

Schwägerl, Christian. Interview with Dennis Meadows, "Copenhagen Is About Doing As Little As Possible." *Spiegel Online*. Accessed December 10, 2009. http://www.spiegel.de/international/world/0,1518,666175,00.html<ref=nlint.

Severn, David. "Vine Watch." *Anderson Valley Advertiser*, April 2, 2003.

Shears, Richard. "Is this the world's most polluted river? " *Daily Mail*, June 5, 2007. Accessed February 19, 2012. http://www.dailymail.co.uk/news/article-460077/Is-worlds-polluted-river.html.

Shepard, Paul. *The Only World We've Got: A Paul Shepard Reader*. San Francisco: Sierra Club Books, 1996.

Sizer, Nigel. "Perverse Habits." *World Resources Institute Forest Notes*, June 2000, 2.

Smith, Sydney. *Works of The Reverend Sydney Smith: Three Volumes Complete in One*. New York: D. Appleton and Company, 1867.

Sniffen, Michael J. "Proposed System Would Use Lots of Data." *The Guardian*, May 19, 2003.

Synder, Ronna. "Faith, Hope And Chastity Jessica Lotze Made A Covenant With God That She Would Remain Chaste Until Her Wedding Day; The Strength Of Her Faith And The Love Of Her New Husband Helped Her Keep That Covenant." *The Spokesman-Review*, April 27, 1997.

Sohel, Robert. *Panic on Wall Street*. New York: E.P. Dutton, 1988.

St. Augustine. *Basic Writings of St. Augustine*. Michigan: Baker Publishing Group, 1993.

Stamets, Paul. *Mycelium Running: How Mushrooms Can Help Save the World*. Berkeley: Ten Speed Press, 2005.

Stannard, David E. *American Holocaust: Columbus and the Conquest of the Nets World*. New York: Oxford University Press, 1992.

Star Wars. Accessed April 23, 2004. http://www.starwars.com/databank/location/deathstar/. No longer available online.

Steamlocomotive.com. "Steam Locomotive Builders." Accessed February 19, 2012. http://www.steamlocomotive.com/builders/.

Stiffler, Lisa. "PDBEs: They are everywhere, they accumulate and they spread: Chemical flame retardants pose threat to humans, environment." *Seattle PI*, March 29, 2007. http://seat-tlepi.nwsource.com/local/309169_pbde28.html.

Taylor, Jessica. "Jessica Taylor talks to veterans of Oak Ridge." *The Guardian*, July 5, 2006. Accessed February 19, 2012. http://www.guardian.co.uk/secondworldwar/story/0,,1812 11,00.html.

Tertullian. *On the Apparel of Women*. Montana: Kessinger Publishing, 2004.

Sedlak, Andrea J., and Broadhurst, Diane D. "Executive Summary of the Third National Incidence Study of Child Abuse and Neglect." US Department of Health and Human Services, 1996. Accessed February 19, 2012. http://www.childwelfare.gov/pubs/statsinfo/nis3.cfm.

Turner, Frederick. *Beyond Geography: The Western Spirit Against The Wilderness*. New Brunswick: Rutgers University Press, 1992.

U.S. Defense Advanced Research Projects Agency (DARPA). "Autonomous Vehicles Grand Challenge." Accessed August 11, 2003. http://www.darpa.mil/grandchallenge/overview.htm. No longer available online.

US Forest Service. *1997 Resources Planning Act Assessment, Final Statistics*. Washington, DC: US Forest Service, July 2000.

US Forest Service. "1998 Report of the Forest Service." Accessed February 19, 2012. http://www.fs.fed.us/pl/pdb/98report/02_stats.html.

United Press International. Android World. August 2, 2003. Accessed February 19, 2012. http://www.androidworld.com/.

Vachs, Andrew. "Stop Child Sex Tourism." Accessed December 3, 2000. http://nightflight.com/dbt/. No longer available online.

Wakefield, Jane. "US Looks to Create Robo-soldier." *BBC News*, April 10, 2002. Accessed February 19, 2012. http://news.bbc.co.uk/1/hi/sci/tech/1908729.stm.

Webster's New Twentieth Century Dictionary of the English Language Unabridged, Second Edition. New York: Simon and Schuster, 1979.

Webster's New Twentieth Century Dictionary of the English Language, 2nd ed. New York: Simon and Schuster, 1979.

Weisman, Alan. "Polymers Are Forever." *Orion*, May/June 2007. Accessed February 19, 2012. http://www.orionmagazine.org/index.php/articles/article/270/.

Wikipedia. "Lozen." Wikimedia Foundation. Accessed February 19, 2012. http://en.wikipedia.org/wiki/Lozen.

Wikipedia. "Tailings." Wikimedia Foundation. Accessed February 19, 2012. http://en.wikipedia.org/wiki/Tailings.

Wikiquote. "Robert Oppenheimer." *Wikiquote*, accessed February 19, 2012. http://en.wikiquote.org/wiki/Robert_Oppenheimer.

Winn, Ira J. "The Psychology of Smog." *The Nation*, March 5, 1973.

Zinn, Howard. *A People's History of the United States*. New York: Harper Perennial, 1980.

Notes

A LANGUAGE OLDER THAN WORDS

1. For the description of a factory slaughterhouse, I unfortunately had to rely on a composite of friends' descriptions and published accounts. My attempts to enter slaughterhouses were met with polite yet insistent refusals. The public relations hacks with whom I spoke did provide me lots of nifty literature, none of which mentions death or killing.
2. Judith Herman, *Trauma and Recovery: The Aftermath of Violence—from Domestic Abuse to Political Terror* (New York: Basic Books, 1992); Diane E. H. Russell, *Sexual Exploitation: Rape, Child Sexual Abuse, and Sexual Harassment* (Beverly Hills: Sage, 1984).
3. Rene Descartes and David Weismann, Ed., *Discourse on the method; and, Meditations on first philosophy* (New Haven: Yale University, 1996), p. 63.
4. 1 Timothy 2: 11.
5. My source for many of the quotes concerning the European hatred of indigenous Africans is Noel Mostert, *Frontiers: The Epic of South Africa's Creation and the Tragedy of the Xhosa People* (Alfred A. New York: Knopf, 1992). Mostert pointed me toward many informative primary documents. The "extremely ugly . . ." quote is from Major R. Raven-Hart, *Before Van Riebeeck: Callers at South Africa from 1488 to 1652* (Cape Town: C. Struik, 1967).
6. Major R. Raven-Hart, *Cape of Good Hope, 1652-1702: The First Fifty Years*, 2 volumes, (Cape Town: A.A. Balkema, 1971).
7. Raven-Hart, *Before Van Riebeeck*.
8. I have two sources for many of my quotes concerning the European hatred of indigenous North Americans. The first is David E. Stannard, *American Holocaust: Columbus and the Conquest of the New World* (New York: Oxford University Press, 1992). The second is Frederick Turner, *Beyond Geography: The Western Spirit Against The Wilderness* (New Brunswick: Rutgers University Press, 1992).

    Both are invaluable references to the primary documents, both provide wonderful analyses, and both are extremely difficult to read because of the atrocities they detail.
9. Stannard, *American Holocaust*.
10. Jeffrey Moussaieff Masson and Susan McCarthy, *When Elephants Weep: The Emotional Lives of Animals* (New York: Delecorte Press, 1995), p. 18.
11. I became aware of the study by Allport, Bruner, and Jandorf through an article someone handed to me at a reading: Ira J. Winn, "The Psychology of Smog," *The Nation*, March 5, 1973.
12. The figure of one hundred and fifty million enslaved children comes from cross-referencing two sources, and then being extremely conservative in my extrapolation. The Anti-Slavery Society finds more than one hundred million enslaved children just in Asia, and the International Programme on the Elimination of Child Labour, which puts the estimate of enslaved children worldwide in "only" the tens of millions, finds that Africa has an incidence of child labor nearly twice that of Asia. The mere fact that I can even consider estimates differing by one hundred million (I think an estimate of two hundred and fifty million enslaved children

worldwide would be defensible) disturbs me greatly. Lost in the numbers— in fact worse than lost, but masked—is the misery inherent in each one of these cases. This little girl enslaved to prostitution in Thailand, this little boy enslaved to rug making in Pakistan. See Associated Press, "Millions Of Children Enslaved Group Reports On Conditions Of Asian Child Labor, Slaves To Western Consumerism," *Spokesman Review*, June 20, 1999, http://www.spokesman.com/stories/1995/sep/19/millions-of-children-enslaved-group-reports-on/, accessed February 19, 2012.

13. Neil Evernden, *The Natural Alien: Humankind and Environment* (Toronto: University of Toronto, 1985).

14. The Okanagan definition of "violation of a woman" was told to me by Jeannette Armstrong.

15. Deuteronomy 21:11.

16. Genesis 3:16.

17. Mary Daly, *Beyond God the Father: Toward a Philosophy of Women's Liberation* (Boston: Beacon, 1985.

18. John Perlin succinctly tells the story of the planet's deforestation in *A Forest Journey: The Role of Wood in the Development of Civilization* (Cambridge: Harvard University Press, 1991).

19. Many books describe the beauty and natural opulence of North America prior to the arrival of civilization. One of the best, and most heartbreaking, is Farley Mowat's *Sea of Slaughter* (Toronto: Seal, 1989).

20. David Beers, Ed. Quinn, *The Voyages and Colonizing Enterprises of Sir Humphrey Gilbert* (London: Hakluyt Society, 1940).

21. Robert Sohel, *Panic on Wall Street* (New York: E.P. Dutton, 1988).

22. The Tertullian quote is from *On the Apparel of Women* (Montana: Kessinger Publishing, 2004).

    The whole quotation is: "You are the devil's gateway. . . . You are the first deserter of the divine law; you are she who persuaded him whom the devil was not valiant enough to attack. You destroyed so easily God's image, man. On account of your desert—that is, death—even the Son of God had to die." I guess what he's saying is that things would be okay for men— who are of course the images of God—if it weren't for those damned women.

23. The story of Origen is not unique. Many early Christians castrated themselves, following Matthew 19.

24. For explorations of the Christian hatred of the body, see Stannard, Daly, Turner, French (*Beyond Power*), Griffin (*Woman and Nature*), or many others. Specifically "What is seen . . ." is from Origen, Selecta in Exodus xviii. 17, Migne, Patrologia Graeca, volume 12, column 296. "I know nothing . . ." is from St. Augustine, *Basic Writings of St. Augustine* (Michigan: Baker Publishing.Group, 1993).

25. Ronna Synder, "Faith, Hope And Chastity Jessica Lotze Made A Covenant With God That She Would Remain Chaste Until Her Wedding Day; The Strength Of Her Faith And The Love Of Her New Husband Helped Her Keep That Covenant," *The Spokesman-Review*, April 27, 1997.

26. D. F. Oliveria, "Act Provocatively and You Provoke; Dumb Decisions; Sometimes Women Lead Men into Temptation," *The Spokesman-Review*, November 7, 1997.

27. The story of the extermination of the great auk is given in Mowat's *Sea of Slaughter*. Here is the destruction of the last egg: "As they clambered up they saw two Geirfugel [great auks] sitting among numberless other sea-birds, and at once gave chase. The Geirfugel showed not the slightest disposition to repel the invaders, but immediately ran along the high cliff, their heads erect, their little wings extended. They uttered no cry of alarm and moved, with their short steps, about as quickly as a man could walk. Jon, with outstretched arms, drove one onto a corner, where he soon had it fast. Sigruder and Ketil pursued the second and seized it close to the edge of the rock. Kertil then returned to the sloping shelf whence the birds had started and saw an egg lying on the lava slab, which he knew to be a Geirfugel's. He took it up, and finding it was broken, dropped it again. All this took place in much less time than it took to tell." I really do hate this culture.

28. The accounts of fecundity prior to the arrival of civilization are from Mowat's *Sea Of Slaughter*.

29. Ruth Benedict wrote up her study for a series of lectures she gave at Bryn Mawr College in 1941. Her notes were lost. But her assistant, Abraham Maslow, was able to assemble fragments. These are presented in *The Farther Reaches of Human Nature* (New York: Viking, 1971). Erich Fromm expanded on these for his necessary book *The Anatomy of Human Destructiveness* (New York: Holt, Rinehart, and Winston, 1973). LaChapelle also does a wonderful job of drawing crucial conclusions from Benedict's study in *Sacred Land, Sacred Sex: Rapture of the Deep: Concerning Deep Ecology And Celebrating Life* (Durango: Kivaki Press, 1988).

30. Lewis Mumford, *The Pentagon of Power* (New York: Harcourt Brace Jovanovich, 1970). This is the second and final volume of *Myth of the Machine*, an exploration that cannot be too highly recommended.

31. Where our dinner came from was derived mostly from George Draffan's phenomenal "Directory of Transnational Corporations," which can be found online at http://www.endgame.org/dtc/directory.html. Better, give George a call. He's got lots more information in his head than in the directory.

THE CULTURE OF MAKE BELIEVE

1. Knights of the Ku Klux Klan National Party Headquarters, "Does the Klan Hate Negroes," http://www.kukluxklan.org/doesthe.htm; no longer available online.

2. Robert E. Conot, *Justice At Nuremberg* (New York: Carrol & Graf, 1983), 384–385.

3. National Criminal Justice Commission, "Key Findings" accessed December 2, 2000, http://www.igc.apc.org/ncia/KEY.HTML; no longer available online.

4. US Department of Justice, Bureau of Justice Statistics, "Correctional Populations in the United States, 1997," November 2000, accessed February 19, 2012, http://bjs.ojp.usdoj.gov/content/pub/pdf/cpus97.pdf, table 1.8.

5. Terry Kupers, *Prison Madness* (San Francisco: Jossey-Bass, 1999), p. 94.

6. Minnesota State Public Defender, "Sources: South Africa's incarceration rates compared to the US," accessed February 19, 2012, http://www.pubdef.state.mn.us/homepages/statepd/south_africa.htm.

7. Kupers, *Prison Madness*, p. 94.

8. Russell, *Sexual Exploitation*, p. 35.

9. "Hate Directory," accessed December 2, 2000, http://www.bcpl.net/-rfrankli/hatedir.htm; no longer available online.

10. Ibid.

11. Noam Chomsky, "The Victors: Part I," *Z Magazine*, September 1990.

12. Andrew Vachs, "Stop Child Sex Tourism," accessed December 3, 2000, http://night-flight.com/dbt/; no longer available online.

13. David Hechler, "Child Sex Tourism," accessed December 3, 2000, ftp://members.aol.com/hechler/tourism.html; no longer available online.

14. Laurie Nicole Robinson, "The Globalization of Female Child Prostitution: A Call For Reintegration and Recovery Measures Via Article 39 of the United Nations Convention on the Rights of the Child," *Indiana Journal of Global Legal Studies* 5, no. 1 (Fall 1997).

15. Ellen Lukas, "Children For Sale: The Stockholm Congress Against the Commercial Exploitation of Children," *Insight on the News*, December 1996.

16. Vachs, "Stop Child Sex Tourism."

17. Gilberto Dimenstein, "Little Girls of the Night," *Report on the Americas*, North American Congress of Latin America, May/June 1994, accessed February 19, 2012, http://pangaea.org/street_children/latin/brzpros.htm.

18. Lukas, "Children for Sale," 1996.

19. Janice Raymond, "Legitimating Prostitution as Sex Work," *Coalition Against Trafficking in Women*, December 1998, Accessed February 19, 2012, http://sisyphe.org/spip.php?article689.

20. Robinson, "The Globilization of Female Prostitution," 1997.
21. Eduardo Galeano, *Upside Down,* translated by Mark Fried (New York: Metropolitan Books, 2000), p. 17.
22. David Edwards, *Burning All Illusions* (Boston: South End Press, 1996), p. 141.
23. Andrea J. Sedlak and Diane D. Broadhurst, "Executive Summary of the Third National Incidence Study of Child Abuse and Neglect," US Department of Health and Human Services, 1996, accessed February 19, 2012, http://www.childwelfare.gov/pubs/statsinfo/nis3.cfm.
24. Lukas, "Children for Sale," 1996.
25. Thomas Roderick Dew, "Abolition of Negro Slavery," *American Quarterly Review,* XII (1832), p. 195.
26. Drew Gilpin Faust, *The Ideology of Slavery* (Baton Rouge: Louisiana State University Press, 1981), p. 92.
27. Dew, "Abolition of Negro Slavery," p. 195.
28. Howard Zinn, *A People's History of the United States* (New York: Harper Perennial, 1980), p. 255.
29. Sydney Smith, *Works of The Reverend Sydney Smith: Three Volumes Complete in One* (New York: D. Appleton and Company, 1867), p. 136.
30. Ibid., p. 131–136.
31. Ibid., p. 136.
32. Brian Inglis, *Poverty and the Industrial Revolution* (London: Hodder and Stoughton, 1971), p. 166.
33. Richard T. Froyen, *Macroeconomics: Theories and Policies* (New York: MacMillan, 1983), p. 36.
34. Mostert, *Frontiers,* p. 108, citing Raven-Hart, *Before Van Riebeeck,* p. 47.
35. Mostert, *Frontiers,* p. 108, citing Cope, 32.
36. Raven-Hart, *Cape of Good Hope,* volume 1, p. 146.
37. Ibid., p. 85.
38. Faust, *The Ideology of Slavery,* p. 116, citing Harper.
39. Ibid.
40. Faust, *The Ideology of Slavery,* p. 293, citing Fitzhugh.
41. Faust, *The Ideology of Slavery,* p. 74, citing Dew.
42. Ibid., p. 60.
43. Faust, *The Ideology of Slavery,* p. 99, citing Harper.
44. Faust, *The Ideology of Slavery,* p. 293, citing Fitzhugh.
45. *Anderson Valley Advertiser,* November 1, 2000.
46. *Webster's New Twentieth Century Dictionary of the English Language Unabridged,* Second Edition, (New York: Simon and Schuster, 1979).
47. Ibid.
48. Mowat, *Sea of Slaughter,* p. 168.
49. Ibid., p. 118.
50. Ibid., p. 29.
51. Carl Ortwin Sauer, *The Early Spanish Main,* (Berkeley: University of California Press, 1966), p. 69.
52. William Brandon, *New Worlds for Old: Reports from the New World and their effect on the development of social thought in Europe, 1500-1800* (Athens: Ohio University Press, 1986), p. 60.
53. Ibid., p. 60.
54. Edmund S. Morgan, *American Slavery, American Freedom: The Ordeal of Colonial Virginia* (New York: W.W. Norton and Company, 1975), p. 39.
55. Ibid., p. 40.
56. Exodus 34: 10-11.
57. Exodus 34: 11-16.
58. Exodus 23: 31-33.
59. For other examples of this same message of God giving power in exchange for disallowing relationships, see, for example, Deuteronomy 7, or Joshua 23.

60. George Draffan, personal communication.
61. Stanley Diamond, *In Search of the Primitive* (New Brunswick: Transaction Publishers, 1993), p. 1.
62. Edmund Burke, Ed. Charles W. Eliot, *Reflections on the Revolution in France* (New York: P. F. Collier & Son, 1937), p. 374.
63. For the Ruth Benedict materials, see Maslow, *The Farther Reaches of Human Nature*, p. 199-211.
64. Kate Eldridge Glaspell, "Incidents in the Life of a Pioneer," *North Dakota Historical Quarterly*, 1941, p. 188.

STRANGELY LIKE WAR

1. Greg Breining, "South China Tiger as Good as Extinct," *San Francisco Chronicle*, January 9, 2003, A1.
2. Public Citizen, "Corporate Welfare Examples in 1999," accessed February 19, 2012, www.citizen.org/congress/welfare/articles.cfm?ID=1053.
3. US Forest Service, "1998 Report of the Forest Service," accessed February 19, 2012, www.fs.fed.us/pl/pdb/98report/02_stats.html.
4. William Wade Keye, "Managing Forests, Protecting Watersheds," *San Francisco Chronicle*, December 1, 2002, D5.
5. Clive Ponting, *A Green History of the World: The Environment and the Collapse of Great Civilizations* (New York: Penguin Books, 1991).
6. Perlin, *A Forest Journey*, p. 35.
7. Ibid., p. 40.
8. Ibid., p. 35–39.
9. Ibid., p. 135–36.
10. International Union for the Conservation of Nature, Species Survival Commission, *1996 IUCN Red List of Threatened Animals* (Gland, Switzerland: IUCN, 1996), p. 36.
11. D. Bryant, D. Nielsen, and L. Tangley, *The Last Frontier Forests: Ecosystems and Economies on the Edge* (Washington, DC: World Resources Institute, 1997), p. 17.
12. Margot Higgins, "Extinction Debt Come Due Long after Deforestation," *Environmental News Network*, October 12, 1999.
13. Rainforest Foundation-US, "Why? Save the Rest," accessed July 25, 2003, www.rainforest-foundation.org/why/html; no longer available online.
14. Richard Drinnon, *Facing West: The Metaphysics of Indian-Hating & Empire-Building* (Norman, OK: University of Oklahoma Press, 1997), p. xiii.
15. Sylvia De Rooy, "Before the Wilderness," *Wild Humboldt* 1, Spring/Summer 2002, p. 12.
16. W. R. Gast, et al., *Blue Mountain Forest Health Report*, US Forest Service, Malheur, Umatilla, and Wallowa-Whitman National Forests, 1991.
17. US Forest Service, *1997 Resources Planning Act Assessment, Final Statistics* (DC: US Forest Service, July 2000), Tables 12–15.
18. John J. Byckowski, "Firm's Plans Still Not in Stone," *Cincinnati Enquirer*, September 5, 1994, D1; Mike Boyer and Cliff Peale, "International Paper to add 400 jobs by mid-'97," *Cincinnati Post*, February 14, 1996, B6; "International Paper to cut 215," *Cincinnati Enquirer*, February 13, 1997, B16.
19. Nigel Sizer, "Perverse Habits," *World Resources Institute Forest Notes*, June 2000, p. 2, citing Lynch.
20. Americas Watch, *Rural Violence in Brazil* (New York: Human Rights Watch, 1991).
21. Tom Barry and Deb Preusch, *Central America Fact Book* (New York: Grove Press, 1986), p.135, citing Jacobo Schatan, *La Agroindustria y el Sistema Centroamericano* (Mexico: CEPAL, 1983), p. 46.
22. Charles C. Geisler, "Ownership: An Overview," *Rural Sociology* 58 (1993), p. 532-546.

WELCOME TO THE MACHINE

1. The "debates" rage around whether or not it's a problem, whether the loot is being fairly distributed, whether the benefits have been worth the destruction, and over what if anything can or should be done to mitigate the damage.

2. Jeremy Bentham, "Panopticon; Or The Inspection-House: Containing The Idea Of A New Principle Of Construction Applicable To Any Sort Of Establishment, In Which Persons Of Any Description Are To Be Kept Under Inspection; And In Particular To Penitentiary-Houses, Prisons, Houses Of Industry, Work-Houses, Poor-Houses, Lazarettos, Manufactories, Hospitals, Mad-Houses, And Schools: With A Plan Of Management," *The Panopticon Writings*, Ed. Miran Bozovic (London: Verso, 1995, p. 29-95), accessed February 19, 2012, http://cartome.org/panopticon2.htm.

3. Bentham, "Panopticon."

4. Ibid.; extravagant italicization in original.

5. Ibid. Bentham's use of the term "Purpose X" is revealing. The purpose of surveillance and exact mandate of the bureaucracy doing it are never made explicit. The rationales are invariably abstract virtues such as taxpayer value, customer service, or the ever-popular "security," which covers everything from happy retirement to preemptive wars. The US government slipped of course when it began to publicly equate national security and economic security; for a discussion of the Carter Doctrine and its informal predecessors, see Michael T. Klare, *Resource Wars: The New Landscape of Global Conflict* (New York: Metropolitan Books, 2001), p. 33ff.

6. Michel Foucault, *Discipline & Punish: The Birth of the Prison,* trans. Alan Sheridan (New York: Vintage Books, 1979), p. 201–202.

7. Foucault, *Discipline & Punish*, p. 202.

8. Jane Wakefield, "US Looks to Create Robo-soldier," *BBC News*, April 10, 2002, accessed February 19, 2012, http://news.bbc.co.uk/1/hi/sci/tech/1908729.stm.

9. Guy Gugliotta, "Rats Turned into Remote-Controlled Robots: Techniques Potential Uses Include Aid to Victims of Disaster or Neural Injuries," *Washington Post*, May 2, 2002, sec. A, accessed February 19, 2012, http://www.washingtonpost.com/ac2/wp-dyn?pagename=article&node=&contentId=A18261-2002May1&notFound=true.

10. Colin A. Ross, *Bluebird: Deliberate Creation of Multiple Personality by Psychiatrists* (Richardson, TX: Manitou Communications, 2000), p. 97.

11. David Hearst, "Sci-fi War Put under the Microscope," *The Guardian*, May 20, 2003, accessed February 19, 2012, http://www.guardian.co.uk/uk_news/story/0,3604,959469,00.html.

12. Hearst, "Sci-fi War," 2003.

13. Associated Press, "Pentagon Hopes to Identify People by the Way They Walk," *News-Star* (Shawnee, Okla.), May 20, 2003.

14. Michael J. Sniffen, "Proposed System Would Use Lots of Data," *The Guardian*, May 19, 2003.

15. Francis Bacon, *The Works*, ed. James Spedding (New York: Garrett Press, 1968), p. 4, 296, quoted in Carolyn Merchant, *The Death of Nature: Women, Ecology and the Scientific Revolution* (San Francisco: Harper SanFrancisco, 1983), 164, 168, and 172.

16. Tim Haynes, "High-tech Robot Gives Glimpse of Future," *Boston Globe*, July 25, 2003.

17. Reuters, "Robot Shows Prime Minister How to Loosen Up," August 22, 2003.

18. United Press International, *Android World*, August 2, 2003, accessed February 19, 2012, http://www.androidworld.com/.

19. U.S. DARPA, "Autonomous Vehicles Grand Challenge," accessed August 11, 2003, http://www.darpa.mil/grandchallenge/overview.htm; no longer available online.

20. Witold Rybcynski, *Taming the Tiger: The Struggle to Control Technology* (New York: Viking Press, 1983), p. vii, 227.

21. Jean-Jacques Rousseau, *On the Social Contract* (Mineola, New York: Dover Publications, 2003), p. 61. Had Rousseau been alive in 2003, he may very well have added, ". . . and even become President of the United States."

22. Rousseau, *On the Social Contract*, p. 3.
23. Éttiene de La Boétie, *Discours de la Servitude Volontaire* (1552), trans. Harry Kurz, *The Politics of Obedience: The Discourse of Voluntary Servitude* (Buffalo, New York: Black Rose Books, 1997). Our quote comes from the abridged and edited version at Personal Empowerment Resources, accessed February 19, 2012, http://www.mind-trek.com/.
24. George Orwell, *1984* (New York: Signet Classics, 1962), p. 210–211.

ENDGAME, VOLUME ONE

1. *Webster's New Twentieth-Century*, s.v. "civilization."
2. *Oxford English Dictionary*, compact ed., s.v. "civilization."
3. Stannard, *American Holocaust* , p. 4.
4. Ibid.
5. Maria Mies, *Patriarchy and Accumulation on a World Scale* (London: Zed Books, 1999), p. 98.
6. Lewis Mumford, *The Myth of the Machine: Technics and Human Development* (New York: Harcourt Brace Jovanovich, 1966), p. 186.
7. Diamond, *In Search of the Primitive*, p. 1.
8. Mumford, *The Myth of the Machine*, 186. There is awkwardness in the original, even though Mumford is normally an exquisite stylist.
9. Diamond, *In Search of the Primitive*, p. 4.
10. Frederick, *Beyond Geography*, p. 182.
11. Faust, *The Ideology of Slavery*, p. 293.
12. Mowat, Stannard, Drinnon, and Turner, for example.
13. R. D. Laing, *The Politics of Experience* (New York: Ballantine Books, 1967), p. 58.
14. I am indebted to Becky Tarbotton for the previous several paragraphs.
15. Just last night I had dinner with a couple of mainstream environmentalists and a bunch of other people. My mom was there, too. The mainstreamers spent much of the dinner putting forward precisely these arguments. They said, and this is a direct quote, "Things will be all right if only we take back the Senate." I don't think they meant by storm, and by *we* I don't think they meant normal human beings but Democrats, the left wing of the corporate party. They then said, "And it would be great if we could take over the White House, too." My mom then said, "It doesn't matter whether the Democrats or Republicans are in the White House. The government would still be run by the big corporations." Wrinkled noses all around. Something stunk. What *was* that awful smell? Then lots of very fast sentences spilling from the mouths of the mainstream environmentalists, anything to make the moment disappear. Earlier in the evening they'd taken a different approach to someone Saying Something That Shouldn't Be Said. I was giving a talk the next day, and someone asked what I would talk about. I said, "How to take down civilization." The same awkward silence. The same wrinkling of noses. But this time the next thing that was said was, "Could you please hand me the hummus? It's awfully good. And this soup is simply delightful." Down the old memory hole.
16. Doesn't that feel good just to admit that? That realization was extraordinarily liberating for me! Now I can just get on with the work.
17. And don't give me any shit about how the wants of most Americans aren't in opposition to the needs of their landbase. That's just crazy. Sure, we can talk about their deep-down desire for connection, but you and I both know that's not what I'm talking about here.
18. Fulcra, for those Latin aficionados keeping score at home.
19. One common way is through amassing money, but this power seeking takes many forms.
20. Or even talk about fighting back.
21. Max Hastings, *Bomber Command* (New York: Touchstone, 1979), p. 227.
22. John Keegan, *The Second World War*, 1st Amer. ed. (New York: Viking Penguin, 1990), p. 430.
23. "The Effects of Strategic Bombing on the German War Economy," The United States Strategic Bombing Survey, Overall Economic Effects Division, October 31, 1945.

24. Keegan, *The Second World War*, p. 430.
25. I've received a large number of letters, by the way, with cogent and radical analysis of the culture from people in the military.
26. And as I did in *A Language Older Than Words*.
27. I need to say something else here that doesn't really fit in the book but is crucial to the discussion, and this is a good place to raise it. I'm often asked if I'm afraid of getting arrested or killed by feds because of my writing. I always answer, "Absolutely. But I'm far more afraid of what this culture is doing to the planet and to all of us. It's as Robert E. Lee said, when asked why he so often attacked even when outnumbered, 'We must decide between the risk of action versus the positive loss of inaction.'"

    I'll tell you my fantasy, which is that as some fed reads this book, perhaps with an increasing sense of outrage, that instead of ordering me arrested or killed, he disproves me. I would like nothing more than to be shown conclusively that my premises are wrong and that we do not have as difficult a path ahead of us as I know we do.

    Show me how a way of life based on the use of nonrenewable resources can be sustainable. Show me how a way of life based on perceiving those living beings around us (and often ourselves) as resources can be sustainable. Show me how civilization can and does benefit landbases. Show me how civilization isn't based on systematic and widespread violence. (As Ursula K. LeGuin writes, "All civilization does is hide the blood and cover up hate with pretty words" [*The Sun*, March 2004, 48].) Convince me. I don't think you can do it.

    I mean, by the way, really convince me. I don't mean throw at me your angry and absurd roadblocks to understanding, tossed at me simply because you are too afraid of the implications not only to allow yourself to examine them but to allow anyone else to examine them either (see R. D. Laing's Jack and Jill above). I get enough of that already. For example, after a recent talk someone emailed me with this question: "If you don't like civilization and all it brings, why don't you and your liberal [*sic*] friends just move someplace else?" I mentioned this at a talk I did a couple of nights later, and a woman in the audience exclaimed, "By Christ, tell me where I can go! The fucking culture is everywhere. I can't get away from it. The poisons are in my cells, and they're in the cells of everyone everywhere. Civilization is killing the planet!"

    That email is one example of what I'm talking about. Here's another. Immediately after another show I did, an older man with a gray ponytail and loose-knit sweater rushed the stage. He demanded, "Do you have a bank account?"

    "Why do you ask?" "Because if you do, I can discount everything you say." I stared at him, eyes wide, dumbstruck.

    "All through your talk, I kept wondering whether you're a hypocrite. If you participate in the system, you're a hypocrite, and then nothing you say matters at all."

    I pointed to his sweater. "Where do you think this was made? And your pants? Your shoes? My shoes? My backpack? Just because we're immersed in this culture that systematically eliminates alternatives doesn't mean—"

    He cut me off, looked smug. "Ah, ha! So you feel defensive. You do have a bank account then."

    I just shook my head and walked away.

    Back to the feds and other cops reading this book. If you don't like what I say, disprove me. I don't think it can be done. And if you can't disprove me, don't simply act out your denial and kill or arrest me. Join me. Do the real work. Protect your landbase. I'm sure we could use your skills.

    I want to be clear, by the way, that this is not a general invitation to debate my life or work in private. I do enough of that in public and have no interest in doing it in private. And frankly, more or less all of the "attempts to disprove me" I've seen have been nothing more than these roadblocks I mentioned or, even more often than that, plain old bursts of anger (and especially passive aggressiveness) because people are afraid, and so they lash out (never once, of course, admitting they're lashing out). This paragraph is instead a very specific invitation for servants of power not to fall back on force to defend that power, but to try real discourse. And for them to seriously examine the premises of this book. If they honestly find errors in my premises or thinking, I'd be willing to reexamine everything I'm saying, on the condition that if they cannot find errors, they not only seriously examine their own role in the ongoing apocalypse that is industrial civilization, but help stop the apocalypse—help bring down civilization.

28. It depends on who "we" are. I don't think members of the French resistance would have included the German occupiers or the French collaborators in a similar statement. Similarly, I'm not in this with Charles Hurwitz or John Stossel. Yes, they're killing the planet they live on, too, but I'm trying to stop them. I'm not on their side.

29. Well, the real point is fear. It's far less scary to not purchase an airline ticket than to blow up a dam. And we still get to say, "Ha! I delivered a blow against the machine!"

30. Just last month I bought a bunch of heirloom apple trees from a very small grower. The trees will eventually pay part of my rent to the bears and deer and birds and insects whose home this was long before I moved in.

31. And why?

32. If I may change this cliché so it finally makes sense.

33. David Barsamian, "Expanding the Floor of the Cage, Part II: An Interview with Noam Chomsky," *Z Magazine*, April 1997.

34. J. Bradford DeLong, "The Corporations as a Command Economy," accessed March 17, 2004, http://www.j-bradford-delong.net/ Econ_Articles/Command_Corporations.html; no longer available online.

35. Don't laugh. It's been done.

36. Too bad, darn it.

37. Star Wars, accessed April 23, 2004, http://www.starwars.com/databank/location/deathstar/; no longer available online.

38. Of course I'm making this up.

39. The draft doesn't exist.

40. They also titled the movie *Star Protest* instead of *Star Wars*.

41. That was to be an example of art imitating life.

42. Star Wars, 2004.

43. It's a joke! There's no script!

ENDGAME, VOLUME TWO

1. I don't think that "using the force" would be a tactic, since it's a spiritual attitude. A tactic would be *how* Luke flies and what approach he chooses to take to the tube itself. "Using the force" is a way of being. A spirituality or a way of being is *not* a tactic.

2. I need to be clear that I am not cynical enough to believe all relationships are this way, nor is that my experience. I was speaking specifically of those I've known or heard about whose specific goal of marriage was more important than either integrity or the quality of the relationship. I am also thinking about the extremely popular (and extremely morally troubling) book from a few years ago called *The Rules™: Time-Tested Secrets for Capturing the Heart of Mr. Right*.

3. I guess this would be a strategy. There isn't really a word for plans to achieve operational goals the way that the words *strategy* and *tactics* exist, and in any case the terms are a bit fuzzier than I'm making them seem, as from a general's standpoint the movement of a brigade might be tactical, but from the perspective of a lieutenant commanding a squad in that particular brigade, such movement would be strategic: it's all about perspective.

4. I told you that you might not believe it.

5. You could also end up on an aircraft carrier or in Washington, DC, or South Carolina.

6. And don't give me any nonsense about how effective we are. If we were effective in the least, the world would not be getting killed.
7. I first wrote "more or less ignoring morality on the larger scale," but that's insane: we are not stopping those in power from killing the planet; this is not "more or less ignoring" larger-scale morality, but ignoring it to an outrageous, unbelievable, unspeakably despicable, and most important of all, unforgivable degree.
8. Men, too, are of course trained to hate women, but we probably shouldn't talk about that, should we?
9. Similarly, women aren't supposed to hate, lest they be called ballbusters.
10. Lundy Bancroft, *Why Does He Do That? Inside the Minds of Angry and Controlling Men*, (New York: Berkley Books, 2002), p. 33.
11. Ibid., p. 34, italics original.
12. Ibid.
13. I have known many women whose husbands beat them only on their bodies, never on their faces, because that would show.
14. Bancroft, *Why Does He Do That?* p. 34.
15. Ibid., p. 34–35, italics and bold in original.
16. Ibid., p. 35.
17. Ibid., p. 54.
18. Ibid., p. 151.
19. Ibid., p. 157.
20. Ibid., p. 152.
21. Faust, *The Ideology of Slavery*, p. 81.
22. Bancroft, *Why Does He Do That?* p. 197, italics original.
23. Ibid., p. 288, italics original.
24. David Edwards, *The Compassionate Revolution: Radical Politics and Buddhism* (Devon, UK: Green Books, 1998), p. 81.
25. Bancroft, *Why Does He Do That?* p. 43.
26. Laing, *The Politics of Experience*, p. 186.
27. With self and other.
28. Bancroft, *Why Does He Do That?* p. 63, italics original.
29. Note that I also disagree with his implication that guilt or empathy are specifically human emotions, and to imply that the abuser distances himself from her humanity suggests that were she not human, there would already be the distance that could enable his abuse. I'm not attacking Bancroft here, who I feel does extraordinary work, but merely trying to point out how easy it is to succumb to this culture's rhetoric of superiority.
30. Bancroft, *Why Does He Do That?* p. 196.
31. Ibid., p. 311.
32. Ibid., p. 361.
33. Ibid., italics original.
34. And I would say most often not even then.
35. Bancroft, *Why Does He Do That?* p. 360.
36. And I would say most often not even then.
37. I owe the term *Selective Law Enforcement Officers* to Remedy, "Mattole Activists Assaulted, Arrested after Serving Subpoena for Pepper Spray Trial," Treesit Blog, August 27, 2004.
38. "Reviving the World's Rivers: Dam Removal," part 4, Technical Challenges, International Rivers Network, accessed July 11, 2004, http://www.irn.org/revival/decom/brochure/rrpt5.html; no longer available online.
39. Gandhi-ites throw this one at me all the time, but it's possible that Gandhi never actually said it. The quote is all over the internet, and it was in the movie *Gandhi*, but David Lean was not known for the historical accuracy of his films.
40. Audre Lorde, "The Master's Tools Will Never Dismantle the Master's House," *In Sister/Outsider* (Trumansburg: The Crossing Press, 1984), p. 112.

41. Ibid., italics original.
42. I am grateful to Lierre Keith for this paragraph.
43. Insofar as we can make a meaningful distinction.
44. I'm embarrassed to admit I made this same assertion in my book *A Language Older Than Words*. I don't know what to say, except that I hadn't thought it through. I was wrong.
45. I am grateful to Lierre Keith for this final story.
46. Bettelheim was a terrible person, far worse than Gandhi. He was accused, most probably accurately, of physical and sexual assault on children. His attitudes on autism were despicable: he blames mothers for it. His attitudes on anti-Semitism were essentially as bad: he once shouted at an audience of Jews, "Anti-Semitism, whose fault is it? Yours! . . . Because you don't assimilate, it is your fault. If you assimilated, there would be no anti-Semitism. Why don't you assimilate?" But the reason I include a few of the despicable Bettelheim's sins in a footnote and a few of Gandhi's in the body of the text is that Bettelheim's position is not one that people claim carries moral high ground (Bettelheim's analysis *here* concerns tactical responses to violence), yet that is, so far as I can tell, the main thing Gandhi really has going for him: his adherents claim (as often and as loudly as they can) that Gandhi's position carries the day because it carries moral weight. This makes an examination of his own morality and the morality of his positions eminently relevant.
47. Bruno, Bettelheim, Introduction to *Auschwitz: A Doctor's Eyewitness Account*, by Miklos Nyiszli (New York: Frederick Fell, 1960), p. vi.
48. *Webster's New Twentieth Century Dictionary*, s.v. "civilization."
49. *Oxford English Dictionary*, compact ed., s.v. "civilization."
50. "Effects of Strategic Bombing," p. 13.
51. This sounds familiar. Hmm, where have we heard it? Ah, "In order to maintain our way of living, we must tell lies to each other and especially to ourselves."
52. Note that I'm not saying it's true in all discourse. Not all discourse is antagonistic.
53. Or rather, some of us are encouraged to vote; the poor, people of color, felons are all either discouraged or prohibited.
54. That's one reason I could not encourage that sixteen-year-old to burn down a factory. I didn't know him well enough to know if he was thinking his own thoughts. To be an adult one must not only know freedom but responsibility. As I say in *Walking on Water*, freedom without responsibility is immaturity, and responsibility without freedom is slavery. We need people mature enough to think for themselves. This young man may have been, or he may not have been. I didn't know, which is one reason I couldn't advise him.
55. If you do not love your landbase, why have you read this far into this book?
56. Which is my big beef with the pacifists: they seem to have a one-size-fits-all prescription. Well, the truth is that one size only fits one.
57. Why do you think the rules of war, written by governments—in other words, those who raise large armies—exclude many non-army combatants from their protection?
58. And I mean, at his *disposal*.
59. Bob, Blaisdell, ed, *Great Speeches by Native Americans* (Mineola, NY: Dover, 2000), p. 80.
60. Samuel G. Drake, *Biography and History of the Indians of North America, from Its First Discovery*. 11th ed. (Boston: Benjamin B. Mussey & Co., 1841), p. 662.
61. H.L. Gordon, *The Feast of the Virgins and Other Poems* (Chicago: Laird and Lee, 1891), p. 343–44.
62. Blaisdell, *Great Speeches by Native Americans*, p. 67.
63. Thomas Jefferson, ed. Andrew A. Lipscomb and Albert Ellery Bergh, *The Writings of Thomas Jefferson*, vol.11 (Washington, DC: Thomas Jefferson Memorial Association, 1903), p. 345.
64. Derrick Jensen and George Draffan, *Welcome to the Machine: Science, Surveillance, and the Culture of Control* (White River Junction: Chelsea Green, 2004), p. 74.
65. B.H. Liddell Hart, ed. *The Rommel Papers*, translated by Paul Findlay (New York: Harcourt, Brace, and Company, 1953), p. 328.

THOUGHT TO EXIST IN THE WILD

1. David Hancocks, *A Different Nature: The Paradoxical World of Zoos and Their Uncertain Future* (Berkeley: University of California Press, 2001), p. 7.

2. He also presumes that all humans are inherently destructive, citing the time and again discredited Pleistocene Overkill Hypothesis.

3. Hancocks, *A Different Nature*, p. 7.

4. R.J. Hoage and William A. Deiss, foreword by Michael H. Robinson, *New Worlds, New Animals: From Menagerie to Zoological Park in the Nineteenth Century* (Baltimore: John Hopkins University Press, 1996), p. vii.

5. For a thorough examination of this definition and of the effects of cities on their host cultures and on the landbases they destroy, see Derrick Jensen, *Endgame, Volume 1: The Problem of Civilization*, and *Endgame, Volume 2: Resistance* (New York: Seven Stories Press, 2006).

6. Hancocks, *A Different Nature*, p. 7.

7. Ibid., p. 8.

8. Vicki Croke, *The Modern Ark: The Story of Zoos: Past, Present and Future* (New York: Scribner, 1997), p. 137.

9. Ibid, p. 136.

10. As if a "human world" exists. It's not a human world, obviously, except in the deranged minds of these narcissists. The degree to which any people believe this is a "human world" is the degree to which they have been inculcated into this culture, and therefore inculpated in the atrocities it commits.

11. And psychopathology.

12. David Montgomery, *King of Fish: The Thousand-Year Run of Salmon* (Boulder, CO: Westview Press, 2003), p. 39.

13. Bob Mullan and Garry Marvin, *Zoo Culture: The Book About Watching People Watch Animals*, second edition (Urbana: University of Illinois Press, 1999), p. 157-158.

14. Ibid., p. 83–84.

15. Rosl Kirchshofer, ed., *The World of Zoos: A Survey and Gazetteer* (New York: Viking, 1968), quoting Hediger, p. 9.

16. Mullan, *Zoo Culture*, p. 84. And note, significantly enough, that even in this example, instead of describing his own experience of being locked in his office, he hypothetically locks *you* up.

17. "Animal Planet: From India's famed camel fair to Indonesia's fierce Komodo dragons—all the world's a zoo," *San Francisco Chronicle*, November 28, 2004, F1, F4.

18. Croke, *The Modern Ark*, p. 252.

19. Frances Densmore, *Teton Sioux Music*, Bulletin 61, Bureau of American Ethnology, Smithsonian Institution, 1918, p. 172.

20. Hancocks, *A Different Nature*, p. 252.

21. Eric Baratay, and Elisabeth Hardouin-Fugier, *Zoo: A History of Zoological Gardens in the West* (London: Reaktion Books, 2002), p. 209.

22. Baratay, *Zoo*, p. 236.

23. Baratay, *Zoo*, p. 115.

24. Ibid., p. 118.

25. Ibid., p. 114.

26. Nigel, Rothfels, *Savages and Beasts: The Birth of the Modern Zoo* (Baltimore: Johns Hopkins, 2002), p. 60.

27. Jeremy Cherfas, *Zoo 2000: A Look Beyond the Bars* (London: British Broadcasting Corp., 1984), p. 19.

28. Rothfels, *Savages and Beasts*, p. 61.

29. Ibid., p. 61.

30. Note that other accounts suggest that only one of the young elephants survived. (Rothfels, *Savages and Beasts*, p. 218.)

31. Rothfels, *Savages and Beasts*, p. 61–62.

32. Ibid., 62.
33. Ibid., 62.
34. Ibid., 64.
35. And just for grins, look up "wisdom" in the index of your average science textbook: I'll bet you a nickel it's not there.
36. "Scientists highlight fish 'intelligence.'" Note that even here, the journalist who came up with the headline is so insecure about human intelligence and so sure that fish can not be intelligent that he or she places quotes around the word intelligence. It's extraordinary how deeply held is this arrogance. (You do know, don't you, that many species of fish form extensive and fluid social groups, and that fish are, as now even scientists have been forced to acknowledge, "regarded as steeped in social intelligence, pursuing Machiavellian strategies of manipulation, punishment and reconciliation, exhibiting stable cultural traditions, and co-operating to inspect predators and catch food".)

WHAT WE LEAVE BEHIND

1. Or rather, one that doesn't rot. And that's the point.
2. Susan Casey, and Gregg Segal, "Our oceans are turning into plastic.... Are we?" *Best Life Magazine*, May 11, 2007.
3. Richard Shears, "Is this the world's most polluted river?" *Daily Mail*, June 5, 2007, accessed February 19, 2012, http://www.dailymail.co.uk/news/article-460077/Is-worlds-polluted-river.html.
4. Casey, "Our Oceans," 2007.
5. Alan Weisman, "Polymers Are Forever," *Orion*, May/June 2007, accessed February 19, 2012, http://www.orionmagazine.org/index.php/articles/article/270/.
6. Casey, "Our Oceans," 2007.
7. Ibid.
8. Ibid.
9. "Tailings," Wikipedia, accessed February 19, 2012, http://en.wikipedia.org/wiki/Tailings.
10. "APP of Mine Overburden," Colorado School of Mines, "CH 126 Experiment #6, Acid Producing Potential of Mine Overburden," accessed July 6, 2007, http://www.mines.edu/fs_home/jhoran/ch126/app.htm; no longer available online.
11. "Microbial Influences: Thiobacillus ferrooxidans," Acid Mine Drain Drainage Experimentsat Colorado School of Mines, accessed July 6, 2007, http://www.mines.edu/fs_home/jhoran/ch126/microbia.htm; no longer available online.
12. Ibid.
13. Ibid.
14. Although the main leak was methyl isocyanate (note the last three syllables), the reaction that led to the release of MIC produced cyanide and other toxic chemicals as well.
15. In this case cyanide is the complexant.
16. All of this is of course from the film *The Treasure of the Sierra Madre*.
17. Mineral Policy Center, "Cyanide Leach Mining Packet," Washington, DC: GPO, 2006.
18. Ibid.
19. Jennifer Bogo, "Crying Rivers (Romania gold mine spills cyanide into rivers) (Brief Article)," *High Beam Encyclopedia*, May 1, 2000.
20. Erin Klauk, "Environmental Impacts at Fort Belknap from Gold Mining," *Impacts of Resource Development on Native American Lands*, accessed February 19, 2012, http://serc.carleton.edu/research_education/nativelands/ftbelknap/environmental.html.
21. "Summitville, the Exxon Valdez of the Mining Industry," *Sprol*, October 17, 2005, http://www.sprol.com/?p=268m site visited July 13, 2007
22. Dr. Greg Laughlin: "The technology is not at all far-fetched.... [W]e just require delicacy of planning and maneuvering" (Robin McKie, "Nasa aims to move Earth," *Guardian Unlimited*,

June 10, 2001, accessed February 19, 2012, http://observer.guardian.co.uk/international/story/0,,504486,00.html).

And what's wrong with these NASA scientists? Did they never see that classic television program *Space 1999* and learn that strange things happen when a celestial body is blown out of orbit? People could end up duplicated and come face to face with their future selves; or they could answer a distress signal and be forced to crash land on a distant moon which will turn out to be a penal colony, where our commander will find himself a prisoner of the beautiful Elizia and her equally beautiful prison guards; or a mysterious power from an alien planet could take control, luring the inhabitants to a paradise of eternal peace but living death (Oh, wait, we already have that one: it's called industrial civilization, except we have neither the paradise nor eternal peace).

23. Please note that he's doing the same damn thing that McDonough does, which is to try to naturalize capitalism, make us think capitalism is somehow natural or compatible with the natural world.

24. Peter Montague, "Rachels' Democracy and Health News #932," November 8, 2007, accessed February 19, 2012, http://www.precaution.org/lib/07/ht071108.htm.

25. AP, "UN chief calls for action on climate change," *San Francisco Chronicle*, November 8, 2007.

26. The journalistic equivalents of biopimpologists: since neither of the words *journopimps* or *pimpalists* really cuts it, how about *pimpwriters*.

27. Jessica Taylor, "Jessica Taylor talks to veterans of Oak Ridge," *The Guardian*, July 5, 2006, accessed February 19, 2012, http://www.guardian.co.uk/secondworldwar/story/0,,1812 11,00.html.

28. Ibid.

29. Steamlocomotive.com, "Steam Locomotive Builders," accessed February 19, 2012, http://www.steamlocomotive.com/builders/.

30. Thank you to Lierre Keith for the first three questions, which come from Alan G. Johnson's book *The Gender Knot*.

DREAMS

1. Likewise, I can and do choose one word over another, one piece of punctuation over another, and I can even determine to write about a particular subject, but I have no say over whether or not the words sing. The words, phrases, sentences, paragraphs: these are as much gifts as are dreams. If they sing, they sing, and if they don't there's not much I can do except ask the muse for more words.

2. Richard Dawkins, "God's Utility Function," in *River Out of Eden: A Darwinian View of Life* (New York: Basic Books, 1995), p. 85.

3. Ibid.

4. "Blind watchmaker" is Dawkins's term.

5. Richard Dawkins, "What is True?" *The Devil's Chaplain* (New York: Houghton Mifflin, 2003), p. 15.

6. It is Assembly Bill 2296. Torturing nonhumans is now a fundamental legal right in the state of California.

7. Dawkins, "What is True?" p. 9.

8. See, for example, Jensen, *Endgame*, vols. 1 and 2. See also Jensen, *The Culture of Make Believe*. See also the murdered oceans.

9. Dawkins, "What is True?" p. 15.

10. Sam Harris, "Mother Nature is Not Our Friend," *Richarddawkins.net*, accessed February 19, 2012. http://richarddawkins.net/article,2096,Mother-Nature-is-Not-Our-Friend,Sam-Harris.

11. Derrick Jensen, *A Language Older Than Words* (New York: Context Books, 2000).

12. Sam Harris, "Science Must Destroy Religion," *MachinesLikeUs.com*, accessed February 19, 2012, http://machineslikeus.com/articles/ScienceMustDestroy.html.

13. Sam Harris, *The End of Faith: Religion, Terror, and the Future of Reason* (New York: W.W. Norton, 2004), p. 225.

14. Wikiquote, "Robert Oppenheimer," accessed February 19, 2012, http://en.wikiquote.org/wiki/Robert_Oppenheimer.

15. It's a train because this culture is a machine, and also because a train runs on tracks: it can only go where it is headed. So, too, with this culture, which will not deviate from its destructive path. The train consisted of one car because this culture reduces all diversity to unity: one God, one culture, one language, one way to know, one meaningful species, and so on.

16. Obviously the big man is this culture, and its members.

17. *Us* represents everyone who is not a member of this culture, at least not anymore.

18. The fact that we had to run along the tracks and couldn't just let the train go represents the fact that at this point the culture has overrun the planet, and there is nowhere anyone can go to get away from it. It forces everyone into this frenetic forced march that can end only with death, or the end of this culture.

19. This culture forces all others to conform to it, to adapt to it, to keep adapting, to keep running, or to die.

20. This culture forces *everyone* to service it.

21. This is one reason the world hasn't fought back more than it has.

22. The tunnel is, I think, the collapse of this culture.

23. I don't know what the other train represents.

24. This is, I think, the voice of the *Saprolegnia*, and other immediate causes of extinction. It seems so clear to me that so many nonhumans are giving up. They'd rather go extinct than continue to put up with this culture's ceaseless rape of the world. I don't blame them.

25. The man and woman together represent the continuation of a species, a people (human people, frog people, toad people, cedar people, salmon people, sturgeon people, crab people, wolf people, mountain people, river people, or anyone else).

26. The world and all its members are being crushed under the weight of humanity.

27. Of course this is also a sad visual pun on humans eating frog legs—which was a major cause of California red-legged frog decline in the nineteenth century.

28. Thomas Mann, *Death in Venice, Tonio Kröger, and Other Writings*, Ed. Frederick A. Lubich (New York: Continuum Publishing Company, 1999), p. 286.

29. Of course you could also read Seneca as suggesting that you can't go against your own will.

30. A more radical New Age version of this holds that not only does one's will influence one's own decisions, but, godlike, influences everything around us. A great articulation of this is by one of the characters in a short story by, oddly enough, thriller writer (and misogynist) Lawrence Block: "There is a metaphysical principle which holds that we choose everything about our lives, that in fact we select the very parents we were born to, that everything which happens in our lives is a manifestation of our will. Thus, there are no accidents, no coincidences." The whole notion is that everything is a response to our own will. Block doesn't seem to espouse this idea, by the way, because later in the story the person to whom this was said throws the speaker out of a high rise window (I wonder if *that* was a manifestation of the speaker's will?), but I include Block's quote because it's such a perfect articulation (Block *is* a good writer) of this attitude put forward by so many. This notion is really the end point for the whole infantile, narcissistic cosmology of this culture. Not only is it narcissistic, but it combines this culture's typically contemptible level of irresponsibility for the abuse it perpetrates with this culture's equally contemptible (and indeed they go hand in hand) habit of blaming the victim: so young children choose to be raped by the parents they chose, for some cosmic lesson? Salmon choose for dams to be erected? It's all bullshit, and it's all about letting abusers not take responsibility for their actions.

31. Barnet, "The Religious System of Santería," in Mararite Fernández Olmos and Lizabeth Para-visini-Gebert, *Sacred Possessions: Voudou, Santería, Obeah, and the Caribbean* (New Brunswick: Rutgers University Press, 2000), p. 86.

32. This bit about choosing our parents is part of what I don't like. It smacks of the blame the victim bigotry that has permeated this culture from the beginning. Even if we drop overtly abusive parents, it's important to note that the Greek philosophers who wrote this were often not the children of slaves forced to work their short miserable lives in mines. It's awfully convenient for someone with the leisure time to write up a bit of cosmology to declare we choose our station in life.

33. Normally the bones used were from sheep or other ruminants, or pigs. I've not read of anyone using human tali. There were other bones used as well, but the talus was one of the main ones.

34. Wikipedia contributors, "Lozen," *Wikipedia*, accessed February 19, 2012, http://en.wikipedia.org/wiki/Lozen.

DERRICK JENSEN is the best-known voice of the growing deep ecology movement. Winner of numerous awards and honors including the Eric Hoffer Book Award, *USA Today*'s Critic's Choice, and Press Action's Person of the Year, Jensen is the author of over twenty books, including *Endgame*, *A Language Older Than Words*, and *Dreams*. Philosopher, teacher, and radical activist, he regularly stirs packed auditoriums across the country with revolutionary spirit. Jensen holds degrees in creative writing and mineral engineering physics. He lives in Crescent City, California.

LIERRE KEITH is a writer, small-scale farmer, and radical feminist activist. She is the author of two novels, as well as *The Vegetarian Myth: Food, Justice, and Sustainability* and *Deep Green Resistance* (with Derrick Jensen). She's been arrested six times. She lives in Humboldt County, California.

About Seven Stories Press

Seven Stories Press is an independent book publisher based in New York City with distribution throughout the world. We publish works of the imagination by such writers as Nelson Algren, Russell Banks, Octavia E. Butler, Assia Djebar, Ariel Dorfman, Coco Fusco, Barry Gifford, Lee Stringer, and Kurt Vonnegut, to name a few, together with political titles by voices of conscience, including the Boston Women's Health Collective, Noam Chomsky, Angela Y. Davis, Human Rights Watch, Ralph Nader, Gary Null, Project Censored, Barbara Seaman, Gary Webb, and Howard Zinn, among many others. Seven Stories Press believes publishers have a special responsibility to defend free speech and human rights, and to celebrate the gifts of the human imagination, wherever we can. For additional information, visit www.sevenstories.com.